MW01388218

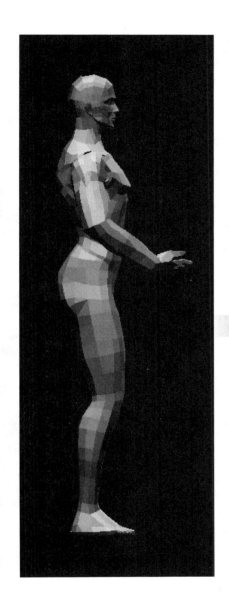

Ergonomics and Musculoskeletal Disorders

RESEARCH ON MANUAL

MATERIALS HANDLING

1983-1996

Edited by
Waldemar Karwowski
Michael S. Wogalter
Patrick G. Dempsey

 HUMAN FACTORS AND ERGONOMICS SOCIETY

ISBN 0-945289-09-X

When citing work from this book, please refer to the **original** proceedings publication, as in the following example:

Gallagher, S., Hamrick. C. A., & Love, A.C. (1990). Biomechanical modeling of asymmetric lifting tasks in constrained lifting postures. In *Proceedings of the Human Factors Society 34th Annual Meeting* (pp. 702–706). Santa Monica, CA: Human Factors and Ergonomics Society.

(Note that in 1993, the Society's name changed to the Human Factors and Ergonomics Society.)

The following HFES proceedings are represented in this book:

1996, 40th Annual Meeting	1991, 35th Annual Meeting
1995, 39th Annual Meeting	1990, 34th Annual Meeting
1994, 38th Annual Meeting	1989, 33rd Annual Meeting
1993, 37th Annual Meeting	1988, 32nd Annual Meeting
1992, 36th Annual Meeting	1987, 31st Annual Meeting

The HFES annual meeting proceedings are indexed or abstracted in the following publications or services: *Applied Mechanics Reviews, Engineering Index Annual, EI Monthly, Cambridge Scientific Abstracts, EI Bioengineering Abstracts, EI Energy Abstracts, Ergonomics Abstracts, ISI Index to Scientific & Technical Proceedings,* and *International Aerospace Abstracts.* This publication is also available on microfilm from University Microfilms International, 300 N. Zeeb Road, Department P.R., Ann Arbor, MI 48106; 18 Bedford Row, Department P.R., London WC1R 4EJ, England.

To obtain copies of papers not included in this book, readers may:

1. purchase back volumes (see back page for ordering information);
2. access documents available through the above-listed indexing/abstracting services;
3. obtain a microfilm/microfiche copy through UMI (see above), or
4. order photocopies from HFES at the cost of $7.50 per paper ($17.50/paper for rush orders).

Additional copies of this book may be purchased from the Human Factors and Ergonomics Society at $35 per copy for HFES members and $50 for nonmembers. Add California sales tax for deliveries in California. Discounts apply on purchases of 5 or more copies. In addition, a 10% discount applies when ordering this book as part of a four-book set of HFES proceedings collections. The other three books are:

Human Factors Perspectives on Warnings: Selections from Human Factors and Ergonomics Society Annual Meetings, 1980–1993 (ISBN 0-945289-02-2, 1994, $35 HFES members, $50 nonmembers)

Human Factors Perspectives on Human-Computer Interaction: Selections from Proceedings of Human Factors and Ergonomics Society Annual Meetings, 1983–1994 (ISBN 0-945289-05-7, 1995, $42 HFES members, $58 nonmembers)

Designing for an Aging Population: Ten Years of Human Factors/Ergonomics Research (ISBN 0-945289-08-1, 1997, $42 HFES members, $58 nonmembers)

To order the four-book set, send $180 plus $15 shipping/handling and sales tax if applicable to HFES at the address above. Orders are accepted with prepayment by check (payable to Human Factors and Ergonomics Society), Master-Card, or VISA.

Library of Congress Cataloging-in-Publication Data

Ergonomics and musculoskeletal disorders : research on manual materials handling, 1983–1996 / edited by Waldemar Karwowski, Michael S. Wogalter, Patrick G. Dempsey.

    p.   cm.

    Includes bibliographical references and index.

    ISBN 0-945289-09 X (pbk.)

    1. Lifting and carrying. 2. Human engineering. 3. Back--Wounds and injuries--Prevention. 4. Musculoskeletal system--Wounds and injuries--prevention. I. Karwowski, Waldemar, 1953– . II. Wogalter, Michael S., 1955– . III. Dempsey, Patrick G., 1967– .

T55.3.L5E74   1997

620.8'2--dc21

97-31460

CIP

Contents

*Note: The original proceedings year and page numbers are
given for each paper in parentheses.*

Note: The original proceedings year and page numbers are given for each paper in parentheses.

Note: The original proceedings year and page numbers are given for each paper in parentheses.

Note: The original proceedings year and page numbers are
given for each paper in parentheses.

Preface

Injuries associated with manual materials handling (MMH) represent the single largest source of losses in many industries, including operations in the manufacturing, service, and distribution sectors. These problems have been researched for more than four decades to provide practitioners with solutions and interventions to stem the associated losses.

This volume contains a selection of the best MMH research that was presented at the annual meetings of the Human Factors and Ergonomics Society between 1983 and 1996. The compilation includes both basic and applied research that can be used by practitioners seeking solutions to MMH problems or seeking insight into the risk factors associated with MMH, by academicians seeking a summary of the last several decades of MMH research, and by students interested in obtaining a fairly broad overview of research and practice in the area of MMH.

Background and Scope of the Problem

The field of industrial ergonomics, which focuses on achieving safety and productivity in the workplace, has grown tremendously in the past two decades. This growth has been spurred, in part, by growing awareness among those in government and industry that focusing on increased productivity without adequate concern for workers' safety and health and the lack of ergonomics considerations in jobs and tasks may result in injury, increased absenteeism, and reduced productivity. This trend continues to expand around the globe as the role of ergonomics in enhancing profitability reaches new audiences.

One of the most important areas of research and application in industrial ergonomics is the prevention of work-related low back disorders (WRLBDs), particularly as related to manual materials handling (which encompasses lifting, lowering, pushing, pulling, carrying, and holding). A significant proportion of WRLBDs is attributed to MMH. Despite increased automation in the workplace, many jobs still require workers to manually handle materials. Considerable research efforts have addressed the responses of the human body to MMH, including acute responses such as energy consumption and tissue stresses, as well as injuries resulting from longer-term exposures to physically demanding tasks. The papers in this collection represent an excellent overview of the research conducted in the past 15 years.

WRLBDs are a major source of economic loss to employers, compensation carriers, and society in general. These losses explain why the research directed toward prevention of injuries associated with manual materials handling in industry has been one of the most active areas in the industrial ergonomics field. In 1980, about one million workers suffered WRLBDs in the United States, and these disorders accounted for almost one-fifth of workplace injuries and illnesses. For eight states providing workers' compensation payment data, almost one-fourth of the expenditures were for injuries to the low back (Bureau of Labor Statistics, 1982). Similarly, Webster and Snook (1994) found that disabling low-back pain cases represented 33% of the costs associated with a large sample of workers' compensation claims costs.

Low back disorders incur costs in the areas of medical care, lost wages, and indirect costs (such as training new workers and production losses; see Snook, 1987). Based on 1980 data from Liberty Mutual Insurance Company, Andersson, Pope, Frymoyer and Snook (1991) estimated that annual workers' compensation costs for WRLBDs in the United States were $4.6 billion for the year 1980. More recently, Webster and Snook (1994) estimated that total annual workers' compensation costs (medical and indemnity) in the United States totaled $11.4 billion, an estimate derived from Liberty Mutual's 1989 experiences. According to Andersson et al. (1991), the total estimated costs (direct plus indirect expenses) for WRLBDs in U.S. industry in 1988 were between $26.8 and $56 billion.

A recent antecedent-oriented analysis of a large sample of work-related injury and illness workers' compensation claims indicated that claims attributed to manual materials handling accounted for 32% of the claims and 36% of the costs (Murphy, Sorock, Courtney, Webster, & Leamon, 1996). MMH claims were the single largest source of claims; the next-largest source of claims (slips and falls—same level) accounted for 15% of the costs. This illustrates that the losses associated with MMH far outweigh any other source of injury in the workplace.

MMH Research

Manual materials handling research has been performed by individuals with diverse backgrounds, ranging from orthopedic surgery to industrial engineering. The papers presented at HFES annual meetings have a strong engineering influence, as is illustrated by the papers in this collection.

The design of MMH tasks typically focuses on ergonomics, job placement, training, or a combination of two or all three approaches. The primary focus should be on permanently reducing the risk of injury through task and workplace design (ergonomics), and supplemented with job placement and/or training when necessary. Strong support for training as an effective means of reducing injuries is lacking, however (Kroemer, 1991). This position is reflected here; the majority of papers focus on the influence of task and workplace design on the biomechanical, physiological, and psychophysical responses of the human body. A secondary focus on job placement is also evident.

Not only have the results of research expanded our knowledge, but technology has expanded the possibilities for research and application. For example, the direct measurement of oxygen consumption was once confined to the laboratory. There are now devices that allow for an almost unobtrusive measurement of oxygen consumption as workers perform their tasks. Likewise, faster computers permit timely processing of biomechanical data, which enables the development of more complex and more anatomically correct biomechanical models.

Researchers have advanced considerably in their ability to model the biomechanical stresses the human body undergoes while performing manual materials handling tasks. Static two-dimensional planar biomechanical analyses were once the norm; today a number of three-dimensional analysis techniques are available which incorporate dynamic stresses as well as accommodate MMH activities not occurring in the sagittal plane. Factors such as the advancement of motion analysis technology and a better understanding of electromyographic responses of the body have resulted in more detailed and representative analyses.

Significant advances have also been made in the understanding of the physiological and psychophysical responses to different task designs. There are extensive psychophysical data for assessing a variety of MMH tasks. There is also an increased understanding of how the physiological and psychophysical approaches may conflict. Such conflicts may present problems to practitioners, but they also serve to increase our understanding of the varied approaches.

The National Institute for Occupational Safety and Health (NIOSH) designed an equation to assess manual two-handed lifting in the sagittal plane (NIOSH, 1981). The 1991 revised equation (Waters, Putz-Anderson, Garg, & Fine, 1993) accommodates lifting and lowering activities not performed in the sagittal plane; in addition, it accounts for tasks performed with less-than-optimal hand-to-object coupling. These equations have motivated considerable research activities, and this trend is likely to continue in the future.

Some of the greatest expansion in the area of the prevention of WRLBDs in the past 15 years has occurred in the area of epidemiology. The reality of economic losses has driven an interest in the determinants of musculoskeletal disorders among occupational populations. Although acute responses of the body to work measured in the laboratory provide a wealth of information, the value of epidemiological studies relating workplace exposures to injury outcomes cannot be understated. This trend is sure to continue as researchers seek valid control measures to reduce economic losses in the workplace.

Content and Use of This Collection

This collection contains 61 papers representing basic and applied MMH research and provides an overview of both the theory and practice of MMH from an ergonomics perspective. The collection shows a strong biomechanics influence, followed by psychophysics and physiology. The majority of the papers focus on the ergonomic approach of fitting jobs to the workforce, with preplacement strength and capacity testing a distant second. Unfortunately, investigations of the determinants of musculoskeletal disorders associated with MMH have been too few, so epidemiological studies are underrepresented here.

There are a number of sources of published work on manual materials handling (journals containing archival material, conference proceedings representing current work, and, recently, Internet sites focusing on ongoing work). However, the *Proceedings of the Human Factors and Ergonomics Society Annual Meeting* represent the best basic and applied research in the field. This selection provides a consolidated source of papers that make the material more accessible for those who cannot regularly attend the HFES Annual Meeting, cannot afford to purchase multiple volumes of proceedings, or cannot access the proceedings in libraries. In addition, the quality and brevity of these papers make them excellent teaching tools for undergraduate and graduate students and for new researchers in the area of MMH.

More important, the variety of perspectives represented by the papers in this compilation will provide practitioners with methods and results that can be applied in numerous workplaces.

Paper Selection Procedure

We used a multistage paper selection process that included peer review. First, using liberal selection criteria, we identified 120 relevant papers from the HFES annual meeting proceedings from the years 1982 to 1996. A 15-year span was judged appropriate based on the length constraints for this book.

We sent the selected papers to three referees, primarily HFES members affiliated with the Society's Industrial Ergonomics Technical Group. The referees were asked to rate papers by considering the quality of the work, the current usefulness of the work, whether or not the work was dated, and the focus on manual materials handling. We asked the reviewers to provide an overall rating of the paper between -3 (very poor quality; must not be included) and +3 (extremely high quality; must be included), as well as a justification for the rating. After we received the reviews, we combined the ratings into a single metric (average rating) for each paper. Papers that received an average rating of +1 or higher were selected for inclusion. (The editors neither reviewed nor selected papers of which they were primary author or coauthor.)

Acknowledgments

We gratefully acknowledge the assistance of Renliu Jang, a graduate student in the Industrial Engineering Department at the University of Louisville, for his help in carrying out the logistics of this project. This collection would not have been possible without the assistance of those that served as referees. In addition to anonymous reviewers, the following individuals served as reviewers:

F. Aghazadeh	J. Jarvinen	G. Mirka
P. C. Champney	W. Karwowski	S. Morrissey
B. Das	C. J. Kerk	M. L. Resnik
P. G. Dempsey	S. Konz	L. Punnett
F. Fathallah	C. Joe Lin	L. J. H. Shulze
S. Gallagher	J. D. McGlothlin	F. Tayyari
R. Goonetilleke	R. McGorry	J. Woldstad
S. Hsiang		

Waldemar Karwowski
Michael S. Wogalter
Patrick G. Dempsey

References

Andersson, G. B. J., Pope, M. H., Frymoyer, J. W., & Snook, S. H. (1991). Epidemiology and cost. In M. H. Pope, G. B. J. Andersson, J. W. Frymoyer, & D. B.Chaffin (Eds.), *Occupational low back pain: Assessment, treatment and prevention* (pp. 95–113). St. Louis, MO: Mosby Year Book.

Bureau of Labor Statistics. (1982). *Back Injuries Associated With Lifting* (Bulletin 2144). Washington, DC: U.S. Government Printing Office.

Kroemer, K. H. E. (1991). Personnel training for safer material handling. *Ergonomics, 35,* 1119–1134.

Murphy, P. L., Sorock, G. S., Courtney, T. K., Webster, B. S., & Leamon, T. B. (1996). Injury and illness in the American workplace: A comparison of data sources. *American Journal of Industrial Medicine, 30,* 130–141.

NIOSH. (1981). *Work practices guide for manual lifting* (HEW(NIOSH) Report No. 81–122). Cincinnati, OH: Author.

Snook, S. H. (1987). The costs of back pain in industry. *Spine: State of the Art Reviews, 2*(1), 1–5.

Waters, T. R., Putz-Anderson, V., Garg, A., & Fine, L. J. (1993). Revised NIOSH equation for the design and evaluation of manual lifting tasks. *Ergonomics, 36,* 749–776.

Webster, B. S., & Snook, S. H. (1994). The cost of 1989 workers compensation low back pain claims. *Spine, 19*(10), 1111–1116.

HANDLE POSITIONS IN A HOLDING TASK AS A FUNCTION OF TASK HEIGHT

J. Deeb, C. G. Drury, K. Begbie

Department of Industrial Engineering
State University of New York at Buffalo
Amherst, New York

ABSTRACT

Six handle positions in a two-handed container holding task were tested with the container at floor, waist and shoulder heights. Fifteen male and fifteen female manual materials handlers participated. Handle position effects on forces exerted, heart rate and psychophysical indicies were large, comparable to a 25 percent change in container weight. As in a previous study (at waist height only) and an industrial survey, handle positions providing both horizontal and vertical stability were better than symmetrical positions. Optimal angles of handle to container changed greatly with task height, giving almost horizontal angles at floor level and almost vertical angles at waist and shoulder level. Implications for the design of handle cutouts on containers are discussed.

INTRODUCTION

Reduction of manual materials handling (MMH) injuries has been shown to be largely a function of reducing loads to within acceptable limits (Snook 1978). One factor greatly affecting the acceptability of the load is the presence or absence of handles (Garg and Saxena, 1980). However, placement of handles on boxes to facilite MMH is by no means an obvious task. Earlier experiments at SUNY Buffalo, both in the field (Drury, Law and Pawenski, 1982) and in the laboratory (Coury and Drury 1982) have shown that handle position has a large effect. For all variables except one in the laboratory task, the size of the effect of handle position exceeded the size of the effect of changing box weight from 10 to 15 kg.

Handle positions on a box-like object can be defined using Figure 1. Both field and laboratory studies agreed that minimum stress (and maximum acceptability) occurred for asymmetric hand positions offering both horizontal and vertical stability. Examples are 3/7, 6/8 and 3/8. Symmetrical hand positions such as 8/8 or 2/2 were used by subjects to lift heavy, compact boxes and were found in the laboratory experiment to minimize forces exerted by the hands.

The previous laboratory experiment was limited to a holding task at waist height. The study presented here extends these results to floor and shoulder heights.

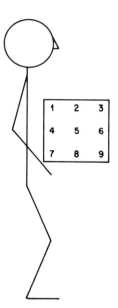

Figure 1. Box handle position nomenclature.

METHODOLOGY

Subjects

Subjects were MMH workers recruited from local industry and comprised fifteen males and fifteen females.

Materials

Two containers, constructed from plywood, measured 400 x 400 x 400 mm and had handle attachment points in all nine positions shown in Figure 1. One container weighed 9 kg and one 13 kg. Handles were Hand Dynamometers so that force exerted on the handle could be measured. Rather than freely pivoting, the handles were arranged to pivot initially but lock when force was applied so that realistic handle use could be achieved. Six handle positions were used based on both laboratory and field data. They were 1/9, 3/7, 3/8, 6/8 (asymmetric) and 2/2 and 8/8 (symmetric). Side view photographs from each side were taken with two 35 mm cameras: the background on each photograph was a black grid on a white wall to allow subsequent measurement of body and container angles. Heart rate and the force exerted by each hand on the hand dynamometer handle were recorded.

Procedure

Each subject was medically examined, briefed and had electrodes attached for heart rate measurement. Each then performed 36 holding trials comprising all combinations of two container weights, three task heights and six handle positions. Each trial lasted 25 seconds with a 95 second break between trials. During each trial, the subject held the container by the handles at either floor height (20-50 mm above the floor) or with the center of the box at waist or shoulder height.

Dependent Variables

Three sets of indicies were used as before.

1. Biomechanical Indicies
   (a) Handle Angle (HA): the angle between the handle and the horizontal

   (b) Box Angle (BA): the angle between the bottom of the box and the horizontal

   (c) Wrist Deviation Angle (WA): the angle between the lower arm (ulna) and the third metacarpal.

   (d) Slippage Angle (SA): the angle between the handle axis and a line at $90°$ to the third metacarpal. (This represents the degree to which the subject fails to grip the handle with the whole width of the hand).

   (e) Elbow Angle (EA): the angle between midlines of lower arm and upper arm.

   (f) Shoulder Angle (SHA): the angle between the midline of the upper arm and the line joining shoulder and hip joints.

   (g) Waist Angle (WA): the angle between the shoulder-hip line and the hip-knee line.

   (h) Knee Angle (KA): the angle between the hip-knee line and the midline of the lower leg.

   (i) Force at the Handle (FH): the force (in kg. wt.) exerted by each hand on the handle.

2. Physiological Index

   (a) Heart Rate (HR): measured during the last fifteen seconds of the trial.

3. Psychophysical Indicies

   (a) Rated Perceived Exertion (RPE): Borg's scale rated after each trial.

   (b) Body Part Discomfort Frequencey (BPDF): Frequency of non zero ratings on Corlett and Bishop's (1976) scale, rated after each trial.

   (c) Body Part Discomfort Severity (BPDS): Mean severity of all non-zero ratings on the above scale.

All of the angular measures were digitized from the side view photographs using a Hewlett-Packard 9823A computer with digitizer. Analyses of variance were performed for each variable using the BMDP package.

RESULTS

Only statistically significant results at $p < 0.05$ will be presented. Task height (H) was significant in all 20 analyses except Slippage Angle (Right Hand) at $p < 0.001$. Handle position (P) was significant for all physiological and psychophysical variables and for 11 of the 14 biomechanical variables. Box weight (W) was significant ($p < 0.01$) for all physiological and psychophysical indicies but only for 4 of the 14 biomechanical indicies. Gender (G) was significant only for two biomechanical indicies. The major interaction present was Task Height x Handle Position (H x P), significant for both hand forces, RPE, heart rate and 12 out of the 14 biomechanical indicies.

Detailed results for selected indicies are presented by group of indicies.

Biomechanical Indicies

The major purpose of this experiment was to find the effect of task height on handle use. Gross body posture changes considerably with task height. At floor level, knee and waist are flexed. At waist level, knees and waist are in an upright posture (extended) while at shoulder height there is even a backward lean. Arm posture changes from moderately extended elbows and moderately flexed shoulders at floor height through increased flexion at each joint at waist height to sharply flexed elbows and less flexed shoulders at shoulder height. The box tilts back progressively as task height increases while the handles move from almost horizontal at floor height to almost vertical at the other two heights. Wrist deviation is non-zero, in contrast to the Coury and Drury (1982) experiment, as handles could not rotate after the subject had achieved a suitable position and applied a force. All wrist deviations were ulnar, and all increased away from waist height. Slippage was small and relatively constant over all task heights.

The upper limbs showed differential effects of left and right. The left hand (always the upper hand on the box) gave more vertical handle angles, higher ulnar deviations of the wrist, a more flexed elbow and a more extended shoulder, all of which would be expected with higher handle positions on the box.

Handle position effects at each task height were tested with Neuman-Keuls tests on the H x P interaction. In general symmetrical positions (2/2 and 8/8) gave very similar angles for each hand. The reversed position (1/9) was often significantly different from other positions, as were two symmetrical positions (2/2 and 8/8).

Forces on the handles are presented as H x P interactions in Figure 2. The forces in the left and right hands are least in the floor position, where the two forces sum to almost exactly the average container weight (11 kg), showing that forces are almost exactly vertical and merely supporting the weight. In the waist position, and to a lesser extent in the shoulder position, forces are higher because much of the force is now directed towards the body with the container supported both by body/container friction and the upward component of the forces at the handles. The force disparity between the hands

was large for the most asymmetrical hand position (1/9) with the lower hand (right) supporting most of the weight.

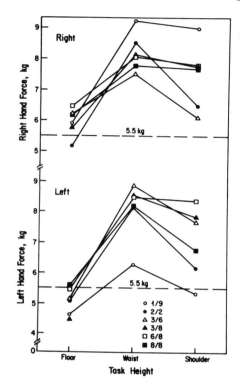

Figure 2. Hand forces at each task height hand handle position.

Physiological and Psychophysical Indicies

Heart rates for comparable handle positions at waist height were slightly higher (96-101 beats/min.) than in the previous study (94-99 beats/min.) despite a reduction in container weight. This may have been due to the changed subject population. Handle positions 3/8 and 6/8 can be seen in Figure 3 to give clearly lower heart rates than the other positions for floor and waist levels while the symmetrical positions (2/2 and 8/8) fared rather poorly. Rated perceived exertion (Figure 4), this time closely comparable in value to the previous study, showed the superiority of 3/8 and 6/8 with 2/2 and 8/8 only poor at the shoulder level. Both heart rate and rated perceived exertion show the waist level as least strenuous but the highest heart rates are found at shoulder level and the highest RPE values at floor level.

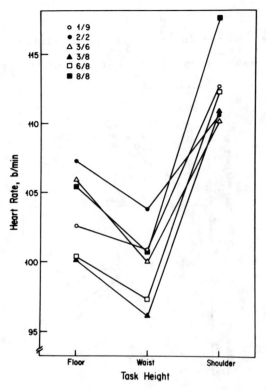

Figure 3. Effects of task height and handle position on heart rate.

Body part discomfort frequency and severity were lowest in the waist position. Handle position effects for these variables are shown in Figure 5 which again emphasize the utility of positions 3/8 and 6/8. It is interesting to note that the body parts rated worst on frequency were in the upper limbs with the order being upper arm (worst), lower arm, shoulder and hand. In severity the four worst were all in the torso with the order being lower back (worst), mid back, upper back and shoulder. A further finding was that buttocks, thighs and legs were stressed only when holding at floor level, presumably due to the squatting posture.

DISCUSSION

As in the previous research, the effect of handle position was large. In the measures of force and stress, the difference between the best and the worst handle positions was closely comparable to the difference between loads of 9 kg and 13 kg. All variables show that handle position can alter the

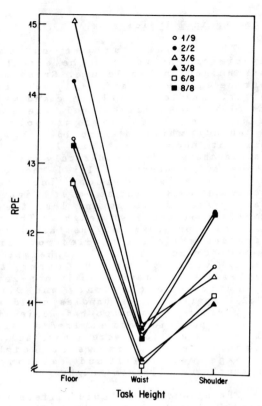

Figure 4. Effects of task height and handle position on R.P.E.

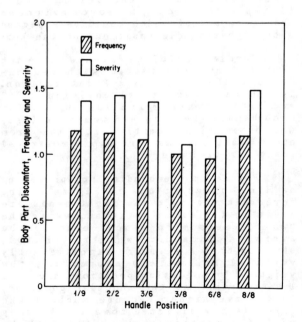

Figure 5. Body part discomfort (Corlett and Bishop scale) as a function of handle position.

cost to the worker as much as a 25%
change in box weight, with half of the
indicies, including heart rate, showing
the equivalent of a 50% change. It
should be remembered that the handle
positions were chosen to be reasonable
ones and not chosen to include obviously
poor positions.

Task height had, as expected, a
very large effect on all indicies. In
particular, the upper limbs and the
container interacted in different ways
at the three heights. At the floor
position arms were extended, wrists
highly deviated in the ulnar direction
and handles almost horizontal on an
almost horizontal box. All of the box
weight was hung from the two hands. At
the higher task heights, the box was
clasped to the body with almost vertical
handles on a tilted box. Wrist
deviation was again present but smaller
in magnitude than at the floor level.
Combined forces were almost 50% higher
than the container weight.

In choice of 'good' handle
positions, it is possible to extend the
results of the earlier experiment at
waist height only. Critical evaluation
variables are heart rate, RPE, BPD and
wrist deviation. The symmetrical
positions (2/2 and 8/8) and the
unbalanced position (3/6) fared poorly
on one or more of the indicies.
Positions 3/8 and 6/8 emerged clearly
from the 'cost to the subject' variables
as being good positions. The reversed
position (1/9) gave no clear evidence of
being good or poor, only of making the
expected changes in arm angles and
forces. Positions 3/8 and 6/8 confirm
the previous findings of horizontal and
vertical stability as being necessary
attributes of a 'good' handle position.
They are also the handle positions
recommended in the Coury and Drury
(1982) paper on the basis of waist
height holding only. As these are
better than the reversed position (1/9),
a tentative conclusion is that in
addition to horizontal and vertical
stability, it is important to have the
upper arm further from the body than the
lower arm.

Finally, it is useful to note that
box weight again had very little effect
on the handle and body angles. Combined
with the previous result of almost no
effect of box size, it means that
recommendations for handle positions and
angles can be given with some confidence
to cover a wide range of manual
materials handling containers.

ACKNOWLEDGEMENTS

This work was supported by a
contract from the National Institute of
Occupational Safety and Health, Division
of Safety Branch in Morgantown, W. Va.
The project officer was T. Pizzatella.

REFERENCES

Borg, G. Physical Performance and
Perceived Exertion (Lund:
GLEERUPS), 1962.

Corlett E.N., and Bishop R.P. A
technique for assessing postural
discomfort. Ergonomics,1976, 175-
182.

Coury B.G., and Drury, C.G. Optimum
handle positions in a box-holding
task. Ergonomics, 1982, 25(7),
645-662.

Drury, C.G., Law, C-H. and Pawenski,
C.S. A survey of industrial box
handling. Human Factors,1982,
24(5), 553-565.

Garg, A., and Saxena, U. Container
characteristics and maximum
acceptable weight of lift. Human
Factors 1980, 22(4), 487-496.

Snook, S.H. The design of manual
handling tasks, Ergonomics,
1978, 21, 963-985.

AN EVALUATION OF TWO METHODS FOR THE
INJURY RISK ASSESSMENT OF LIFTING JOBS[1]

D. H. Liles   P. Mahajan   M. M. Ayoub
University of Texas   California State University   Texas Tech
at Arlington   Fresno   University

ABSTRACT

This paper compares two methods of evaluating the risk potential of manual materials handling jobs. This comparison is based upon two large field studies. A total of 101 different lifting jobs were analyzed using each of the two methods. Injury profiles, representing a total of over one million hours of worker exposure, were also compiled. The results indicate that the two methods tend to agree in their risk assessment of most jobs. The results also indicate that the assessments tend to be correct when compared to observed injury statistics. There are, however, certain differences between the two methods. These and other points are also discussed.

INTRODUCTION

Manual Materials Handling injuries, particularly back injuries associated with the lifting of loads, account for a significant part of the total occupational injury problem. Direct and indirect economic losses are enormous.

In recent years, much research has been undertaken to find the causes and potential solutions to the problem. Two principles have emerged. First, it is apparent that some workers are physically more capable than others and are, therefore, less likely to be injured. Secondly, research has indicated that injury incidence and severity increase as the relative job demands increase beyond certain limits. (Chaffin and Park, 1973; Ayoub et al., 1978 and 1983; Keyserling, et al.,1980). It seems critical, therefore, that any program designed to reduce lifting injury include some system for the evaluation of job demands as compared to the worker or worker population working in the job.

This paper reports the findings of recent research which compared the effectiveness of two methods that have been proposed as ways to identify and redesign high risk lifting jobs. The first of the two methods is based upon the Job Severity Index (JSI) which was developed by Ayoub et al. (1978). In basic terms, the JSI is the ratio of job demands to predicted worker capacity. The second method was reported by the National Institute for Occupational Safety and Health (1981). This method uses an Action Limit (AL), which is defined as the weight that can be safely handled by 75% of the female population and 99% of the male population.

The results of the research are based upon data collected during two large field studies. The combined field studies investigated the injury performance of 453 industrial subjects working in one of 101 different lifting jobs. In excess of one million exposure hours of injury data were collected. Job requirement data such as weight of lift, frequency of lift, task geometry, etc. were determined as required for complete analysis of each job using both methods. In addition, human data such as strength and body size were collected as required for the JSI method. During the study period, each injury incident was analyzed to determine the type of injury, the cause of injury, the associated lost time, and the total direct cost of injury.

THE TWO METHODS

Complete descriptions of each method can be found in the original documents (Ayoub et al., 1978 and 1983; and NIOSH, 1981). A brief summary of each is, however, provided below.

The Job Severity Index, or JSI, is a ratio of job demands to predicted worker capacity. The use of the JSI requires that the job be described, in detail, in terms of the required weights of lift, frequencies of lift, exposure times, lifting weights, and load centers of gravity. In addition, lifting capacities either for individual workers or for a worker population must be predicted. These capacity predictions are based upon human data including age, sex, body weight, stature, abdominal depth, arm strength, back strength, and dynamic endurance.

[1]This research was supported in part by NIOSH Grant No. 5RO1OH00798-04.

The JSI can be expressed algebraically as shown below.

$$JSI = \sum_{i=1}^{n} \left[\frac{Hours_i \times Days_i}{Hours_t \times Days_t} \sum_{j=1}^{m_i} \frac{F_j}{F_i} \times \frac{WT_j}{CAP_j} \right]$$

where:

n = number of task groups.

$Hours_i$ = exposure hours/day for task group i.

$Days_i$ = exposure days/week for task group i.

$Hours_t$ = total hours/day for job.

$Days_t$ = total days/week for job.

m_i = number of tasks in task group i.

WT_j = maximum required weight of lift for task j.

CAP_j = the adjusted capacity of the person (or worker population) working at task j.

F_j = lifting frequency for task j.

F_i = total lifting frequency for task group i.

Lifting capacity for an individual is the predicted maximum acceptable weight of lift. Maximum acceptable weight of lift is predicted using one of six regression equations developed by Ayoub et al. (1978). Population capacities are estimated using the predicted capacities of all 453 industrial subjects tested during the two studies. In either case, the capacities are adjusted for lifting frequency and load center of gravity.

It has been found that a value of JSI in excess of 1.5, either for an individual or for a large percentage of the population, is indicative of a stressful or high risk job (Ayoub et al., 1983).

As discussed in the Work Practices Guide (NIOSH, 1981), the Action Limit, or AL, is defined as that weight of lift which can be safely handled by 75% of the female population and 99% of the male population. A second limit called the Maximum Permissible Limit (MPL) is defined as the weight of lift which is safe for only 25% of the male population and virtually none of the female population. If the required weight of lift is greater than the AL but less than or equal to the MPL, it is recommended that administrative or engineering action be taken. If the required weight of lift is greater than MPL, it is recommended that engineering changes be required. The job data required for the AL method are similar to those required by the JSI method. The AL method, however, requires no human strength or other data.

The AL can be expressed algebraically as shown below.

$$AL(Kg) = 40(15/H)(1-0.004|V-75|)(0.7 + 7.5/D)(1-F/F_{max})$$

where:

H = horizontal location of hands at origin of lift. 15cm. $\leq$ H $\leq$ 80 cm. measured from the ankles.

V = vertical location of hands at origin of lift. 0cm. $\leq$ V $\leq$ 175 cm. measured from the floor.

D = vertical travel distance of lift from start to end of lift. 25cm. $\leq$ D $\leq$ (200-V)cm.

F = average frequency in lifts/ minute. $0.2 \leq F \leq F_{max}$, F = 0 if < 0.2.

F_{max} = maximum frequency that can be sustained and is determined as follows:

For periods $\leq$ 1 hour,
F_{max} = 18 for V > 75cm,
F_{max} = 15, otherwise.

For periods of >1 hour,
F_{max} = 15 for V > 75cm.
F_{max} = 12 otherwise.

ANALYSIS

The objective of the analysis was to test two hypotheses. The first hypothesis had to do with how well the two methods agree in their assessment of risk potential. The second was concerned with the correctness of the risk assessments when compared to actual injury statistics.

The first step in the analysis was to evaluate each of the jobs using both methods. To facilitate comparison, a new ratio called the Action Limit Ratio (ALR) was defined. The ALR is simply the ratio of the equivalent or averaged weight of lift to the calculated Action Limit. The use of an averaged weight of lift is recommended by the Guide (NIOSH, 1981) to properly account for varying weights of lift. An ALR of less than or equal to one corresponds to a required weight of lift less than or equal to the AL. An ALR of more than three corresponds to a required weight of lift greater than the MPL. This ratio was devised so that a single number could be used to rate each job.

Using the Job Severity Index, an estimate was made of the percent of the population that would be overstressed if working on a particular job. This estimate was made using the 385 male subjects measured during the two studies. The estimated percentage overstressed is defined as that percentage of the sample population having a JSI of 1.5 or greater if the population is assumed to be working in a given job. This facilitated the grouping of the jobs into three JSI categories consistent with the three ALR categories. The three ALR categories are: ALR ≤ 1 (≤ 1% overstressed), 1 < ALR ≤ 3 (between 1 and 75% overstressed), and ALR > 3 (> 75% overstressed). The three JSI categories are: ≤ 5% overstressed, between 5 and 75% overstressed, and > 75% overstressed. It should be noted that the boundaries between the low and moderate stress categories are not identical. It was concluded, during the JSI analysis, that basing the 99 percentile capacity estimates upon a sample as small as 385 males was not justified. The 95 percentile values (≤ 5% overstressed) were used, instead, to represent the accommodation of "almost all" of the male population.

Given the preliminary calculations discussed above, the remainder of the analysis was relatively straight forward. Various correlation analyses were performed to determine the similarity of the risk assessments produced by the two methods. Secondly, the correctness of the assessments was evaluated by calculating and comparing the observed injury statistics for each risk classification.

RESULTS

The classifications of the 101 jobs by the two methods are shown in Table 1. Had the two methods been in complete agreement, the off-diagonal cells would have been empty. It can be seen that 19 jobs were given higher risk ratings by the AL method than by the JSI method. Ten jobs were given lower risk ratings by the AL method than by the JSI method. It may also be observed that the disagreement seems to be slightly greater in the high risk category. Possible causes of this disagreement are discussed later.

Basing a statistical correlation upon the estimated percent overstressed was not possible. This is due to the fact that the AL method does not provide estimates of the percent overstressed for all risk levels. Such estimates are given only for the AL and MPL levels (for ALR values other than one or three, the estimated percentages overstressed are not available.) To overcome this problem, the jobs were ranked by each method from the least severe job to the most severe. The two sets of rankings were then analyzed and found to have a correlation of 0.68.

The second statistical procedure performed was a Chi-squared contingency test. This was done using Table 1 as a contingency table. The hypothesis for this test is that the two methods are statistically independent and do not agree in their assessments. This hypothesis was rejected with a large statistical significance ($(1-\alpha) > 0.9995$). From this, it can be inferred that the two methods generally agree. It should be noted that the validity of this test may be questioned due to the relatively small number of jobs (13 and 10) given high risk evaluations by the two methods. In this particular case, however, the error introduced is not large enough to change the inference stated above.

Tables 2 and 3 give the injury and cost statistics for both methods and each of the three risk categories. These statistics describe only injuries to the back caused by lifting. Data on injuries of other types and causes were collected but are not presented here. Table 3 includes only those back injury costs observed during the second study.

TABLE 1

NUMBER OF JOBS OBSERVED IN EACH
RISK CATEGORY* FOR BOTH METHODS

	JSI			
ALR	L	M	H	TOTAL
L	38	7	0	45
M	13	27	3	43
H	1	5	7	13
TOTAL	52	39	10	101

TABLE 2

BACK INJURY STATISTICS FOR EACH
JSI AND ALR CATEGORY (COMBINED STUDIES)

		Exposure Hours	# Inj/ 100 FTE	# LTI/ 100 FTE	Days Lost per LTI
JSI	L	305 333	5.24	2.62	39.75[a]
	M	510 485	10.97	7.05	15.11
	H	242 063	15.70	11.57	51.07
ALR	L	308 387	3.89	1.95	52.67[b]
	M	406 071	13.79	9.36	41.68[c]
	H	343 423	12.23	8.15	14.00

a. Excluding one serious injury this
   statistic is 1.75.
b. Excluding one serious injury this
   statistic is 2.0.
c. Excluding one serious injury this
   statistic is 13.53.

TABLE 3

INJURY COST STATISTICS FOR EACH
JSI AND ALR CATEGORY (SECOND STUDY)

		Exposure Hours	Medical $ per 100 FTE	Total $ per 100 FTE
JSI	L	192 781	5 014	9 208
	M	238 123	16 956	35 092
	H	58 727	15 686	36 338
ALR	L	212 891	4 522	8 320
	M	237 667	19 868	41 041
	H	39 063	6 159	18 934

*L = Low Risk Category, M = Moderate
Risk Category, H = High Risk Category

100 FTE = 100 Full Time Employees
        = 200,000 exposure hours
LTI = Lost Time Injury (Injuries)

Cost information is not available from
the first study.

Table 2 shows the number of back injuries per 100 FTE (Full Time Employees),
the number of lost time back injuries
per 100 FTE, and the number of days lost
per lost time injury. It can be easily
seen, based upon injuries per 100 FTE
and lost time injuries per 100 FTE, that
both methods seem to be equally effective at isolating low risk jobs. The
JSI method, however, seems more effective at differentiating between high and
moderate risk jobs. This may be due to
the previously mentioned fact that the
two methods disagree on some of the
higher risk jobs or it may be due to
the relatively small sample size. The
days lost statistic appears to be somewhat erratic. This was caused by two
high loss injuries. If these two injuries are excluded, the days lost
statistic behaves more like the other
statistics.

Table 3 shows the medical expense
per 100 FTE and the total expense per
100 FTE, including worker's compensation, for injuries falling in each
category. Again, it is indicated that
the two methods seem equally able to
correctly isolate low risk jobs. Using
the cost statistics, however, neither
method seems able to differentiate
between moderate and high risk jobs.

CONCLUSIONS

As job analysis and assessment tools,
the two methods seem to have equivalent
potential, based upon the data analyzed.
Both provide relatively effective measures of the injury risk associated with
lifting jobs and can be used to place a
job into one of at least two risk categories. No conclusive statement can be
made regarding the effectiveness of
either method as a job design or redesign tool because no data of the type
required were collected. Such a statement would require the longitudinal
study of the effects of actual job redesign upon injury rates. It can reasonably be assumed, however, that back
injury incidence and severity rates
could be significantly reduced through
the use of one or both of the two methods as job design tools.

The JSI method can also be used to
select or screen workers. This involves the evaluation of a single worker
as compared to specific job demands.
Worker selection using the JSI method
is discussed in Ayoub et al. (1983).

The AL method has no direct provisions for worker selection.

As stated earlier, the two methods produce different evaluations of some jobs. There are several possible reasons for this. The first has to do with the lowering of loads. The JSI method ignores lowering, whereas, the AL method assumes that there is no difference between lifting and lowering. For tasks that require significant lowering, the JSI method consistantly underestimates the risk. To the extent that lowering is less stressful than lifting, the AL method consistantly overestimates the risk. Neither method considers other materials handling tasks such as pushing, pulling, and carrying. A second difference has to do with the definition of the required weight of lift. On this point, the JSI method is the more conservative of the two. The third difference is related to how each method defines task geometry factors such as a load center of gravity and lifting heights. The exact effects of these differences are difficult to determine. Both methods, however, account for the increased stress associated with remote centers of gravity and certain lifting heights and distances. Other differences include the procedures used to measure the effect of lifting frequency and task exposure time. If the two methods are used concurrently, the user must resolve assessment differences on a case by case basis.

REFERENCES

Ayoub, M. M., Bethea, N. J., Deivanayagam, S., Asfour, S. S., Bakken, G. M., Liles, D. H., Mital A., and Sherif, M. Determination and Modeling of Lifting Capacity. Final Report, NIOSH Grant No. 5RO1OH00945-02, September 1978.

Ayoub, M. M., Asfour, S. S., Bakken, G., Bethea, N., Liles, D. H., Selan, J., and others. Effects of Task Variables on Lifting Capacity. Final Report, NIOSH Grant No. 5RO1OH00798-04, January 1983.

Chaffin, D. B. and Park, J. S. A Longitudinal Study of Low-Back Pain as Associated with Occupational Weight Lifting Factors. *American Industrial Hygiene Journal*, December 1973, 573-725.

Herrin, G. D. A New Method for Evaluating Manual Lifting Tasks. Presented at Annual Conference of American Institute of Industrial Engineers in Detroit, Michigan, May 1981.

Keyserling, W. M., Herrin, G. D. and Chaffin, D. B. Isometric Strength Testing as a Means of Controlling Medical Incidents of Strenous Jobs. *Journal of Occupational Medicine*, 1980, 22(5), 332-337.

National Institute of Occupational Safety and Health. *Work Practices Guide for Manual Lifting*. Department of Health and Human Services (NIOSH) Publication No. 81-122, March 1981.

ISOMETRIC STRENGTH TEST CONSISTENCY OVER TIME

Terrence J. Stobbe
Ralph W. Plummer
Donald P. Shreves

West Virginia University
Morgantown, WV

ABSTRACT

Workmen's Compensation costs have become a major financial burden on industry. A significant part of these costs are the result of musculoskeletal injuries. One method of controlling these injuries is matching employees to jobs based on strength. Isometric strength testing has been shown to be an effective method of matching employee strength capability to job strength requirements. The use of screening test raises the question of test-retest consistency. This study was designed to determine the consistency of isometric strength test results over time. Thirteen student volunteer subjects (10m, 3F) participated in a testing protocol consisting of five strength tests performed a minimum of two times each at two week intervals. Four test sessions were held. An analysis of variance was used to identify week to week strength differences in the test population. Each of the five tests was analyzed separately and no significant week to week strength differences were found. This result further supports the validity of using isometric strength testing as a selection tool.

INTRODUCTION

One of the major cost containment problems currently facing industry is the prevention and control of worker's compensation costs. These costs vary widely, and depend on the industry and individual employer experience with claims. One example of the impact of these costs comes from West Virginia where for some companies, the compensation premiums exceed the salary costs (Greenwood, 1982). A significant reduction in these costs has the same dollar impact on a company as a reduction in personnel but traditionally, much more effort has been focused on manpower control than on injury and illness control. It is perhaps, time to adjust.

A large portion of these injuries and costs are the result of musculoskeletal injuries. Specific percentages again vary across companies and industries, with musculoskeletal injuries accounting for 20 to 60 percent of the injuries and more than 30 percent of the costs (NIOSH 1981). Nationally, these injuries cost billions of dollars per year.

Control of these costs and injuries has been approached from four perspectives. They are employee lifting training, medical diagnosis and rehabilitation improvements, use of basic human factors principles in job design, and employee selection based on the matching of employee strength capability and job strength requirements, low back x-rays, or some other basis. This paper focuses on the last approach, and is concerned with potential validity problems associated with the use of strength tests.

Strength assessment based on either the psychophysical approach or the isometric approach has been shown by a number of authors to be a reliable method of assessing employee strength capability. In essence, the method consists of simulating the highly stressful tasks associated with a job with some type of strength test. The subject's test score is then compared to the job's strength requirements which have been assessed separately. Those persons who demonstrate the required strength are at less risk of injury, and thus are preferable to hire or place on the job.

The use of any testing methodology such as this immediately raises the question of test-related consistency since it would be quite unfair to mislabel or disqualify a person on the basis of a test if the test is inherently unreliable--i.e., gives widely divergent results over a series of tests. This question has been address by Stobbe (1982) and Chaffin (1978). They found that the COV for isometric testing to be in the .05 to .10 range for laboratory studies and .05 to .15 for in-plant field studies.

These results were obtained using a protocol in which each subject was tested a number of times on a given test in a single test session. While this result is clearly important, it begs the question of day to day measured strength consistency. Briefly stated, the problem is that people undergo employment screening on a given day, and if they happen to be "weak" on that day, they may be misclassified. Clearly this problem exists in all types of testing---people may have a bad day---but it would be a major problem if day to day variation is generally large compared to within day variation. This study was designed to evaluate the day to day variation in isometric strength.

METHODS

Subjects

Thirteen unpaid student volunteer subjects were recruited from the West Virginia University student body. Ten males and three females ranging in age from 20 to 27 participated. All potential subjects were medically screened prior to participation and those with a history of medically relevant problems (including back injuries and hypertension) were eliminated. Subjects were also screened with respect to their involvement with activities likely to produce strength changes during the study. The only volunteers excluded were those actively engaged in weight-training programs. They were eliminated because there was no way to distinguish between training related strength changes, and random day to day strength changes.

Equipment

The equipment used in this study was standard isometric strength testing equipment available from Prototype Design and Fabrication Co. of Ann Arobr, Michigan. It consisted of a set of handles with the grips spread 18 inches apart mounted on a horizontal piece of box section. The horizontal section produced the vertical position adjustment by sliding up and down on a vertical piece of box section. The vertical section was bolted to a platform constructed from welded I-beams. A sheet of plywood was laid down on top of the I-beams to provide a platform for subjects to stand on.

The data was collected using a Force Monitor (also built by Prototype Design and Fabrication) designed to meet the criteria established in the "Ergonomics Guide to the Assessment of Human Static Strength" (AIHA, 1975). The force monitor received its input signal from an Interface model SM-500 force transducer.

The force monitor was calibrated and its sensitivity matched to that of transducer. The recording system accuracy was checked periodically with dead weight calibration of 20 to 50 pounds.

The force monitor provided both the peak strength value attained during a given test, and a four second average strength value obtained during the central portion of each test. Only the average value was used in the subsequent analysis

Testing Protocol

In this study, subjects performed a series of trials on each of five tests on four different days. The test days were two weeks apart to assure that the test subjects experienced no strength training as a result of test participation. The two week interval was selected based on the work of Muller and Hettinger which showed that the muscle growth and strength increase following a strength training session had returned to the pre-training level after two weeks (Muller, 1970), and that a fourteen day interval between training sessions resulted in no relative weekly increase in strength (Muller and Hettinger, 1956).

Five strength tests were used in the study. They were a torso lift, an arm lift, an overhead lift, a push and a pull. A description of each test is provided in Table 1. The test battery was chosen to represent a range of test conditions. In the arm lift, the subject was in a highly structured posture with the handle position determined by the subject's anthropometry. In the torso and overhead lifts, the handle and foot positions were fixed and the subjects' posture was determined by their anthropometry. In the push and pull tests, the handle location was fixed, but the subjects were free to assume the foot position and posture they felt would give them the best performance. (This posture and foot position were recorded and used during each test session.) This combination of test types covers the range of testing conditions likely to be encountered in isometric selection testing.

During each testing session, each subject performed a number of trials on each test, with the number determined using the ten percent criterion proposed by Stobbe (1982). In essence this criterion states that on a given test, the series of trials should be continued until the two largest test values are within ten percent of each other. Local muscle fatigue was avoided by providing a two minute rest period between trials (Schanne, 1972).

Subject motivation in a strength testing situation is always somewhat controlled. In this study, the guidelines of the "Ergonomics Guide" were followed. Standardized instructions with no quantitative performance feedback were provided. Positive qualitative feedback was always given, but was limited to comments such as "that was good", or "you are doing fine".

RESULTS

The results of this study show that while there was week to week variability in subject strength, the variability was not significantly greater than the within day variability reported previously by Stobbe and Chaffin. In addition, it was found that the less structured tests had the greatest variability. The data is summarized in Tables 2 and 3 and Figure 1. In the analysis, only the maximum value achieved on each test during each weeks test session was used.

It is evident from Table 2, which presents the group means for each trial of each test, that there is very little week to week variability on any of the tests. The percent

Table 1

Description of Strength Tests Used

Test Name	Test Description
Torso Lift	Subjects stand facing handles, feet spread a comfortable distance apart and equidistant from the handles. Horizontal distance from center of ankles to center of handles is 15 inches. Vertical distance from platform to center of handles is 15 inches. Subjects are free to choose between bent and straight knee postures. Lift is vertical.
Overhead Lift	Subjects stand facing handles, feet spread a comfortable distance apart. Front foot positioned with ankle 20 inches from handle centerline. Other foot can be equidistant or farther away. Vertical distance from platform to center of handles is 70 inches. Subjects free to choose posture subject to those constraints. Lift is vertical.
Arm Lift	Subjects stand facing handles, feet spread a comfortable distance apart and equidistant from the handles. Subjects upper arm is parallel to and against torso, with forearm extending forward making a 90 degree angle at the elbow. Subjects feet are positioned so hands grasp handles comfortably while standing erect. Vertical distance to floor adjusted so handles are easily grasped in this position. Lift is vertical..
Push	Subjects stand facing handles with the ankle of the forward foot 14 inches from the handle centerline. The other foot is positioned equidistant or farther away. Vertical distance from platform to the centerline of the handles is 49 inches. Subjects are free to choose this pushing posture. Push is horizontal.
Pull	Subjects stand facing handles with the ankle of the forward foot 13 inches from the handle centerline. The other foot is positioned equidistant or farther away. Vertical distance from the platform to the handle centerline is 62 inches. Subjects are free to choose their pulling posture. Pull is horizontal.

difference between the highest and lowest mean trial values is consistently less than 10 percent. An analysis of variance was performed on this data comparing the weekly (trial) means on each test. No significant inter-trial difference was found on any test (α = .05).

Further inspection of Table 2 reveals another interesting fact. For all of the tests except the torso lift, the first trial has the greatest mean value. On the torso test the four trials are less than one percent different so the differences are not meaningful. This result is contrary to what was expected namely that some learning or training would occur causing some increases above the initial week's result. This result further supports the use of the 10 percent test-retest criterion to determine the number of trials to use for each test each week.

Table 2

Group Means by Trial by Test
(data in kg)

Test: Torso Lift		Test: Overhead Lift	
Week	Group Mean	Week	Group Mean
1	41.1	1	57.3
2	41.5	2	54.5
3	41.3	3	56.0
4	41.3	4	55.9

Test: Arm Lift		Test: Pull	
Week	Group Mean	Week	Group Mean
1	41.0	1	22.1
2	38.2	2	20.3
3	38.9	3	20.0
4	38.1	4	20.6

Test: Push	
Week	Group Mean
1	43.7
2	41.5
3	42.6
4	41.9

Both of these results are more easily seen in Figure 1 which presents them graphically. This clearly shows the relatively flat curves as well as the relatively large initial week values.

Previous studies have reported strength consistency results in terms of the COV. Table 3 presents a summary of the COV data collected in this study. The overall COV of .102 across both subjects and tests is quite comparable to the previously reported range of .05 to .10. The previous studies did not report the within test subject to subject COV range, but the low overall average COV in this study suggests this data is consistent with that previously collected.

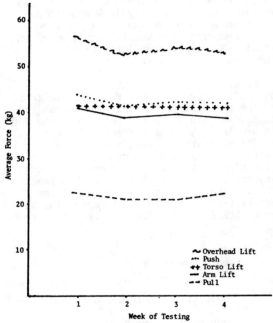

Figure 1. Average Force Versus Week of Training

Table 3

Summary of Coefficient of Variation Data

Test	Average COV	Range of COV Across Subjects
Torso Lift	.086	.025 - .217
Overhead Lift	.097	.031 - .172
Arm Lift	.105	.027 - .215
Pull	.097	.044 - .196
Push	.126	.040 - .184
All Test Combined	.102	.025 - .217

DISCUSSION

If isometric strength testing is to be used effectively as a screening and/or placement tool, it is critical that it produce repeatable results. The repeatability should hold both for tests completed in a single day, and for tests completed over a period of time---that is a strength test performed at noon today should be a good measure of strength at 5:00 p.m. today, and 10:00 a.m. two weeks from today.

Previous studies have addressed the question of within day isometric strength measurement consistency and found it to be a highly reliable

measure. Our purpose in conducting this study was to evaluate the second question. Our results showed that week to week isometric strength measurement consistency was about equivalent to within day consistency. This was true of both group and individual data.

This result further strengthens the scientific basis of underlying the use of isometric strength testing as a method of evaluating people's strength for employment and selection purposes. When used properly, the method clearly provides repeatable results both within and between days, and as such meets the reliability criterion for human capability assessment.

REFERENCES

AIHA, Ergonomics Guide for the Assessment of Human Static Strength, monograph edited by D. B. Chaffin and published by the American Industrial Hygiene Association, Akron, Ohio, 1975.

Chaffin, D. B., and Stobbe, T. J., "New Developments in Strength Testing," presented at the American Industrial Hygiene Association Annual Conference, Los Angeles, CA, May, 1978.

Greenwood, J., Personal Communication, September, 1982.

Muller, E. A. and Hettinger, T., "Der Zerlaus Der Zunahanlder Muskelkraft mach einem Einmaligen Maximalen Trainings Reiz" Int Z Angew Physiol, fbed 184-191, 1956.

Muller, E. A., "Influence of Training and of Inactivity on Muscle Strength", Arch of Phys Med & Rehab., Aug., 1970.

NIOSH, Work Practices Guide for Manual Lifting, Technical Report, March, 1981.

Schanne, F. J., Three Dimensional Hand Force Capability Models for a Seated Person, unpublished PhD Dissertation, The University of Michigan, Ann Arbor, 1972.

Stobbe, T. J., The Development of a Practical Strength Testing Program for Industry, University Microfilm, Ann Arbor, Michigan, July, 1982.

HANDLE POSITIONS AND ANGLES IN A DYNAMIC LIFTING TASK

Colin G. Drury and J.M. Deeb

Department of Industrial Engineering
State University of New York at Buffalo

ABSTRACT

Thirty industrial subjects took part in a manual lifting task, using different handle positions on a container and different angles between handle and container. Lifts were from floor to waist, waist to shoulder and floor to shoulder. Upper extremity body angles were measured, with heart rate and rated perceived exertion. As in previous static holding experiments, it was found that handle positions with both horizontal and vertical stability gave good results. As a result of this work, handle positions are recommended in the middle of the front edge of a box (at $60°$) and in the middle of the lower edge (at $50°$). Such an arrangement will minimize wrist deviation and slippage angle between handle and hand.

INTRODUCTION

Over the last several years NIOSH has supported a research program to understand the role of handles as an aid to manual materials handling injury reduction. This work has covered an industrial survey, laboratory experiments on one-handed holding tasks and, most recently, experiments on a two-handed box-holding task. All are summarized in Drury and Pizatella, 1983. The work reported here extends the two handed holding task to one requiring dynamic movement of the box between different heights.

From the previous experiments, it had been found that a number of factors were of minor importance. Box size did not affect body angles and neither did subject gender, height or weight. There were, of course, effects of box weight and subject gender on heart rate and psychophysical measures of workload. Factors such as handle position and handle angle did have a large effect on body angles, physiological and psychological measures. Figure 1 shows the convention used throughout these studies to specify handle position. The previous work had shown that hand positions contributing to both horizontal and vertical stability (eg 3/7, 3/8 or 6/8) were both frequently chosen by industrial users and gave low heart rates and psychophysical ratings. However, symetrical positions, such as (2/2 or 3/8) gave the lowest hand forces, as the box was held away from the body, hanging freely from the hands. In contrast, the assymetrical positions made considerable use of the exterior surface of the body to carry part of the load.

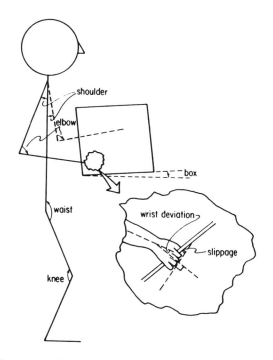

Figure 1. Definition of Handle Positions and Biomechanical Measures.

Angles of the handle to the box were found by allowing the handles to pivot and finding that angle which gave a neutral angle to the wrist. Generally, handle positions at the front of the box gave optimum angles near to vertical, while positions along the bottom gave more horizontal angles. The height at which the box was held above the floor had a large effect on handle angle, so that no single angle was optimum at all heights.

From the previous work, it was decided to examine four handle positions (3/7, 3/8, 6/8 and 2/2) under dynamic conditions. It was also important to determine whether the compromise handle angle of 70° to the horizontal would be confirmed with a fixed handle under dynamic conditions. For comparison a handle angle midway between 70° and the usual horizontal handle was used. Again male and female industrial subjects were employed to increase the validity of the results.

METHODOLOGY

Thirty local industrial manual materials handling workers participated in the study. The fifteen males and fifteen females had mean ages of 24.6 and 24.9 years respectively, with standard deviation of 4.9 and 8.4 years.

Plywood containers, with their centers of gravity in the center of the container and weighing 11kg were used for the lifting task. Each measured 400 mm on each dimension and had handhold cutouts in positions appropriate to the four hand positions (3/7, 3/8, 6/8, 2/2). Each cutout was angled at 70° to the horizontal on one box and 35° on the other.

After a medical examination and briefing, each subject performed eight lifts for each combination of lift height (floor to waist, waist to shoulder, floor to shoulder), handle position and handle angle. These 24 combinations were given in a random order with one lift each 15 seconds and a two-minute rest period between combinations. Subjects both lifted and lowered the box but only lifting results are reported here for brevity.

Dependent Variables

Side view movie films were taken of each lift, using a large, single-surfaced mirror to observe both sides of the body. Angles of body segments were digitized from the movie film at five heights between floor (1) waist (3) and shoulder (5). The following angles were measured at each stage of each lift for the 5th and 7th of the eights lifts:

(a) Box Angle (BA): the angle between the bottom of the box and the horizontal.

(b) Wrist Deviation Angle (WA): the angle between the lower arm (ulna) and the third metacarpal.

(c) Slippage Angle (SA): the angle between the handle axis and a line at 90° to the third metacarpal. (This represents the degree to which the subject fails to grip with the whole width of the hand).

(d) Elbow Angle (EA): the angle between midlines of lower arm and upper arm.

Mean heart rate was measured over the duration of the 5th and 7th lifts. After each condition, subjects gave their Rated Perceived Exertion (Borg 1962) and their Body Part Discomfort on the Corlett and Bishop (1976) five point scale.

RESULTS

Biomechanical Measures

For brevity, only the biomechanical results from the floor-shoulder lift are presented here. In a similar vein, only the major measures of job cost, heart rate and RPE are given. Table 1 shows the summary of the ANOVA's of the various biomechanical measures. No gender main effects and no trials main effects were found, showing that results are generalizable and repeatable. Stages of lift, handle angle and handle position all have highly significant effects on most variables. In all cases there was a highly significant stage of lift x handle position interaction, but few other interactions proved significant.

Handle angle effects are shown in Table 2. Only the slippage angles showed large changes from a 35° to a 70° handle angle. The difference was almost 35° (i.e. 70°-35°) for each hand, indicating that, as in previous work, slippage between hand and handle was a major mode by which subjects accommodated to changing demands of the task.

The Stage x Handle Position interaction, while highly significant, showed more consistency of stage effects at each handle position then it showed dissimilarities. For box angle, all handle positions were grouped within 1° to 2° showing on average an initially forward-tilted box at floor to waist level and a backward-tilted box at shoulder level. Both elbow angles started at 120°-150° at floor level, dropped to 80-120° at waist level and rose again to 85-140° at shoulder level, except for the right hand in the 3/7 position which continued to decrease up

Source	df	Box Angle	Elbow Angle R	Elbow Angle L	Waist Angle R	Waist Angle L	Slippage Angle R	Slippage Angle L
Gender (G)	1	–	–	–	–	–	–	–
Stage of Lift (S)	4	***	***	***	***	***	***	***
Handle Angle (A)	1	–	***	***	***	***	***	***
Handle Position (P)	3	***	***	***	***	–	***	***
Trial Number (T)	1	–	–	–	–	–	–	–
GxS	4	*	–	–	**	–	–	–
GxA	1	–	–	**	–	–	–	–
SxP	12	***	***	***	***	***	***	***
AxP	3	–	–	***	–	–	–	–

Table 1. Summary of Analysis of Variance of Biomechanical Measures ($* = p < 0.05$; $** = p < 0.01$; $*** = p < 0.001$).

Handle Angle	Box Angle	Elbow Angle R	Elbow Angle L	Wrist Angle R	Wrist Angle L	Slippage Angle R	Slippage Angle L
35°	(not	104°	108°	-13°	-11°	35°	50°
70°	significant)	100°	114°	-10°	-8°	5°	16°

Table 2. Mean Values of Biomechanical Measures at Each Handle Angle.

to shoulder level. Wrist angles were all grouped within a 5° range at each stage of lift. They changed smoothly from 5° radial at floor level, through 15° ulnar at waist level to about 20° ulnar at shoulder level.

Figures 2 and 3 show the effects of handle position and stage of lift on slippage angle for right and left hands. The mean slippage angles shown are the averages across both Handle Angles (35° and 70°). For Slippage Angle, the Handle Angle effect was quite large so that a neutral slippage angle on Figures 2 and 3 would be $+16^{\circ}$ for the 70° Handle Angle and -16° for the 35° Handle Angle. The neutral position for the 70° Handle Angle is the main one of interest and is marked on Figures 2 and 3. The upper (left) arm was neutral near the floor level and gradually increased, with position 6/8 remaining most neutral except at the floor level where 3/8 was the most neutral. For the lower (right) arm, neutral position was reached at about waist level, with positions 3/8 and 6/8 being closest to neutral throughout. As the handle positions were closely grouped for wrist angle, the implication was that position 6/8, and to a lesser extent 3/8, minimized the accommodation of the body to the box for the better of the two Handle Angles (70°).

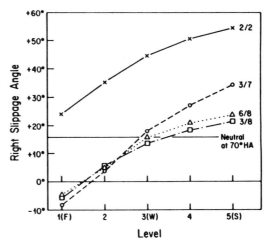

Figure 2. Effects of Handle Position and Stage of Lift on Right Slippage Angle.

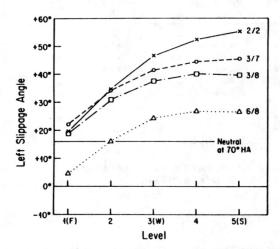

Figure 3. Effects of Handle Position and Stage of Lift on Left Slippage Angle.

It was found that where the same hand position occurred more than once (2/2 on both hands, 3/7 and 3/8 on the left hand, 3/8 and 6/8 on the right hand), the results were closely comparable for all biomechanical indices. This finding enables handle angle recommendations to be made with some confidence.

Heart Rate and RPE

For both heart rate and RPE ANOVA's, Gender and Height of Lift were significant at p < 0.01. For Heart rate there was an additional effect of Trail 5 to Trial 7 with heart rate increasing by about 3 beats/min. between the two trials. Small, and practically insignificant even if statistically significant, interactions were found for heart rate between Trials and Gender, Height of Lift and Handle Angle. No handle position effects were found.

Figure 4 summarizes the Gender and Height of lift effects for both heart rate and RPE. Females were more highly stressed than males and both were least stressed for the waist-shoulder lift. The heart rates were close to ten times the RPE scores, as is expected in a dynamic task involving large muscle masses. For females in the two shorter lifts, ten times RPE underestimated heart rate by about 10 beats/min.

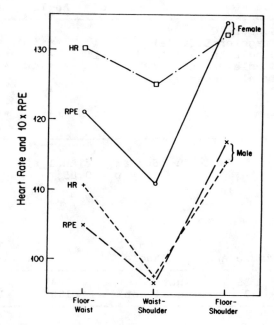

Figure 4. Effects of Lift Height on Heart Rate and 10xRPE for Each Gender.

DISCUSSION AND CONCLUSION

There was little to choose between the four handle positions and the two handle angles in terms of Heart Rate and RPE scores, but large differences were seen in terms of biomechanical measures. Slippage was the main accommodating mechanism by which the human operator adapted to the changing task demands. For both hands, slippage was minimized, with a 70° handle angle, by handles positions of 3/8 and, to a lesser extent, 6/8. The only exception was at floor level itself, where a 2/2 position gave less slippage.

The optimum handle angle of 70° proposed on the basis of static experiments may require some changes as a result of dynamic lifting. Taking the various body angles together, one can define a box-to-handle angle which would give neutral wrist and slippage angles. Averaged over all stages of the lift, these are as follows:

Handle Position	2	3	6	7	8
Optimum Angle	83°	76°	60°	56°	50°

Close agreement overall was obtained between these results and the 70° recommended earlier.

From this discussion, a good set of handle recommendations can be made. Have handles, on hand held cutouts, in positions 6 and 8 with angles of 60° and 50° to the horizontal respectively.

REFERENCES

Borg, G.Z.V. 1962. *Physical Performance and Perceived Exertion*, Ejnar Munksgaard, Copenhagen.

Corlett, E.N., and Bishop, R.P. 1976, A Technique for Assessing Postural Discomfort. *Ergonomics*, 175-182.

Drury, C.G. and Pizatella, T., Handle Placement in Manual Materials Handling. *Human Factors*, 1983, *25(5)*, 551-562.

ACKNOWLEDGEMENTS

This work was supported by the National Institute for Occupational Safety and Health, Division of Safety Research. The NIOSH project officer was T. Pizatella.

VALIDITY OF ISOMETRIC STRENGTH TESTS FOR PREDICTING
PERFORMANCE IN PHYSICALLY DEMANDING TASKS

Andrew S. Jackson   H. G. Osburn
University of Houston
Houston, Texas

Kenneth R. Laughery
Rice University
Houston, Texas

ABSTRACT

Job analysis has shown that many tasks in the energy industry are physically demanding. This study examined the validity of an isometric strength test battery for predicting performance in work sample tests that simulated physically demanding tasks in coal mining and oil production. Since two of the work sample tests involved an endurance component, metabolically determined VO(2) (L/Min) was used to measure arm endurance. All tests were administered to 25 male and 25 female physically fit subjects. Correlations between the isometric strength tests and the work sample tests ranged from .67 to .93. In addition, the isometric strength tests were more highly correlated with the endurance work sample tests than metabolically measured arm endurance. Slopes and intercepts of the male and female regression lines were homogeneous. These results support the validity of isometric strength tests for predicting performance in physically demanding jobs.

INTRODUCTION

This study investigated the validity of isometric strength tests for predicting performance in physically demanding tasks in the energy industry. However, instead of predicting performance in actual work tasks simulated work sample tests were used as criteria of performance. These simulated work sample tests closely reproduced actual work tasks and provided improved standardization and reliability.

The purpose of the study was to examine the predictive validity of a three test isometric strength battery. The study sought answers to the following three questions: (1) To what extent are isometric strength tests valid for predicting performance in physically demanding work tasks? (2) Does a measure of arm endurance increase the accuracy of predicting performance on physically demanding endurance tasks over and above the accuracy afforded by the isometric strength tests alone? (3) Is the regression of job performance on isometric strength the same for males and females?

METHOD

Two separate studies were conducted. Study 1 investigated physically demanding tasks required of underground coal mining technicians (Jackson & Osburn, 1983). Study 2 involved tasks performed by roustabouts on off-shore oil production facilities (Laughery & Jackson, 1984). The same isometric strength tests were used in both studies.

Study 1: Underground Coal Mining

Work Sample Tests. Based on a job analysis of work performed by underground mining technicians four physically demanding tasks were identified: (1) roof bolting, (2) lifting and carrying 50-pound bags, (3) lifting and stacking 82-pound concrete blocks, and (4) shoveling coal. Roof Bolting. Roof bolting was measured with an isokinetic leg press machine modified to simulate the movements required to straighten a roof bolt that has been bent in order to be installed in low coal conditions. A strip chart recorder provided a record of the torque generated during the movement.

Bag Carry. The bag carry test was the number of bags lifted and transported nine feet in five minutes.

Block Carry. The block carry test involved lifting an 82-pound concrete block, transporting the block three feet, and placing it in a defined position. A total of six blocks were transported. Total score was blocks moved per unit time.

Shoveling. This test involved shoveling polyvinyl chloride from the floor over a three foot wall. The measure was the time required to shovel 800 pounds.

Isometric Tests. Isometric tests included grip strength, arm lift and back lift. The tests were administered using specially designed equipment manufactured by Lafayette Instrument Company (Model 32528). The equipment used a load cell and a digital recorder to measure isometric strength. The unit was designed to start and end the strength trial with an audio signal. The trial duration was three seconds and the average force in the final two seconds was used as the strength score. A detailed description of the tests can be found in Jackson and Osburn (1983).

Arm Endurance Test. A Monarch bicycle ergometer (Model GC1) was mounted on a table thirty-one inches high and the pedals were modified for arm use. The subject sat on a chair behind the ergometer and pedaled it with his/her arms. The pedaling rate was 60 rpm and a metronome was used to help the subject keep the correct pace. The test was started by pedaling without any resistance on the ergometer. This initial stage lasted for one minute and was designed to help the subject learn the work rate and provide a warm-up trial. An incremental test was used in which the work intensity was systematically increased by 0.5 KP every two minutes. The subject continued to pedal until the point of physical exhaustion. Oxygen uptake (VO(2)L/Min) was used to measure arm work capacity. An open circuit system was used during the entire test. Standard methods were used to calibrate (VO(2)L/Min) for peak work and during the final minute of each work state (Consolazio, Johnson & Pecora, 1963).

Subjects. All tests were administered to 25 male and 25 female college students who were paid to participate in the study. The subjects were either physical education majors or intercollegiate athletes. All were physically fit and motivated. The testing was completed over a three-day period. In addition to the physical tests basic anthropometric measurements (height, weight, and percent fat) were obtained.

Study 2: Oil Production

Work Sample Test. A pipe transport task was found to be the most physically demanding roustabout task (Laughery & Jackson, 1984). This task require two people to lift a pipe (182 cm long and 6.3 cm in diameter) from a height of 10.5 inches and place it on a table 21.5 inches above the floor. The pipe was lifted again from 21.5 inches and maintained for about 2 seconds at a height of 26.5 inches. The pipe was carried by a steel bar inserted into the pipe. The total weight lifted by two people is 300 pounds. This task was simulated by mounting a pipe to a fixed base. The Lafayette load cell and digital recorder was used to measure lift force. A system of chains and "S" hooks were used to stabilize the load cell to the floor and pipe. The chain system was used to place the pipe at the proper height. The simulated lifting positions were low (10.5), medium (21.5) and high (26.5). Each subject was given three lifts at each height. For the lowest two heights peak force exerted during a three-second trial was used. For the 26.5" height average force was recorded.

Subjects. A second group of 25 male and 25 female college students were recruited and paid for participation. These subjects were also physically fit physical education majors and/or intercollegiate athletes. In addition to the physical tests anthropometric measures were obtained on all subjects. Testing was completed in a single session.

RESULTS

Table 1 shows anthropometric characteristics, strength and age of the subjects used in each study. The total strength measure is the sum of grip, arm and back strength. Both studies show that males were taller, heavier,

stronger and leaner than females. These are typical gender differences that have been consistently reported in the literature (Well & Plowman, 1983).

Table 1. Demographic and Physical Characteristics Data on Study Participants Means, Standard Deviations

Coal Mining Study 1

Variable	Females		Males	
	Mean	SD	Mean	SD
Age(yrs)	23.3	3.5	26.2	4.8
Height(ins)	65.4	3.1	70.4	3.0
Weight(lbs)	132.2	19.5	174.6	30.4
Percent Fat(%)	17.9	3.9	9.8	5.2
Total Strength (lbs)	242.6	46.8	417.5	77.3

Roustabout Study 2

Age(yrs)	23.9	3.3	23.8	4.7
Height(ins)	63.9	2.5	69.0	3.0
Weight(lbs)	129.2	22.1	176.8	35.3
Percent Fat(%)	19.2	6.9	12.5	6.3
Total Strength (lbs)	230.7	44.9	350.3	83.0

Study 1: Underground Coal Mining

Table 2 presents correlations between total isometric strength and the four work sample test, as well as correlations between arm endurance and the two endurance work sample tests. These data are presented separately for males, females and the total sample. The correlations within gender groups were somewhat lower than for the combined sample due to the large differences between males and females in upper body strength. The percentage of female performance relative to male performance on the work sample tests were as follows: Roof Bolting-56%, Bag Carry-79%, Block Carry-28%, Shoveling-66%.

Table 2. Correlations Between Strength and Arm Endurance Tests and Work Sample Tests

Test	Isometric Strength	Arm Endurance
Roof Bolting		
Total	.91	--
Females	.74	--
Males	.77	--
Block Carry		
Total	.87	--
Females	.72	--
Males	.64	--
Shoveling Rate		
Total	.71	.68
Females	.56	.46
Males	.38	.58
Bag Carry Rate		
Total	.63	.46
Females	.65	.46
Males	.37	.13a
Total Work		
Total	.88	.66 (a=
Females	.79	.56
Males	.68	.33 p>.05)

Curvilinearity of Regression. Tests for linearity of regression of the work sample tests on total isometric strength showed that the quadratic component was significant for all work samples with the exception of roof bolting. Inspection of the plots showed that the regression tended to flatten out at higher strength levels. For this reason a log transformation was made on total isometric strength. The correlations for the log transformed models (Block Carry 0.83, Bag Carry 0.67, Shoveling 0.75) were nearly identical to those of the quadratic models indicating that the log transformed model provided a good fit to the data.

Contribution of Arm Endurance. Hierarchical regression analysis was used to determine if arm endurance accounted for additional variance in the work sample tests beyond that accounted for by total isometric strength. It was found that arm endurance did account for additional variance in shoveling rate. However, the increase in the multiple correlation from 0.75 to 0.80 was modest. No significant contribution was found for bag carry.

Gender Differences in Validity. Slopes and intercepts for regression of the work sample tests on total isometric strength were examined separately for males and females. Neither gender or gender by strength interaction yielded a significant increase in the squared multiple correlation indicating that the prediction system was equally valid and fair for males and females.

Study 2: Oil Production

Table 3 presents the correlations between total isometric strength and the three work sample lift tests. These data show that total isometric strength was highly predictive of work sample lift tests. In addition, there were no significant differences in the slopes of the regression lines for males and females. Tests for gender differences were not significant for the low and medium height work sample tests. However, gender was significant for the high lift position. Thus, it appeared that differences in height between men and women may have influenced performance on this task.

Using hierarchical regression the influence of height and gender on work sample performance was examined. For comparative purposes all three work samples were studied. Height did not account for a significant proportion of the variance in the low lift test. However, height did account for a significant proportion of the variance in the medium and high lift tests. Taller people were able to bend their legs and utilize lower body strength more effectively. The high lift position was especially difficult for shorter subjects because they had to execute the lift entirely with their upper body. After height was accounted for gender was not statistically significant.

Table 3. Correlations Between Isometric Strength Tests and Work Sample Tests

| | Work Sample Test | | |
	Low Peak	Medium Peak	High Average
Total Strength			
Total	.93*	.92*	.88*
Females	.80*	.75*	.68*
Males	.91*	.88*	.79*

*$p < .01$

DISCUSSION

The results of both studies support the use of the three isometric strength tests for pre-employment screening for physically demanding jobs. These tests sample muscular strength for major muscle groups used in the performance of many industrial tasks. Although the types of lifts and carries involved in industrial jobs are biomechanically different, they elicit action from the same muscle groups.

The substantial correlations between isometric strength and the endurance work sample tests were not unexpected. Muscular strength and absolute muscular endurance are highly correlated. Absolute endurance involves moving a weight load that is constant for all subjects. In this situation weaker subjects must utilize a higher percentage of their maximum strength in order to complete a given amount of work. Working at higher percentage of maximum strength will more quickly lead to fatigue and exhaustion. Our data suggest that the use of time consuming and costly endurance tests does not substantially increase the accuracy of predicting performance on tasks that involve an absolute endurance component.

SIMULATION OF CART PUSHING AND PULLING

Kwan S. Lee

Department of Industrial and Systems Engineering
Ohio University, Athens, Ohio

and

Don B. Chaffin and Gary D. Herrin

Center for Ergonomics
University of Michigan, Ann Arbor, Michigan

ABSTRACT

A dynamic biomechanical model is formulated for cart pushing and pulling. The model is used to simulate different task and human parameters to determine the potential effects of handle height, cart speed, body weight and the cart load that minimize slip potential during cart pushing or pulling. In addition, predicted vertebral column loadings as a function of task and subject parameters were also studied. Low handles were found to be better for pulling while high handles were better for pushing to reduce slip potential. As expected, a heavy person, combined with a light task force and a slow speed of movement proved to reduce the slip potential.

INTRODUCTION

Several researchers (Chaffin et al., 1983; Ayoub et al., 1974; Snook et al., 1969; Kroemer, 1968; Gaughran et al., 1956) have studied pushing and pulling, but their studies did not include dynamic pushing and pulling. Others have studied the dynamic nature of industrial work (Ayoub et al., 1978; Fisher, 1967) but their work was confined to the study of lifting. These studies offer little comprehensive information on dynamic pushing and pulling tasks. The limitation of the above mentioned research suggests the necessity of a study which would yield a more thorough modelling and experimental approach to dynamic pushing and pulling tasks.

More analyses should be performed for conditions where the possibility of work hazards exists. Indirect experimental methods are often required because of the limitations inherent in experimenting with human subjects in these potentially hazadous situations, such as when the subject is about to slip. The proposed model can simulate cart pushing and pulling (Figure 1) and predict the required friction without placing human subjects in potentially hazardous conditions. This prediction could also assist in reducing the confusion caused by many different recommendations (which range from 0.4 to 0.6) for the coefficient of friction required for a working floor to be "safe" (Day and Sharmburger, 1965; Sigler, 1943; Hunter, 1930).

In addition, the model can estimate gross low-back loads caused by dynamic motion during pushing and pulling.

GENERAL MODEL DESCRIPTION

The entire simulation is flowcharted in Figure 2. The general idea is to iterate the simulated cart pushing and pulling through a biomechanical model by changing velocity, body weight, hand force and handle height. Then the maximum compressive force at L5/S1 disc and the maximum coefficient of friction between shoe sole and floor which is required to keep the body balanced can be obtained from this simulation. The model is a sagittal plane model and assumes that all external hand forces acting on the body do so at the center of the grip of hands, and all external foot forces acting on the body (at the center of the contact area) do so at one contact point of the heel or sole of each foot.

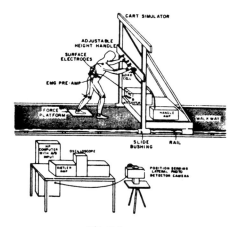

Figure I
A Cart Simulator and Instruments for the Dynamic Experiments

The simulated human body is assumed to be made of eleven solid links (hand, lower arm, upper arm, trunk and neck, right and left upper leg, right and left lower leg, right and left foot). The mass of each link has been assumed to be proportional to the total body mass as indicated by the distribution of Dempster et al. (1964). A principal moment of inertia of each link has been determined according to Dempster's Data (1955). The radius of gyration has been determined as a percentage of segment of link length, according to Plagenhoef's data. More details are explained in other paper by Lee et al. (1984).

The model requires the body weight, displacements of the joints of the body, foot contact point and forces exerted on the cart handle (hand forces) as inputs and produces vertical and horizontal foot forces and compressive forces at the L5/S1 disc as outputs.

METHOD

Body weight, handle height, hand force and velocity of the exertion (push or pull) were varied (Table 1) to determine the sensitivity of the model coefficients of friction and back forces. To get the movement pattern of a subject, the vertical and horizontal positions of the wrist, elbow, shoulder, L5/S1 disc, hip, knees and ankles are calculated from photographic reference points obtained during each cycle.

Body Weight (kg)	Hand* Force (N)	Handle Height (cm)	Velocity (km/hr)
42	90	66	1.8
52	196	109	2.7
62	294	152	3.6
72	392		
82	490		

* Horizontal

Table 1.
Different Values of Variables used
in the Simulation

RESULT

The Effect of Body Weight

It was found that μ showed a non-linear and interactive effect with handle height (Figure 3). This tendency was the same for all subjects, regardless of sex, or direction of exertion (pushing or pulling). The predicted coefficient of friction decreased from .30 to .22 for high handle pushing and from .96 to .46 for low handle pushing as body weight changed from 52 kg to 82 kg. The opposite result was shown in pulling. The predicted coefficient of friction decreased from .92 to .49 for high handle pulling and from .32 to .20 for low handle pulling as body weight changed from 42kg to 82

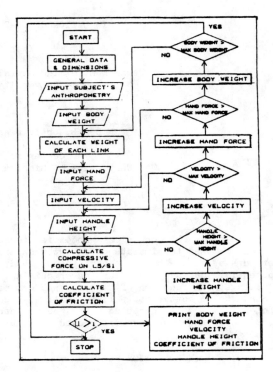

Figure 2

Flowchart of the Cart Simulation

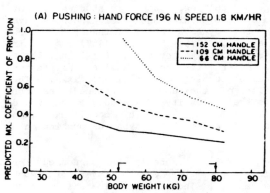

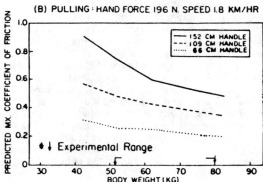

Figure 3
Predicted Coefficient of Friction vs. Body Weight

kg. As depicted in Figure 4, the predicted compressive force increased linearly as the body weight increased. However, the slopes were higher in pulling than in pushing.

The Effect of Hand Force

The effect of hand force on the predicted coefficient of friction (μ) depended on the handle height and direction of exertion (pushing or pulling). This was true for all subjects. The predicted coefficient of friction increased from .33 to .98 in low handle pushing and from .17 to .31 in high handle pushing as the horizontal hand force changed from 98 newtons to 196 newtons while it increased from .3 to .41 in low handle pulling and from .41 to .76 in high handle pulling as the horizontal hand force changed from 98 newtons to 196 newtons (Figure 5).

In pulling, the maximum compressive force increased approximately linearly (Figure 6) as the horizontal hand force increased, but in pushing it had a saddle point in the neighborhood of 196-294 newtons (20-30 kg). This may be due to the fact that, until the saddle point, the vertical hand force was reducing the body weight effect on the L5/S1 disc. It was found that low handle pulling (66 cm from the floor) showed the highest compressive force which increased at a steeper rate as the horizontal hand force increased. The same was true for high handle pushing (154 cm from the floor).

The Effect of Handle Height

In pushing, the high handle showed smaller μ (required coefficient of friction) while the low handle showed greater μ. Nonetheless, the high handle showed greater μ in pulling while the low handle showed smaller μ (Figure 7).

Pulling showed significantly greater compressive force on the L5/S1 disc than pushing (Figure 8) for all subjects regardless of the handle height and the hand force. When the hand force was increased to more than 392 newtons, the hand force became the limiting factor in pushing or pulling, since the subject could not perform at a constant velocity in the experiment. The smaller compressive forces with the 109 cm handle height may be due to different vertical hand force and velocity which may have affected the amount of compressive force greatly.

The Effect of Velocity

The use of motion velocity faster than standard velocity was not reasonable because the motion required great horizontal foot forces (Y) and thus required the coefficient of friction of greater than 1.0 which is not realistic in industrial situations. This condition happened mostly in the low handle (66 cm from the floor)

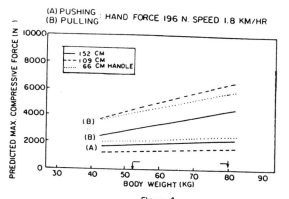

Figure 4
Predicted Maximum Compressive Force vs. Body Weight

* ↓ Experimental Range

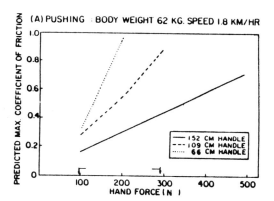

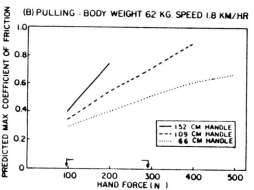

Figure 5
Predicted Coefficient of Friction vs Hand Force

* ↓ Experimental Range

pushing and the high handle (154 cm from the floor) pulling (Figure 9). The requirement for the coefficient of friction almost doubled as the velocity increased from 1.8 KM/HR to 3.6 KM/HR. This was expected since all subjects could not push against a heavy resistive force with the low handle and pull with the high handle against a heavy resistive force during the cart experiment. In low handle pushing, the vertical foot force decreased to a great extent since the vertical hand force increased as the velocity increased and supported some of the body weight. The vertical foot force remained the same or decreased somewhat in pushing the middle handle height and increased when pushing the high handle height. This means that in pushing, the horizontal foot force as a function of velocity may increase at the same rate in spite of the different handle height since the horizontal hand force increases as the velocity increases. However, the angle for the hand force causes the opposite effect on the vertical foot force since an increase in vertical hand force decreases the vertical foot force. This explanation is also applicable for pulling.

Figure 10 shows the effect of velocity on the compressive force at the L5/S1 disc. Generally, as the velocity increased, the maximum compressive forces at the L5/S1 disc increased as expected. However, the higher the velocity the greater the compressive forces and the coefficient of friction. This may be due to the interactive effect of the vertical hand force which became greater as the velocity increased.

The effect of an increase in velocity was different for different handle heights. Middle handle height in pushing and high handle height in pulling showed the least increase.

SUMMARY

A dynamic biomechanical model is formulated and used to simulate cart pushing and pulling. Heavy body weight, light hand force and slow velocity were found to be better to reduce slip potential. Light body weight and slow velocity were also found to be better to reduce the compressive force at the L5/S1 disc. Handle height and hand force showed opposite effects in pushing and pulling.

------- * ------- * -------

Acknowledgement --- The authors would like to express their gratitude to Mr. James A. Foulke and Dr. Robert A. Andres at the Center for Ergonomics for their help. This research was partially supported by research grants from NIOSH (No. 210-81-3104) and NASA (No. NAS 915244).

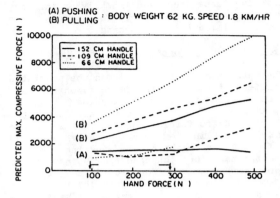

Figure 6
Predicted Maximum Compressive Force vs. Hand Force
* ↓ Experimental Range

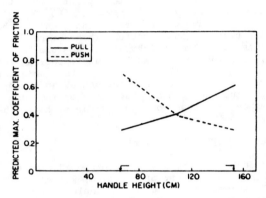

Figure 7
Predicted Coefficient of Friction vs. Handle Height
Averaged over All Conditions Studied
* ↓ Experimental Range

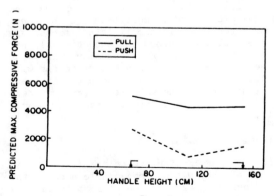

Figure 8
Maximum Compressive Force vs. Handle Height
Averaged over All Conditions Studied
* ↓ Experimental Range

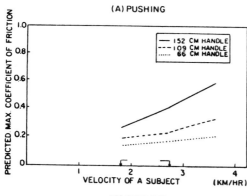

(A) PUSHING

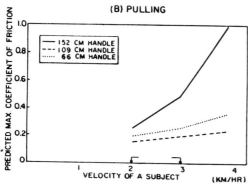

(B) PULLING

Figure 9
Predicted Coefficient of Friction vs. Velocity
Averaged Over All Conditions Studied
(Averaged Hand Force for a Subject)

✦ ↓ Experimental Range

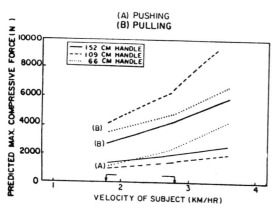

(A) PUSHING
(B) PULLING

Figure 10
Predicted Compressive Force vs. Velocity
Averaged Over All Conditions Studied
(Averaged Hand Force for a Subject)
✦ ↓ Experimental Range

REFERENCES

Ayoub, M. and J. McDaniel "Effects of Operators Stance on Pushing and Pulling Tasks." *AIIE Transactions,* 6(3):185-195, September, 1974.

Ayoub, M. and M. ElBassoussi "Dynamic Biomechanical Model for Sagittal Plane Lifting Activities." *Safty in Manual Materials Handling,* ed. C. Drury, USDHEW, PHS, CDC, NIOSH, DBBS, July, 1978.

Chaffin, D. B. , R. O. Andres and A. Garg "Voltional Postures during Maximal Push/Pull Exertions in the Sagittal Plane", *Human Factors,* 1983, 25(5), 541-550.

Day, S.S. and E Shamburger "Factors Controlling Skill Resistance of Resident Floor Coverings" *Hospitals, J.A.H.A.* vol, 39, April 16, 1965.

Dempster, W. T., L.A. Sherr, and J.G. Priest "Conversion Scales for Estimating Humeral and Femoral Lengths and the Lengths of Functional Segments in the Limbs of American Caucasoid Males." *Human Biology* 36(3) : 246-261, September, 1964.

Fisher,B.J."A Biochemical Model for the Analysis of Dynamic Activities." *MSE Thesis,* The University of Michigan, Department of Industrial and Operations Engineering, Ann Arbor, MI, 1967.

Gaughran, G.R.L. and W.T. Dempster "Force Analysis of Horizontal Two-Handed Pushes and Pulls in the Sagittal Plane." *Human Biology,* 28:69-92, 1956.

Kroemer, K.H.E. "Horizontal Push and Pull Forces."*Appl. Ergonomics,* 5:94-102, 1974.

Lee K.S., D.B. Chaffin and M.H. Zarrugh "Biomechanical Prediction of Slip Potential During Pushing and Pulling" submitted to *J. Biomech.* (Manu. No. 84-050) , 1984.

Plagenhoef, S.C. "Methods for Obtaining Data to Analyze Human Motion." *Res. Q. Am. Assoc. Health Phys. Ed.* 37:103-112, 1966.

Sigler, P.A."Relative Slipperiness of Floor and Deck Surfaces." US National Bureau of Standards, Building Materials and Structures Reports, B.H.S. 100.,1943.

Snook, S.H., C.H. Irvine and S.F. Bass "Maximum Weights and Workloads Acceptable to Male Industrial Workers While Performing Lifting, Lowering, Pushing, Pulling, Carrying and Walking Tasks." *Proc.* American Industrial Hygiene Conference, Denver, Colorado, May, 1969.

BIOMECHANICS AND WORK PHYSIOLOGY IN UNDERGROUND MINING

J.L. Selan*, M.M. Ayoub*, J.L. Smith*, J. Peay**

*Texas Tech University
Lubbock, Texas 79409

**U.S. Bureau of Mines
Pittsburgh, PA 15236

ABSTRACT

A biomechanical and work physiology study in underground mining was conducted. Anthropometric, isometric strength, and physical work capacity (PWC) data were collected. Female miners were significantly heavier and larger in circumferential measurements than comparison groups (males did not differ from other groups in terms of anthropometry). Underground miners were significantly stronger than comparison groups, and were rated low in terms of PWC. Energy expenditure measurements were made on 35 underground mining tasks; the majority falling under the categories of moderate to heavy work. Postural stresses were assessed using a modified version of the Ovako Working Posture Analysis System (OWAS; developed by Karhu, et al. 1977). Several tasks appeared to cause substantial postural stress. Postural stress was correlated significantly with physiological demand. Recommendations based on the findings of the study included: increased use of powered equipment in place of manual labor for several tasks, modifications to existing equipment designs, and modifications to the work clothing design standards developed by Caddel, et al. (1981) for low coal miners.

INTRODUCTION

Underground mining has traditionally been considered a hazardous and physically demanding occupation. In 1981, miners suffered 37,017 accidents; the accidents stemming from various underground mining operations (Perry, et al. 1981). Peters (1981) estimated that 79% of all back injuries suffered by coal miners were the result of overexertion. Despite this, very little research has been conducted to quantify some of the capacities of the underground mining population and the demands associated with underground mining tasks.

Because of this, a biomechanical and work physiology study of underground metal and non-metal mining (excluding coal mines with seam heights of 48 inches or less) was conducted. Two major areas were emphasized:

1. defining the mining population in terms of their physical and physiological characteristics; and
2. determining the physiological and biomechanical stresses imposed on the mining population by underground mining tasks.

CHARACTERISTICS OF UNDERGROUND MINERS

Over 360 male and 80 female underground miners were measured. Forty-two anthropometric measurements were made using standard anthropometric protocols and instruments. Isometric strength measurements were made using equipment and protocols described by Ayoub, et al. (1978). Arm, shoulder, back, standing leg, sitting leg, and composite strength were measured. PWC was predicted from submaximal

techniques using a bicycle ergometer. A nine minute protocol was used with the work load increased after the third and sixth minute. To obtain PWC, heart rate and oxygen consumption during the third minute of each work load were plotted. A straight line of best fit was drawn through the points and extended to intersect the age-predicted maximum heart rate of 220 minus age (Astrand and Rodahl, 1977) to determine the predicted maximum. A complete summary of these data is given in Ayoub, et al. (1984).

Anthropometry

Anthropometric data were gathered on 364 male and 86 female underground miners. The data were compared to anthropometric data collected by Ayoub, et al. (1981) on miners working in low coal (seam height < 48 inches). Male underground miners did not differ significantly from the comparison population. Female underground miners were larger than low coal miners for all common circumferential body measurements and were significantly heavier.

Isometric Strength

Isometric strength data were gathered on 313 male and 82 female underground miners. The strength data are summarized, and compared with isometric strength data gathered on low coal miners by Ayoub, et al. (1981), in Table 1.

As Table 1 indicates, both male and female underground miners were significantly stronger than their low coal counterparts for the majority of the strength measures. It is hypothesized that the heavy lifting and generally high level of physical exertion required in underground mining, and the lack of postural constraints placed on underground miners relative to low coal miners, may in part explain the higher strength scores obtained by the underground miners.

Physical Work Capacity

Physical work capacity data (ml/kg min) were collected on 250 male and 43 female underground miners. The PWC data are summarized, and compared with PWC data from low coal miners (Ayoub, et al. 1981), in Table 2.

Based on classifications of cardiovascular fitness from Astrand (1960) and Hodgson (1971), the underground mining population was generally rated low. However, the rate of decline in PWC with advancing age displayed a tendency to be slower for the underground miners compared to the reference populations. This may indicate that physical conditioning of the miners is occurring as a result of performing underground mining tasks.

UNDERGROUND MINING TASK DEMANDS

Underground mining tasks were analyzed to determine their physiological and biomechanical demands. The energy expenditure of mining tasks was estimated by measuring ventilation volume (V_E) using a Max Planck respirometer. Oxygen consumption was predicted from V_E using equations developed at Texas Tech University, and converted to kilocalories of energy assuming 5.00 kcal per liter of oxygen.

Postural stresses were determined using a modified version of the Ovako Working Posture Analysis System (OWAS; developed by Karhu, et al. 1977). The procedure allowed for the identification of 96 distinct postures consisting of a back-arms-legs configuration. A weighting system was developed such that each posture could be represented by a total postural stress score, along with an arm stress, back stress, and leg stress score.

Task Physiological Demands

Ventilation volume data were collected on 35 underground mining tasks. Some of these data are shown in Table 4. Using the categorization system for grades of work developed by Christiansen (1964), the majority of tasks sampled fell under the category of moderate to heavy work.

The energy expenditure requirements of low coal mining tasks were compared to comparable underground mining tasks. Energy expenditures for miner helpers were significantly lower in underground mining than low coal. Ayoub, et al. (1981) combined miner helper and bolter helper data in determining energy expenditure, which may explain the difference. Timbering had

significantly higher energy expenditure requirements in underground mining; due in part to larger, heavier timbers being used.

Based on job analyses, 24 hour energy expenditures were estimated for several underground mining jobs. The average 8 hour work day energy expenditure was 2247.7 Kcal. Based on one third of the male underground miners' PWC, it is reasonable to expect that the miners could expend approximately 2400 Kcal per work day. Thus, underground mining jobs appear on the average to be reasonable as a percentage of the miners' PWC. However, the work day energy expenditure estimates include large periods of idle/down time (approx. 3 hours per work day). If efficiency of mining operations was increased, and thus down time decreased, the ability of miners to handle the additional stress would be questioned.

Task Postural Stress

Postural stress data were collected on 24 mining tasks. Table 4 summarizes the total postural stress, arm stress, back stress, leg stress, and energy expenditure for these tasks.

The highest rated tasks in terms of total postural stress (hand mucking, timbering, track maintenance) were similar in that they all involved a large amount of work at or near ground level, and hence the back was frequently flexed. Those tasks receiving the highest arm stress ratings (scaling, pipe/vent installation) all involved overhead work. Kneeling on the part of miners was responsible for the high leg stress scores obtained for tasks like surveying, timbering, and track installation.

Total postural stress, back stress, and leg stress were positively and significantly correlated with energy expenditure.

CONCLUSIONS/RECOMMENDATIONS

Based on the results of this study, a number of recommendations were made:

1. Equipment used and work clothing worn by underground miners should be based on their anthropometric characteristics. Given the similarity in anthropometric characteristics between male underground and low coal miners, the suggestion made by Ayoub, et al. (1981), that anthropometric data bases available from the Public Health Service or the various military branches be used in mine hardware design, is applicable. Work clothing sizing standards developed by Caddel, et al. (1981) for female low coal miners may need to be revised based on the anthropometric differences between female underground and low coal miners.

As an example of use of anthropometric data in equipment design, it was observed that several operators of roof bolting machines had to bend at the waist in order to operate boom controls (the low height of the boom controls was due to the design of these machines for low coal). Making the boom controls of adjustable length would eliminate this unnecessary stress.

2. Increased use of powered equipment in place of manual labor where appropriate is recommended. This would include: 1) use of jumbo drills in place of jackleg or stoper drills; 2) use of roof bolting machines in place of jackleg or stoper drills when inserting roof bolts; 3) use of mucking machines in place of hand mucking; and 4) use of motorized transport (e.g., loaders) when moving pipe, vent tubing, etc., to a work location.

In addition to the above methods of decreasing job demands, use of lighter weight materials where appropriate is recommended. An example of this is the use of fiberglass support posts for ground control rather than timbers.

REFERENCES

1. Astrand, I. Aerobic Capacity in Men and Women, With Special References to Age. Aeta. Physiol. Scand. 49 (Suppl. 169): 67, 1960.

2. Astrand, P.O. and K. Rodahl. Textbook of Work Physiology, McGraw-Hill Book Co., Inc., New York, 2nd ed., 1977.

3. Ayoub, M.M.; Bethea, N.J.; Bobo, M.; Burford, C.L.; Caddel, K.; Intaranont, K.; Morrissey, S.; and Selan, J.L. Mining in Low Coal Volume I: Biomechanics and Work Physiology, Final Report, U.S. Bureau of Mines Contract No. H0387022, September, 1981.

4. Ayoub, M.M.; N.J. Bethea, S. Deivanayagam, S.S. Asfour, G.M. Bakken, P. Liles, A Mital, and M. Sherif. Determination and Modeling of Lifting Capacity. Final Report, Grant #5R01OH-0054502, HEW, NIOSH, September, 1978.

5. Ayoub, M.M.; Selan, J.L.; Burford, C.L.; Intaranont, K.; Rao, H.P.R.; Smith, J.L.; Caddel, D.K.; Bobo, W.M.; Bethea, N.J.; and Chang, H. Biomechanical and Work Physiology Study in Underground Mining Excluding Low Coal. Final Report, U.S. Bureau of Mines Contract No. J0308058, July, 1984.

6. Caddel, D.K.; Selan, J.L.; Smith, J.L.; and Ayoub, M.M. Mining in Low Coal Volume III: Clothing Design Study, Final Report, U.S. Bureau of Mines Contract No. H0387022, November, 1981.

7. Christensen, E.H., L. Homme au travail, Serie "Securite, hygiene et medecine du travail" No. 4, Bureaua International du Travail, Geneve, 1964.

8. Damon, A., Stroudt, H.W. and McFarland, R.A. The Human Body in Equipment Design. Harvard University Press, Cambridge, Mass., 1966.

9. Gary, R.C., R. Passmore, G.M. Warnock, and J.V.G.a. Durnin. Studies on Expenditure of Energy and Consumption of Food by Miners and Clerks, fife, Scotland, 1952. Medical Research Council Special Report Series, 1955.

10. Humphreys, P.W. and A.R. Lind. The Energy Expenditure of Coal Miners at Work. British Journal of Industrial Medicine, v. 19, 1964, pp. 264-275.

11. Hodgson, J.S. Age and Aerobic Capacity of Urban Midwestern Males. Ph.D. Dissertation, University of Minnesota, Minneapolis, Minnesota, 1971.

12. Karhu, O., P. Kansi, and I. Kuorinka. Correcting Working Postures in Industry: A Practical Method for Analysis. Applied Ergonomics, V. 8, No. 4, 1977, pp. 199-201.

13. Perry, T.J.; N.D. Schwalm; and W.H. Crooks. Human Factors Analysis of Underground Work Areas and Tasks in Metal and Nonmetal Mines. U.S. Department of the Interior Bureau of Mines. Contract No. J0387230. Final Report, March 11, 1981, pp. 150.

14. Peters, R.H. Activities and Objects Most Commonly Associated with Underground Coal Miners' Back Injuries. Proceedings: Bureau of Mines Technology Transfer Symposia. August, 1983, pp. 23-31.

This research was sponsored by the U.S. Bureau of Mines, Contract No. J0308058.

TABLE 1. Strength characteristics of underground miners vs.
low coal miners from Ayoub, et al. (1981)

	Measurement	Underground Miners			Low Coal Miners		
		N	Mean	SD	N	Mean	SD
MALE	Back Strength	294	185.06	49.47	46	138.50*	35.62
	Arm Strength	311	107.10	22.64	53	94.54*	17.71
	Shoulder Strength	307	152.28	35.61	53	115.64*	27.55
	Standing Leg Strength	309	373.28	93.60	49	350.98*	73.82
	Sitting Leg Strength	313	68.28	20.19	47	78.65*	27.40
	Composite Strength	286	332.24	86.00	-	-	-
FEMALE	Back Strength	80	123.25	27.82	30	100.07*	28.3
	Arm Strength	81	58.90	9.48	39	52.04	10.78
	Shoulder Strength	83	75.70	15.38	39	59.71*	14.12
	Standing Leg Strength	79	209.01	42.43	30	186.05*	40.33
	Sitting Leg Strength	82	55.15	16.33	40	52.79	20.46
	Composite Strength	58	202.28	38.65	-	-	-

Note: Composite strength measure not taken on low coal miners.

*Significant at .05 level using students t-test.

TABLE 2. Maximum oxygen uptake of male underground miners
compared to sample populations by age group

MALES

	N	Mean Age	Mean WT (KG)	VO$_2$ MAX	
				MEAN	SD
Underground Miners	144	25.33	79.59	39.15	8.73
Ayoub, et al. (1981)	32	25.59	81.52	36.41	9.71
Underground Miners	73	34.26	81.48	36.66	8.48
Ayoub, et al. (1981)	15	33.00	81.55	32.36	7.11
Underground Miners	21	44.52	86.94	35.91	12.72
Ayoub, et al. (1981)	3	44.00	85.27	26.15*	4.39
Underground Miners	12	56.75	81.29	31.17	6.48
Ayoub, et al. (1981)	5	55.40	77.12	30.72	13.05

FEMALES

	N	Mean Age	Mean WT (KG)	VO$_2$ MAX	
				MEAN	SD
Underground Miners	24	27.13	61.70	27.82	5.92
Ayoub, et al. (1981)	21	25.29	61.71	32.19*	7.21
Underground Miners	15	33.87	70.00	30.93	8.43
Ayoub, et al. (1981)	10	33.00	65.42	28.54	8.62
Underground Miners	4	45.25	62.84	21.16	6.22
Ayoub, et al. (1981)	4	43.00	66.90	25.72	4.27

*Significantly different from underground miners at .05 level.

TABLE 3. Energy Expenditures (kcal/min) of
Underground Mining Tasks

RANK	TASK	MEAN KCAL	STD	TIME*	MAX	MIN	N**
1	Chute Tapping	9.42	2.23	494	14.08	3.17	32
2	Hand Mucking	9.25	1.98	532	13.76	3.91	30
3	Scaling	8.47	1.99	393	12.95	3.19	21
4	Loading Rounds	7.17	1.43	451	11.35	3.65	26
5	Track Maintenance	7.15	1.78	171	10.56	2.56	10
6	Timbering	7.14	1.96	1050	14.09	1.34	60
7	Moving Cables	7.11	1.27	110	12.19	3.73	6
8	Jackleg Drilling	7.09	1.94	437	12.99	2.29	26
9	Tugger Operation	6.86	1.49	50	10.59	3.46	3
10	Concrete Construction	6.47	1.19	429	10.31	2.92	23
11	Machine Maintenance	6.17	1.65	190	10.67	2.24	11
12	Pipe Installation	6.01	2.20	391	10.74	1.55	22
13	Ventilation Installation	5.99	1.49	237	11.31	2.23	14
14	Stoper Drilling	5.69	0.79	85	8.10	3.99	5
15	Miner Helper	5.68	2.14	71	10.62	2.77	5
16	Roof Bolting	5.62	1.46	270	9.68	1.87	17
17	Mucking Machine Operation	5.20	1.52	104	8.41	1.73	6
18	Slusher Operation	4.94	1.89	110	11.83	1.84	6
19	Jumbo Drilling	4.68	1.22	84	9.49	2.65	5
20	Truck Operation	4.21	1.08	91	8.13	2.28	5

*number of minutes for which data were collected on task
**number of miners on which data were collected

Ergonomics and Musculoskeletal Disorders

TABLE 4. Means and rankings for postural stresses and
energy expenditure for underground mining
tasks

Task	Total Postural Stress	Rank	Back Stress	Rank	Arm Stress	Rank	Leg Stress	Rank	Energy Expenditure (Kcal/min)	Rank
CHUTE TAP	43.23	10	26.88	9	3.85	14	12.50	7	8.38	5
HAND MUCK	50.01	3	33.71	3	3.13	24	13.17	5	10.42	1
SCALING	44.29	9	20.11	15	12.64	1	11.55	10	8.73	2
F HAND MUCK	57.32	1	41.80	1	3.13	24	12.40	8	7.78	8
F TIMBERING	46.94	6	27.88	8	6.07	5	12.98	6	7.95	7
F TRACK MAINT	48.13	5	30.63	5	3.13	24	14.38	3	8.09	6
ROCK DUST	31.60	19	20.83	13	3.13	24	7.64	22	8.63	3
LOAD HOLES	34.42	15	18.71	18	5.35	6	10.37	13	7.34	10
TRACK MAINT	49.64	4	33.17	4	3.13	24	13.34	4	7.64	9
TIMBERING	53.46	2	33.92	2	4.75	10	14.79	2	8.51	4
JACKLEG DRILL	29.32	21	14.34	23	5.06	9	9.93	16	4.59	22
LOAD POWDER BAG	33.11	18	19.71	16	4.39	12	9.01	19	6.99	11
CONCRETE CON	44.52	8	29.00	6	3.92	13	11.59	9	6.56	13
PIPE INSTALL	41.52	11	22.06	11	9.09	2	10.38	12	6.70	12
VENT INSTALL	37.67	13	20.65	14	8.35	3	8.67	20	5.72	19
SURVEYING	44.97	7	27.03	8	3.13	24	14.82	1	5.93	15
STOPER DRILL	33.11	18	16.48	20	6.80	4	9.83	17	5.90	16
MINER HELPER	25.54	22	12.78	24	4.46	11	8.30	21	5.77	18
ROOF BOLTER	39.90	12	24.33	10	5.17	8	10.40	11	5.65	20
F VENT INSTALL	34.27	16	19.58	17	5.31	7	9.38	18	6.06	14
MUCK MACHINE	34.54	14	21.20	12	3.23	15	10.11	15	5.83	17
SLUSHER OPER	25.18	23	15.81	21	3.13	24	6.25	24	3.08	23
JUMBO DRILL	30.55	20	17.24	19	3.13	24	10.18	14	5.57	21
MINER OPER	24.38	24	15.00	22	3.13	24	6.25	24	2.83	24

HUMAN STRENGTH AS A PREDICTOR OF LIFTING CAPACITY

M.M. Ayoub, J.L. Selan, H.C. Chen
Texas Tech University
Lubbock, Texas

ABSTRACT

A sampling of lifting capacity prediction models utilizing human strength measurements as input are presented. The models utilize isometric, isokinetic, and/or isoinertial strength measurements. Comparison of such predictive models appears to indicate that lifting capacity is better predicted using dynamic rather than static strength tests.

INTRODUCTION

The physical strengths of individuals, alone or in conjunction with other individual/task variables, have been commonly used to predict lifting capacity for purposes of job design and employee placement. The purpose of this paper is to provide an overview of some of these prediction models.

HUMAN STRENGTH ASSESSMENT

The measurement of human strength is performed using either static (isometric) or dynamic (isokinetic, isoinertial) strength tests. Isometric refers to a test in which the length of the muscle does not change during performance (i.e., no movement occurs). The subject exerts a maximal voluntary effort against a fixed resistance. Isokinetic tests measure muscle strength while a body segment(s) moves at a constant speed using devices such as a CYBEX isokinetic dynamometer, with maximum torque constituting the output data. Isoinertial strength tests involve the movement of a constant mass (weight) on the part of the subject. The weight is increased in a prescribed manner until the maximum weight that can be handled is reached. Lifting capacity prediction models utilizing each of these strength testing methodologies will be reviewed.

DETERMINING LIFTING CAPACITY

The 3 primary approaches used to determine lifting capacity are 1) the biomechanical approach, 2) the physiological approach, and 3) the psychophysical approach. The biomechanical approach attempts to establish the physical stresses imposed on the musculoskeletal system during lifting, which then serve as the criteria upon which capacity of lift is based. The physiological approach uses metabolic energy expenditure criteria (e.g., oxygen consumption) to determine lifting capacity limits. The psychophysical approach requires the subject to adjust the weight of load according to his/her perception of effort such that the lifting task does not result in overexertion or excessive fatigue. The final weight decided on by the subject represents the maximum acceptable weight of lift. The majority of lifting capacity prediction models utilizing strength characteristics as predictors employed the psychophysical approach.

LIFTING CAPACITY PREDICTION MODELS

Table 1 presents a sample of lifting capacity prediction models using a measurement of human strength as a predictor variable. The models are listed in chronological order and indicate that, while earlier prediction models almost exclusively employed isometric strength tests, more recent efforts have utilized isokinetic or isoinertial tests. Also, in many models the strength variables are used in conjunction with other classes of variables (e.g., anthropometric characteristics) to predict lifting capacity. Finally, all of the models presented in Table 1 utilized the psychophysical approach to determine lifting capacity.

The lifting capacity prediction models developed by Ayoub, et al. (1978) will be used to exemplify the use of such models. These models, developed for six ranges of lift and employing strength and anthropometric data as input, are presented in Table 2. For an individual with the following characteristics:

Static arm strength = 40 kg
Shoulder height   = 150 cm
Back strength   = 100 kg
Abdominal depth   = 20 cm
Dynamic endurance = 1 minute
Body weight   = 73 kg

computation of lifting capacity for a floor-knuckle height lift (F-K) would proceed as follows:

Table 1. Lifting Capacity Prediction Models Using Strength
Measurements As Input Variables

SOURCE	MODELS	COMMENTS
McDaniel, 1972	Cap F-K (lb) = 11.9339 - 1.1204 x BS + 0.1581 x RPI^2 + 0.0046 x BS^2 - 8.8072 x SE^2 - 0.0955 x SEX x FI + 0.0601 x HT x RPI + 0.0321 x RPI x LS - 0.0002 x BS x LS - 0.0271 x LS x SE + 0.1109 x SE x FI	BS = isometric back strength RPI = reciprocal ponderal index SE = static endurance FI = fitness index (step test) HT = stature LS = isometric leg strength
Dryden, 1973	Cap K-S (lb) = 25.1212 + 0.3719 x SEX x DE	DE = dynamic endurance

Knipfer, 1974	Cap S-R (lb) = 5.225 x SEX + 0.0049 x SS
	
	+ 0.1944 x HP
		SS = isometric shoulder
		    strength
		HP = isometric horizontal
		    pushing strength

Ayoub, et. al., 1978	See Table 2	

Garg, et al., 1980	T1 = 22.6 + 1.02 x T2
	
	T1 = 27.5 + 0.299 x T3
	
	T1 = -2.1 + 1.49 x T4
	
	T5 = 20.0 + 0.473 x T2
		T1 = psychophysical capacity
		    F-K (kg)
		T2 = isometric strength at
		    origin of lift
		T3 = isometric strength close
		    to body
		T4 = max weight hold for 3 sec
		T5 = psychophysical capacity
		    with slow vertical lift

Pytel and Kamon, 1981	Cap F-S (n) = 102 + 0.74 x DLS
	= 230 + 0.58 x DBES
	= 139 + 1.29 x DEFS
		DLS = isokinetic lifting
		    strength
		DBES = isokinetic back
		    extension
		DEFS = isokinetic elbow flexion
		Mini-Gym set at 0.73 was used
		to determine isokinetic
		strength

Ayoub, et al., 1982	L2 = 0.6489 X1 + 12.7477	L2 = capacity 1 hand tool box
		    lift (lb)
		X1 = isoinertial 6 ft
		    incremental lift
		Models using X1 developed for
		13 simulated USAF handling
		activities

Aghazadeh, 1983	BXB = 54.72 - 9.68C - 9.68LT
	- 2.21F + 0.27DS
	BXB = 4.94 - 9.68C - 0.11LT
	- 2.21F + 0.24SS + 0.12LS
		BXB = capacity for box and bag
		    lifting (lb)
		C = container type
		LT = lift range
		F = frequency of lift
		DS = isokinetic strength using
		    CYBEX dynamometer
		SS = isometric shoulder
		    strength
		LS = isometric leg strength

Ayoub, et al., 1985a	U = 10.248 + 1.166L	U = capacity supine lift using
		    2 hands (lb)
		L = isoinertial 6 foot
		    incremental lift
		Models using L as predictor
		developed for 11 simulated USAF
		maintenance activities

Ayoub, et al., 1985b	A1 = 6.457 + 0.323M	A1 = capacity K-S lift with
		    twisting motion
		M = isoinertial 6 foot
		    incremental lift
		Models developed using M as
		Predictor for 10 lifting/
		lowering tasks involving
		twisting/carrying

```
F-K = -101.17 + 0.232(AS) + 0.804(SH)
      + 0.266(BS) + 2.707(AD) + 0.640(DE)
      - Body Weight
    = -101.17 + 0.232(40) + 0.804(150)
      + 0.266(100) + 2.707(20) + 0.640(1)
      - 73
    = 36.45 kg.
```

As noted in Table 2, the F-K lifting capacity of 36.45 kg represents a frequency of 1 lift/min, and a box size of 46 cm (sagittal plane). Adjustment factors for various frequencies of lift and box sizes have been provided elsewhere (e.g., Ayoub, et al. 1983).

COMPARISON OF STRENGTH TESTS AS PREDICTORS OF LIFTING CAPACITY

Given that different strength testing methodologies are used in the development of lifting capacity prediction models, a major issue is their relative efficacy as predictors. Although available data are not conclusive, it would appear that dynamic strength tests may better predict lifting capacity than static tests. Ayoub, et al. (1982, 1985a, b) utilized both isometric and isoinertial strength tests in predicting handling capacity. Results indicated that an isoinertial lift to 6 foot level was the best single predictor of performance on the various handling activities. Aghazadeh (1983) and Aghazadeh and Ayoub (1985) predicted lifting capacity based on isometric and isokinetic strength tests. Although both isometric and isokinetic models predicted lifting capacity with a reasonable degree of accuracy, the isokinetic model produced less absolute error between actual and predicted weights of load. Garg, et al. (1980), Kamon, et al. (1982), and Mital and Karwowski (1985) have all suggested the superiority of isokinetic strength tests relative to static strength tests in predicting lifting capacity. Finally, Kroemer (1983) reported that test-retest reliability was higher for isoinertial strength tests than isometric tests.

REFERENCES

1. Aghazadeh, F. "Simulated Dynamic Lifting Strength Models for Manual Lifting." Unpublished Ph.D. Dissertation, Texas Tech University, Lubbock, Texas, 1983.

2. Ayoub, M.M.; Bethea, N.J.; Deivanayagum, S.; Asfour, S.S.; and M. Sherif. "Determination and Modeling of Lifting Capacity." Final Report HEW (NIOSH) Grant No. 5R010H-00545-02, September 1978(a).

3. Ayoub, M.M.; Denardo, J.D.; Smith, J.L.; Bethea, N.J.; Lambert, B.K.; Alley, L.R.; and Duran, B.S. Establishing Physical Criteria for Assigning Personnel to Air Force Jobs. Final Report Air Force Office of Scientific Research Contract No. F49620-79-C-0006, September, 1982.

4. Ayoub, M.M.; Gidcumb, C.F.; Hafez, H.; Intaranont, K.; Jiang, B.C.; and Selan, J.L. "A Design Guide for Manual lifting". OSHA, 1983.

5. Ayoub, M.M.; Smith, J.L.; and Selan, J.L. Development of an Employee Screening Program for Emery Air Freight. Final Report prepared for Emery Air Freight. Final Report prepared for Emery Air Freight, 1985.

6. Ayoub, M.M.; Smith, J.L.; Selan, J.L.; Chen, H.C.; Fernandez, J.E.; Lee, Y.H.; and Kim, H.K. Manual Material Handling in Unusual Positions Phase II. Final Report prepared for University of Dayton Research Institute, 1985.

7. Dryden, R.D. A Predictive Model for the Maximum Permissible Weight of Lift from Knuckle to Shoulder Height. Unpublished Ph.D. Dissertation, Texas Tech University, Lubbock, Texas, 1973.

8. Garg, A.; Mital, A.; Asfour, S.S. A Comparison of Isometric Strength and Dynamic Lifting Capability. *Ergonomics*, 1980, 23(1), 13-27.

9. Kamon, E.; Kiser, D.; Pytel, J.L. Dynamic and Static Lifting Capacity and Muscular Strength of Steelmill Workers. *American Industrial Hygiene Association Journal*, 1982, 43(11), 853-857.

10. Knipfer, R.E. Predictive Models for the Maximum Acceptable Weight of Lift. Unpublished Ph.D. Dissertation, Texas Tech University, Lubbock, Texas, 1974.

11. Kroemer, K.H.E. An Isoinertial Technique to Assess Induvidual Lifting Capability. *Human Factors*, 1983, 25(3), 493-506.

12. McDaniel, J.W. Prediction of Acceptable Lift Capability. Ph.D. Dissertation, Texas Tech University, Lubbock, Texas, 1972.

13. Mital, A. and Karwowski, W. Use of Simulated Job Dynamic Strength (SJDS) in Screening Workers for Manual Lifting Tasks. *Proceedings of the Human Factors Soceity*, 1985, 513-515.

14. Pytel, J.L. and Kamon, E. Dynamic Strength Test As A Predictor for Maximal and Acceptable Lifting. *Ergonomics*, 1981, 24(9), 663-672.

Table 2.   Regression Coefficients for Maximum Acceptable Weight of
Lift Plus Body Weight (kg) for Both Males and Females*

Regression Terms	Lifting Ranges					
	F-K	F-S	F-R	K-S	K-R	S-R
Constant Term	−101.17	−115.83	−96.62	−83.47	−94.30	−96.59
Arm Strength	0.232	0.097	0.014	0.208	0.253	0.011
Shoulder Height	0.804	0.863	0.710	0.620	0.730	0.794
Back Strength	0.266	0.220	0.268	0.244	0.172	0.221
Abdominal Depth	2.707	3.138	3.264	3.077	2.854	2.746
Dynamic Endurance	0.640	1.370	0.134	0.164	0.541	0.246

*Adjusted to 1 lift/min, box size of 46 cm in sagittal plane

EFFECTS OF POSTURE ON BACK STRENGTH AND LIFTING CAPACITY

Sean Gallagher and Thomas G. Bobick
U.S. Department of the Interior - Bureau of Mines
Pittsburgh Research Center
Pittsburgh, Pennsylvania

ABSTRACT

The Bureau of Mines performed a pilot study examining the effects of posture on back strength and Maximum Acceptable Weight of Lift (MAWL) on six healthy male subjects ($\underline{M}$ = 32 years $\pm$ 4 $\underline{SD}$). Six back strength measurements (3 static and 3 dynamic) were made while the subjects were kneeling and standing. In addition, these subjects (who were unaccustomed to lifting in these postures) volunteered to participate in a study of psychophysically determined MAWL in both postures. Results of the back strength tests showed a significantly lower peak torque per body weight output in kneeling versus standing back strength measurements for five out of six test comparisons ($\underline{p}$ < .05). Subjective estimates of lifting capacity in the kneeling posture were significantly lower than those for the stooped posture ($\underline{p}$ < .05). The results of tests of back strength and lifting capacity in these two postures provide useful information to consider in determining the physiological and psychophysical stresses imposed by these work postures.

INTRODUCTION

Miners who work in low-seam coal mines (roof height $\leq$ 48 inches) often must handle materials in severely constrained postures (Peay, 1983). Supplies that must be manually lifted in low coal are typically lifted while kneeling or stooped (Gallagher, 1985). Despite the abundance of manual materials-handling literature, relatively little is known about the physiological and psychophysical responses to materials handling in these positions. Body posture has been shown to affect muscular strength capabilities (Ayoub, et al., 1981). However, a review of the literature provided no information on the back strength capabilities of persons while kneeling.

Back strength measurements have been shown to be positively correlated with the ability to lift (Poulsen, 1981). Typically, back strength has been measured while the subject is standing (Ayoub, et al., 1981; Poulsen, 1981; Marras, et al., 1984). This posture may involve the measurement of muscular forces other than those of the low back region, especially the strong muscles of the posterior leg region (i.e. biceps femoris, semimembranosus, semitendinosus, gastrocnemius, and soleus). While these measurements may be an indication of the total muscular force that workers may utilize when lifting in a standing posture, they may not accurately reflect the muscular strength available when they must perform lifting tasks while kneeling. In addition, it is possible that back strength measurements taken in the kneeling position may correlate better with the ability to lift during kneeling materials-handling activities. One aim of the present study was to examine the hypothesis that back strength measurements (both isometric and isokinetic)

taken in a kneeling posture may be less than those obtained when standing.

The position or bearing of the body has been shown to have an important effect on biomechanical, physiological, and psychophysical parameters during the performance of work tasks (Astrand and Rodahl, 1974; Adams and Hutton, 1981; Westgaard and Aaras, 1984). A stooped posture causes the weight of the torso to be added to the stress imposed on the low back during a lifting task. Forward flexion of the vertebral column causes an increase in electrical activity of the back muscles until flexion is extreme (Floyd and Silver, 1955). In the extremely flexed posture, electrical discharge from the back muscles ceases and the load is assumed by the ligamentary structure of the back (Basmajian and DeLuca, 1985). This may lead to increased incidence of muscle strain or sprain. Deviation from the erect posture also causes a decrease in total lung volume and oxygen consumption and results in higher ventilation rates (Moreno and Lyons, 1961). Finally, changes in posture have been demonstrated to increase loadings on the small muscle groups of the upper limbs and torso, as well as causing circulatory changes such as higher blood pressures and heart rates, and redistribution (pooling) of the blood supply (Ayoub, et al., 1981). All of these changes increase the physiological responses of workers that have to assume these postures, thus increasing fatigue and the corresponding risk of injury they may experience. The second purpose of the present investigation was to examine the effects of posture on psychophysically determined lifting capacity. The results of the present study and future Bureau studies will be used to develop lifting guidelines for low coal mines.

METHOD AND PROCEDURE

Six healthy males (M = 32 ± 4 SD) participated in a study examining the effects of posture on back extensor strength and lifting capacity. Subjects were volunteers from the U.S. Bureau of Mines Research Center in Pittsburgh, PA, and had no prior experience in handling materials in restricted work postures. Each subject was required to undergo a thorough physical examination and graded exercise tolerance test prior to their participation in the experiment, to ensure that no health problems were present that would put the subjects at an increased risk of injury. Informed consent was obtained from all participants in the study. Prior to the start of testing, each subject warmed-up by exercising for five minutes on a bicycle ergometer and then performed a series of five back and trunk stretching exercises prior to testing.

Back strength was measured in standing and kneeling postures using a CYBEX Isokinetic Dynamometer (LUMEX. Inc.)[1]. A total of 12 conditions were studied in this experiment: six kneeling and six standing. In each posture, three back strength measurements were taken using an isometric contraction (22.5°, 45.0°, and 67.5° from the vertical) and three measurements were made using a dynamic contraction (30°/sec, 60°/sec, and 90°/sec). Figure 1 shows the device used to measure back strength during tests in the standing and kneeling postures. The subject was secured by two pelvic stabilization straps in each posture. All back strength test conditions were conducted in a randomized order.

The Maximum Voluntary Contraction (MVC) for each test condition was obtained using a test-retest procedure whereby peak torque measurements (kilogram-meters) of two maximal exertions were required to be within 10% of one another. The higher of these two values was taken as the MVC for that test condition (Stobbe and Plummer, 1984). Two minutes rest was given between exertions, and consistent verbal encouragement was given to the subject in order to facilitate maximal exertions from the participants. One subject did not complete the back strength testing portion of the experiment due to equipment problems and was excluded from the analysis of these data.

In the study of lifting capacity, each subject was asked to adjust the weight in a 50.8- by 33.0- by 17.8-cm (20- by 13- by 7-in) lifting box according to his estimate of lifting capacity for each posture (stooped or kneeling). The lifting tasks were performed

[1]Reference to specific brands, equipment, or trade names in this report is made to facilitate understanding and does not imply endorsement by the Bureau of Mines.

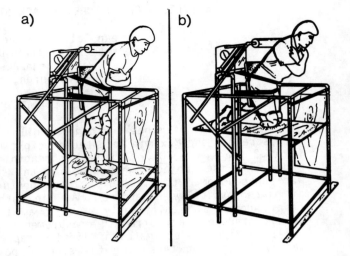

Figure 1. Subject performing back strength measurement in a) standing, and b) kneeling posture.

under an adjustable-height mine simulator that restricted the subject's posture. The height of the simulator was set at 121.9 cm (48 in) for this study. The test set-up is shown in figure 2. Lifting instructions were given to the subject before the experiment started. In this study, the subjects were told to adjust the weight in the box so the load could be handled for a 20-minute period (the actual lifting period) and to assume that this 20 minutes of lifting would have to be performed four times during a workday.

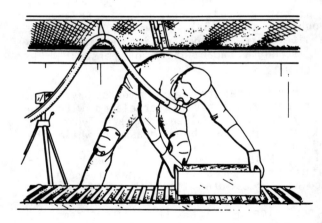

Figure 2. Subject performing stooped MAWL test.

The subject lifted the box at a frequency of 10 lifts/min for two 20-minute periods in each posture. One period started with a heavy box, weighing approximately 38.6 kg (85 lb) and the other with a light box, weighing approximately 6.8 kg (15 lb) in order to control for bias due to initial starting weight of the box. A ten-minute rest break was provided between tests so that subjects could rest and/or attend to personal needs. The average subjectively determined weight chosen for the two test conditions in a posture was taken as the maximum acceptable weight of lift (MAWL) for that posture.

The primary dependent measures for the lifting study were the MAWL for the kneeling (KMAWL) and stooped (SMAWL) postures. Secondary dependent measures included heart rate (HR), oxygen utilization ($\dot{V}O_2$), and ventilation volume ($\dot{V}_E$). Heart rate was obtained during the last ten seconds of every minute using a Beckman Dynograph Recorder, Model 511-A. The average heart rate for each condition was taken as the average of the final 15 values obtained. $\dot{V}O_2$ and $\dot{V}_E$ values were obtained approximately every 30 seconds during the final five minutes of lifting using a Beckman Metabolic Measurement Cart. The data were averaged by the number of values acquired during this five-minute period. One subject was not able to finish the lifting portion of the experiment due to other work commitments, and was excluded from the data analysis.

RESULTS

Tables 1 and 2 show the results of the static and dynamic back strength tests, respectively. The results showed a significantly lower peak torque/body weight output in kneeling versus standing back strength measurements for five out of six test comparisons ($p < .05$). The only posture where statistical significance was not achieved was the comparison between standing and kneeling static back strength in the fully flexed (67.5° from the vertical) position ($p = .08$).

Table 3 gives the results for the psychophysical lifting study. Subjective estimates of lifting capacity were significantly lower for the kneeling posture than for the stooped posture ($p < .05$). However, despite the fact that less weight was lifted in the kneeling posture, the secondary dependent measures of HR, $\dot{V}O_2$, and $\dot{V}_E$ were all higher than those in the stooped posture. These differences, however, were not statistically significant ($p > .05$).

Table 1. Static back strength (N=5) expressed in torque produced (kg-m)/body weight (kg).

	Standing	Kneeling	Significance
22.5°	0.34 (± 0.08)	0.23 (± 0.05)	$p < .01$
45.0°	0.40 (± 0.07)	0.21 (± 0.04)	$p < .01$
67.5°	0.26 (± 0.09)	0.17 (± 0.03)	n.s. ($p = .08$)

n.s. - not significant

Table 2. Dynamic back strength (N=5) expressed in torque produced (kg-m)/body weight (kg).

	Standing	Kneeling	Significance
30°/sec	0.28 (± 0.08)	0.22 (± 0.07)	$p < .05$
60°/sec	0.26 (± 0.04)	0.20 (± 0.03)	$p < .01$
90°/sec	0.19 (± 0.06)	0.13 (± 0.02)	$p < .05$

Table 3. Results of maximum acceptable weight of lift test (N=5).

	Stooped	Kneeling	Significance
MAWL (kg)	27.2 (± 9.3)	22.5 (± 6.9)	$p < .05$
HR (bpm)	133 (± 18)	139 (± 27)	n.s.
$\dot{V}O_2$ (ml/kg/min)	15.9 (± 2.2)	17.1 (± 4.0)	n.s.
$\dot{V}_E$ (1/min)	38.7 (± 6.1)	40.1 (± 9.1)	n.s.

n.s. - not significant

Table 4. Pearson correlation coefficient and statistical significance between stooped MAWL and standing static and dynamic back strength measurements (N=4).

Stooped MAWL	Standing Static 22.5°	Standing Static 45°	Standing Static 67.5°	Standing Dynamic 30°/sec	Standing Dynamic 60°/sec	Standing Dynamic 90°/sec
Correlation Coefficient	.60	.88	.81	.79	.91	.35
Statistical Significance ($p=$)	.40	.12	.19	.21	.09	.65

Table 5. Pearson correlation coefficient and statistical significance between stooped MAWL and kneeling static and dynamic back strength measurements (N=4).

Stooped MAWL	Kneeling Static 22.5°	Kneeling Static 45°	Kneeling Static 67.5°	Kneeling Dynamic 30°/sec	Kneeling Dynamic 60°/sec	Kneeling Dynamic 90°/sec
Correlation Coefficient	.86	.97	-.15	.83	.96	.58
Statistical Significance ($p=$)	.14	<.05	.85	.17	<.05	.42

Table 6. Pearson correlation coefficient and statistical significance between kneeling MAWL and standing static and dynamic back strength measurements (N=4).

Kneeling MAWL	Standing Static 22.5°	Standing Static 45°	Standing Static 67.5°	Standing Dynamic 30°/sec	Standing Dynamic 60°/sec	Standing Dynamic 90°/sec
Correlation Coefficient	.52	.75	.74	.81	.80	.42
Statistical Significance ($p=$)	.48	.25	.25	.19	.20	.58

Table 7. Pearson correlation coefficient and statistical significance between kneeling MAWL and kneeling static and dynamic back strength measurements (N=4).

Kneeling MAWL	Kneeling Static 22.5°	Kneeling Static 45°	Kneeling Static 67.5°	Kneeling Dynamic 30°/sec	Kneeling Dynamic 60°/sec	Kneeling Dynamic 90°/sec
Correlation Coefficient	.79	.98	-.38	.89	.99	.78
Statistical Significance ($p=$)	.21	<.05	.62	.11	<.01	.22

Four subjects performed both back strength and psychophysical lifting tests. In an effort to determine whether any of the back strength measurements related to the psychophysically determined maximum acceptable weight of lift, a post hoc correlation analysis was performed. Results of this analysis are given in Tables 4 through 7.

DISCUSSION

The results of the back strength tests supported the hypothesis that a significant amount of muscular force is generated by the posterior leg muscles when back strength is measured in the standing posture. Measurement of back strength in the kneeling posture apparently reduces the contribution of the leg muscles to the back strength measurements. In this case, the back muscles may be better isolated. In the present study, it is likely that there is still a considerable contribution from muscles other than those directly supporting the vertebral column (especially the glutei and hamstrings) during the kneeling back strength measurements. However, the input from these muscles is somewhat decreased and the contribution from the gastrocnemius and soleus is probably diminished to a greater extent. The findings of the present investigation suggest that further research may be necessary to determine the best posture for examination of back musculature function.

The data from the psychophysical lifting study indicate that subjects unaccustomed to performing materials-handling tasks in restricted work postures find it more difficult to lift weight in the kneeling posture than stooped. This is probably attributable to the size of the muscle mass used during the lifting procedure. It seems clear that in the kneeling posture, the muscular mass available to accomplish a lifting task is a good deal smaller than that available in the stooped posture, thus less weight is subjectively chosen in this posture by the subjects during the lifting capacity tests. Although the leg muscles have a limited utility in the stooped posture, apparently they still are able to contribute somewhat to the lifting process in this position. It should be noted that some subjects found the stooped posture to be more uncomfortable than the kneeling posture, although none of the subjects terminated their participation due to discomfort.

One unanticipated finding of this study is that the physiological measures of heart rate, oxygen consumption, and ventilation volume are all higher in the kneeling posture despite the fact that significantly less weight is lifted in this position. Although this difference does not achieve statistical significance, it is probably due to the difference in workload handled in the two postures. In other words, it may be that in order for the subject to achieve the same physiological workload in each posture, more weight must be handled by the subjects in the stooped position.

There are two possible physiological principles that may explain why the kneeling posture may lead to elevated heart rate, oxygen consumption, and ventilation volume values. The primary reason deals with the size of the muscle mass available to perform a given workload. The smaller the muscle mass used to accomplish a workload, the higher the heart rate and respiratory adjustment will be, and the sooner the work will have to be interrupted due to exhaustion (Stegemann, 1981). The second reason that kneeling may contribute to higher physiological values is that blood flow to the lower extremities may be partially inhibited. Diminishing the blood flow to the legs has also been shown to increase the heart rate (Stegemann, 1981).

Analysis of the relationship between back strength and lifting capacity shows that two back strength tests (i.e. kneeling static 45°, and kneeling dynamic 60°/sec) correlate well with MAWL in both stooped and kneeling positions. None of the standing back strength measurements are found to correlate significantly with lifting capacity in either posture. These data suggest that certain kneeling back strength measurements may be better predictors of lifting capacity in restricted work postures than back strength measurements taken in the fully erect posture. Further research is necessary to assess this method of predicting lifting capacity.

REFERENCES

Adams, M.A., and Hutton, W.C. The effect of posture on the strength of the lumbar spine. *English Medicine*, 1981, 10, 199-202.

Astrand, P.O., and Rodahl, K. Textbook of work physiology. McGraw-Hill Book Co., New York, 1977.

Ayoub, M.M., Bethea, N.J., Bobo, M., Burford, C.L., Caddel, K., Intaranont, K., Morrissey, S. and Selan, J. Mining in Low Coal: Volume I - Biomechanics and Work Physiology (Final Report -U.S. Bureau of Mines contract H0387022), 1981.

Basmajian, J. V., and DeLuca, C.J. Muscles alive. Williams and Wilkins, Baltimore, 1985.

Floyd, W.F., and Silver, P.H.S. The function of the erectores spinae muscles in certain movements and postures. *Journal of Physiology*, 1955, 129, 184-203.

Gallagher, S. Back injury research at the U.S. Bureau of Mines ergonomics laboratory. Paper in Proceedings, Fourth Annual Meeting of the Collegiate Association for Mining Education (Rolla, MO, October 3-4, 1985), 170-191.

Marras, W.S., King, A.I., and Joynt, R.L. Measurements of loads on the lumbar spine under isometric and isokinetic conditions. *Spine*, 1984, 9, 176-198.

Moreno, F., and Lyons, H.A. Effect of body posture on lung volumes. *Journal of Applied Physiology*, 1961, 16, 27-29.

Peay, J.M. (comp.) Back injuries, U.S. Bureau of Mines Information Circular 8948, 1983, 110 pp.

Poulsen, E. Back muscle strength and weight lifting limits in lifting burdens. *Spine*, 1981, 6, 73-75.

Stegemann, J. Exercise physiology. Yearbook Medical Publishers, Chicago, 1981.

Stobbe, T.J., and Plummer, R.W. A test-retest criterion for isometric strength testing, *Proceedings of the Human Factors Society Annual Meeting*, 1984, 455-459.

Westgaard, R.H., and Aaras, A. Postural muscle strain as a causal factor in the development of musculo-skeletal illnesses. *Applied Ergonomics*, 1984, 15, 162-174.

THE EFFECTS OF EXPECTATION ON TRUNK LOADING

W. S. Marras, S. A. Lavender, and S. L. Rangarajulu
Department of Industrial and Systems Engineering
The Ohio State University
Columbus, OH 43210

ABSTRACT

The literature has reported that there is a link between sudden unexpected load handling and the risk of a low back injury. It was hypothesized that sudden unexpected loads would create excessive forces upon the trunk due to the overcompensation of the trunk muscles. An experiment was performed to test this theory and quantify the degree of overcompensation. Subjects were asked to hold a box in a static lift position while weights ranging from 5 to 20 pounds were dropped into the box from a constant height. Under some conditions (expected) the subjects were permitted to observe the weight drop while under other conditions (unexpected) the subjects were deprived of visual and auditory cues during the weight drop. Generally, it was found that during sudden unexpected loading the trunk response resembled an expected loading of twice the weight value. These findings may provide guidelines for work situations where unexpected loading conditions are common.

INTRODUCTION

Several researchers have pointed out that the risk of a back injury increases in environments where workers are exposed to sudden unexpected loadings (Magora 1973, Andersson 1981). Epidemiologist have observed such occurences where loads may shift suddenly or when a worker attempts to lift an object whose weight he underestimates. We also know that slips and falls also account for a significant portion of low back disorder (LBD) injuries. Slips and falls also contain an unexpected element which is not well understood.

It is hypothesized that the increase in LBD risk due to unexpected loading of the trunk may be due to an overcompensation of the trunk muscles to the unexpected load. Thus, greater loading may be imposed upon the spine compared to loadings which occur under similar expected conditions. This trunk muscle over response is believe to overload the muscles in two ways. Both the magnitude of trunk muscle forces as well as the temporal characteristics to trunk muscle loading may change significantly during the overcompensation event. If overcompensation occurs, the mean and peak force magnitudes generated by a trunk muscle would be expected to be significantly greater under unexpected conditions. The rate of muscle force onset as well as the time of muscle force duration would also be expected to increase under unexpected conditions. Studies have shown that when greater levels of force are exerted, the onset rate of muscle force increases (Kroemer and Marras, 1981).

Knowledge of expectation may be valuable for the design of the work environment. If the manner in which the trunk is overloaded is understood, it may be possible to specify load handling limits so the load on the spine, even during sudden unexpected loading, does not exceed an acceptable level. This knowledge could then be applied to unstable load handling situations, such as liquids handling, where it is common to experience unexpected or shifting loads.

The objective of this study was to determine the degree to which the spine is overloaded using unexpected loading compared with expected conditions. In order to quantify this over response, the key components of the trunk muscle magnitude and temporal components were investigated. These components include: 1) the average relative magnitude of muscle force, 2) the peak relative force exerted by the muscle, 3) the onset rate of muscle force development during the exertion, and 4) the time duration of the muscle force. These components were investigated as a function of both sudden expected and sudden unexpected load handling and as a function of load weight.

METHOD

Subjects

Subjects in this experiment consisted of 12 male volunteers ages 21 to 32. All subjects were in good physical shape and none had a history of LBD. Mean subject height was 71.7 inches and mean weight was 175.8 lb. Subjects were informed of the experimental conditions but were not told the experimental hypothesis.

Design

Independent variables in this experiment consisted of the expectation condition and the weight level of the load handled. Expectation conditions consisted of both expected and unexpected loadings of the trunk. Under both conditions a load was suddenly imposed upon the subject, by dropping a weight into a box which the subject was holding. Under the expected condition, the subject was allowed to view the weight dropping into the box. However, under unexpected conditions the subject was deprived of visual or auditory cues which would indicate when the weight would be dropped into the box. Weight conditions consisted of four levels: 5, 10, 15 and 20 lbs.

The dependent variables consisted of the electromyographic (EMG) activity of the trunk muscles. These muscles include the right latissimus dorsi (LATR), left latissimus dorsi (LATL), right erector spinae (ERSR), left erector spinae (ERSL), right rectus abdominus (RCAR) and left rectus abdominus (RCAL). These muscles represent the main supporting structures of the trunk. These muscles are also primarily responsible for loading the trunk. The EMG activity was "integrated" (RMS integration) so that the signal represented the relative force produced by the muscle.

Procedure

Subject muscle activity was monitored with surface electrodes. These electrodes were glued to the muscles of interest according to standard procedures (Basmajian 1978). These electrodes were connected to small lightweight preamplifiers which were attached to a belt worn by the subject. The signals were then amplified, filtered, and "integrated".

The subject was asked to hold a box in a static position. The box was 12 x 12 x 9 inches in size and was used to catch the experimental weights. The box contained a micro switch which indicated the point in time at which the weight was dropped into the box.

The 6 EMG signals and the micro switch signal were fed into an analog-to-digital converter and then monitored by a micro computer.

Subjects stood in a static position with the box held in a position such that the lower arms were horizontal and the box was positioned 17.5 inches from the spine. With the body in this position all experimental weight conditions fall within the action limit as defined by NIOSH (1981). During the experimental trials weights were dropped from a height of 33 inches above the box. During unexpected conditions the subject was blindfolded and wore earplugs so that he would not be aware of exactly when the weight would be dropped. Under the expected condition the subject was permitted to view the weight drop. Subjects were permitted at least a two minute rest period between trials.

Data Treatment

In order to quantify the muscle activity occuring during the experimental conditions a measure of peak and average magnitude was defined in terms of the percent increase over static EMG activity. This measure is determined for each peak or average muscle signal by computing the sudden EMG component minus the static component needed to support the weight under a holding condition divided by the maximum EMG activity for a given muscle minus the static hold EMG component.

Temporal rate of signal onset was defined according to Kroemer and Marras (1980, 1981). The duration of the EMG signal was defined as the length of time the EMG signal magnitude was above the static holding value for a given weight.

RESULTS

A summary of the experimental results appears in Table 1. Relative trunk force magnitudes revealed significant differences for the expectation and weight conditions but not for their interaction. When average muscle force was considered as a function of expectation, four of the muscles exhibited significant differences. In each of these cases the unexpected condition resulted in a greater relative muscle force. Among these four muscles the unexpected condition produced muscle force increases which averaged 2.7 times those of the expected

conditon. Three of the muscles tested also displayed significantly greater muscle force as the weight condition increased.

When peak EMG activities were evaluated all muscles indicated a significant difference between expectation conditions. In this case the unexpected condition also produced greater peak forces than the expected conditions. The unexpected condition increases averaged 1.7 times those of the expected conditions. Five of these muscles also indicated an increase in activity when the weight conditions were increased.

Table 1 also indicated that significantly greater rates of signal onset were observed for all muscles under unexpected conditions. The greatest of these diffences were seen in the rectus abdominus muscles.

Muscle force duration was also significantly different between expectation conditions (F(1,88) = 6.52, p<.05) and weight conditions F(3,88) = 19.77, p<.05). Unexpected conditions produced muscle force durations which were 12% greater than expected conditions, whereas, load increases of 5 lb. increased duration by .09 sec.

Table 1: ANOVA Summary of F Values.
Summary of Significant Effects

	Expectation			Weight			Expectation X Weight		
	Avg.	Peaks	Onset	Avg.	Peaks	Onset	Avg.	Peaks	Onset
LATR	0.58	4.83*	19.21*	1.31	4.35*	0.61	1.42	0.59	0.78
LATL	7.46*	25.7*	6.26*	6.11*	8.80*	2.46*	0.68	2.16	0.62
ERSR	19.10*	18.20*	7.72*	5.36*	7.12*	0.39	0.96	0.61	0.41
ERSL	4.99*	7.22*	7.07*	1.96	5.54*	0.46	0.85	2.01	0.46
RCAR	2.19	4.36*	16.22*	5.51*	2.82*	0.67	1.41	1.07	0.91
RCAL	7.69*	13.10*	26.42*	2.03	2.48	2.29	0.50	0.85	0.44

* = Significant at .05 level

DISCUSSION

These findings suggest that there is indeed an overreaction of trunk loading that occurs when sudden unexpected loads are imposed upon the trunk during manual materials handling. In order to understand the nature of this FMC magnitude overreaction, the expected and unexpected reactions were observed simultaneously for all muscles as a function of load weight. The peak components of muscular force displayed significant differences to both expectation and weight for all muscles reactions compared. This comparison is shown in Figure 1. This figure indicates that the effect of unexpected loading essentially makes the muscles respond as if they were doubling the weight of an expected load. For example, Figure 1 indicates that a 5 lbs. unexpected load elicits peak muscle responses which resemble a 10 lb. expected load. Similarly, a 10 lb. unexpected load elicits peak muscle responses which are similar to a 20 lb. expected load. Further research is needed to determine how this relationship behaves throughout a larger weight range.

The rate of muscular force onset is also exaggerated in an unexpected exertion. This may lead to a particularly dangerous situation since greater trunk forces are occurring very rapidly. This suggest that the occurence of great acceleration or jerk forces may occur more simultaneously among muscles and lead to a traumatic low back disorder. These mechanisms may also represent the basis for LBD injury during slips and falls. Increased muscle loading in conjuction with an impact may intensify trunk trauma.

The data presented in this study indicates that differences in response to expected and unexpected loading are quantifiable. Future research is needed to fully understand the implications of these findings. This information represents a method by which guidelines could sents a method by which guidelines could be developed for safe loading limits applicable to unstable load handling. This information may also provide a useful starting point for a biomechanical model which describes the powerful cognitive-biasing effects of expectation upon the musculo-skeletal system.

REFERENCES

Andersson, C. P. (1981). Epidemologic aspects of low back pain in industry. Spine, 6 (1),53-60.

Basmajian, J. V. (1978). Muscles Alive: Their functions revealed by electromyography, (4th ed.), Williams and Wilkins, Baltimore, MD.

Kroemer, K. H. and W. S. Marras (1980). Towards an objective assessment of the "maximal voluntary contraction" component in routine muscle strenght measurements. Eur J Appl Physiol, 45, 1-9.

Kroemer, K. H. and W. S. Marras (1981). Evaluation of maximal and submaximal static muscle exertions. Human Factors, 23 (6), 643-653.

Magora, A. (1973). Investigaton of the Relation between low back pain and occupation 4. Physical requirements: bending, rotation, reaching and sudden maximal effort. Scand J Rehab Med, 5, 191-196.

NIOSH technical report (1981). Work practice guide for manual lifting. DHHS (NIOSH) Publication No. 81-122, March.

MODELING OF MANUAL HANDLING IN UNUSUAL POSTURES

Selan, J.L., Ayoub, M.M., Smith, J.L., Chen, H.C., Lee, Y.H., Kim, H.K.
Texas Tech University
Lubbock, Texas

ABSTRACT

Subjects performed a series of psychophysical lifting tasks simulating USAF aircrew maintenance activities. The tasks involved lifting while in unusual postures: lying prone, supine, sideways, and standing. Maximum acceptable weights on the tasks were then related to performance on a series of isoinertial and isometric strength tests. High correlations were obtained between lifting task and strength test performance. Prediction models developed indicated that an isoinertial 1.8 m lift and isometric one hand pull were the best predictors of task performance. The prediction models can be used in the establishment of design weights for aircrew maintenance tasks.

INTRODUCTION

The modeling of manual materials handling (MMH) activities has been utilized by numerous researchers (e.g., Ayoub, et al., 1982) for purposes of ergonomic job design and personnel placement. In general, these activities consisted of lifting within the sagittal plane of the subject. The purpose of the present study was to determine the relationship between a series of simulated USAF aircrew maintenance activities requiring lifting in unusual postures and a series of strength tests for purposes of modeling. This study is part of a large research and development program being conducted for the USAF to determine the relationships between the physical attributed and capabilities of aircraft maintenance technicians to enhance the maintainability of large systems.

METHODS

Subjects

50 male and 50 female students ranging from 18 to 25 years of age participated in the study. Subjects were selected based on height and weight to be representative of the USAF population. All subjects underwent a physical examination prior to participation in the study.

Lifting Tasks

Based on typical MMH activities performed by aircrew maintenance personnel, a series of eleven lifting tasks involving unusual postures were simulated. The following lifting tasks were performed:

- Face down lift (right, left hand)
- Face up lift (one, two hands; box starting position at chest, elbow level)
- Lying sideways (one, two hands; box starting position against body, at elbow distance)
- One hand standing lift.

The psychophysical approach was used for all lifting tasks. Subjects adjusted the weight in the container by adding/subtracting lead pieces of varying weights. The final weight selected represented the maximum acceptable weight of load for the task.

Strength Tests

A series of strength tests were conducted to relate to lifting task performance. An isoinertial incremental 1.8 m lift was performed using a modified Air Force Factor-X weight machine. The subject lifted weights in 44.6 n increments until the maximum weight that could be lifted to 1.8 m was determined. Three isometric strength tests were performed: right hand static pull, right hand sideways static exertion, and 38 cm vertical pull. Grip strength was measured using a hand dynamometer.

RESULTS

Lifting Tasks

Descriptive statistics for the various simulated aircrew maintenance tasks are presented in Table 1.

Strength Tests

Descriptive statistics for the various strength tests are presented in Table 2.

Prediction Models

The first step in the development of prediction models was determining whether a strong relationship existed between lifting task and strength test performance. This was done using correlation analysis, and is summarized in Table 3. Correlations ranged from 0.74 to 0.93. The right hand isometric pull and incremental lift consistently showed the highest correlations with lifting task performance, although all the strength tests correlated highly with the tasks.

A step-wise regression program was used to develop regression equations based on data from 70 randomly-selected subjects (30 subjects used for model validation). Variables entered the model based on maximum improvement in R-square. The right hand isometric pull and incremental lift entered the equations as the first two or three variables. Based on this, regression equations were developed using only the incremental lift (Table 4) or the right hand isometric pull (Table 5). R-squares ranged from 0.720 to 0.853 using the incremental lift, 0.665 to 0.813 using the isometric pull.

Since the incremental lift predicted as well or better than any other single strength test, and the incremental lift is a standard test performed in USAF entry stations, only the incremental lift model was validated using 30 subjects selected randomly. Table 6 indicates average and average absolute errors for predicted versus actual lifting task performance.

CONCLUSIONS

The prediction mdoels were considered adequate to predict individual lifting task performance based on performance on the isoinertial 1.8 m lift. These data will be used to establish design weights to enhance the maintainability of USAF systems.

REFERENCES

Ayoub, M.M., Denardo, J.D., Smith, J.L., Bethea, N.J., Lambert, B.K., Alley, L.R., and Duran, B.S., Establishing Physical Criteria for Assigning Personnel to Air Force Jobs, AF Contract No. F-49620-79-C-0006, Institute for Ergonomics Research, Texas Tech University, 1982.

Table 1. Lifting Task Data (Newtons)

Variable	Males $\overline{X}$	SD	Min	Max	Females $\overline{X}$	SD	Min	Max
A	188.65	43.16	129.18	320.73	89.05	19.15	62.36	135.86
B	193.02	45.57	122.50	322.95	90.07	18.13	66.82	143.55
C	136.71	31.05	89.09	213.82	64.28	13.94	37.86	104.68
D	132.52	27.22	84.64	204.91	65.35	14.25	37.86	102.45
E	156.67	36.93	100.23	236.09	79.69	18.71	44.55	113.59
F	178.14	39.56	115.82	267.27	87.71	20.98	42.32	131.41
G	195.11	56.88	86.86	358.59	93.63	23.83	57.91	158.14
H	244.20	60.27	131.41	394.23	127.36	31.98	80.18	196.00
I	545.64	135.55	347.45	917.64	239.08	45.93	164.82	383.09
J	656.96	156.18	414.27	957.73	289.28	50.83	191.55	454.36
K	432.36	78.17	249.45	603.59	197.96	36.93	135.86	296.23

A = Face Down Left Hand
B = Face Down Right Hand
C = Side Lift 1 Hand Close
D = Side Lift 1 Hand Elbow
E = Side Lift 2 Hand Close
F = Side Lift 2 Hand Elbow
G = Face Up 1 Hand Close
H = Face Up 1 Hand Elbow
I = Face Up 2 Hand Close
J = Face Up 2 Hand Elbow
K = 1 Hand Stand Lift

Table 2. Strength Test Data (Newtons)

Variable	$\overline{X}$	SD	Min	Max	$\overline{X}$	SD	Min	Max
	Males				Females			
M	553.43	105.97	343.00	797.36	306.21	59.91	173.73	454.36
N	102.45	23.34	57.91	200.45	52.30	9.84	31.18	71.27
P	1370.66	281.22	735.00	1982.27	731.08	194.17	343.00	1260.64
Q	390.66	70.25	264.60	539.00	194.35	48.64	107.80	362.60
R	522.96	109.80	311.82	846.36	212.04	58.71	0.00	311.82

M = Right Hand Pull
N = Sideways Pull
P = 38 cm Vertical Pull
Q = Grip Strength
R = 1.8 m Incremental Lift

Table 3. Lifting Task/Strength Test Correlation Matrix
(Male and Female Data Combined)

	M	N	P	Q	R
A	0.88	0.82	0.85	0.81	0.85
B	0.89	0.81	0.83	0.83	0.87
C	0.90	0.86	0.79	0.80	0.91
D	0.87	0.87	0.79	0.81	0.87
E	0.85	0.85	0.82	0.80	0.83
F	0.85	0.84	0.82	0.83	0.86
G	0.83	0.83	0.74	0.75	0.87
H	0.90	0.82	0.81	0.82	0.88
I	0.88	0.88	0.82	0.79	0.90
J	0.89	0.86	0.83	0.79	0.93
K	0.88	0.86	0.83	0.86	0.91

A = Face Down Left Hand
B = Face Down Right Hand
C = Side Lift 1 Hand Close
D = Side Lift 1 Hand Elbow
E = Side Lift 2 Hand Close
F = Side Lift 2 Hand Elbow
G = Face Up 1 Hand Close
H = Face Up 1 Hand Elbow

I = Face Up 2 Hand Close
J = Face Up 2 Hand Elbow
K = 1 Hand Stand Lift
M = Right Hand Pull
N = Sideways Pull
P = 38 cm Vertical Pull
Q = Grip Strength
R = Incremental Lift

Table 4. Regression Equations Using Incremental Lift
As Single Predictor (Newtons)

Equation   R-Square

A = 32.84 + 1.31 * R   0.734
B = 36.16 + 1.25 * R   0.738
C = 20.64 + 0.94 * R   0.791
D = 25.27 + 0.88 * R   0.755
E = 32.38 + 1.06 * R   0.720
F = 29.99 + 1.29 * R   0.796
G = 28.81 + 1.40 * R   0.719
H = 43.10 + 1.71 * R   0.760
I = 38.86 + 4.30 * R   0.798
J = 45.65 + 5.19 * R   0.853
K = 71.54 + 2.99 * R   0.835

A = Face Down Left Hand   G = Face Up 1 Hand Close
B = Face Down Right Hand   H = Face Up 1 Hand Elbow
C = Side Lift 1 Hand Close   I = Face Up 2 Hand Close
D = Side Lift 1 Hand Elbow   J = Face Up 2 Hand Elbow
E = Side Lift 2 Hand Close   K = 1 Hand Stand Lift
F = Side Lift 2 Hand Elbow   R = Incremental Lift

Table 5. Regression Equations Using Right Hand Isometric
Pull As Predictor (Newtons)

Equation   R-Square

A = -8.51 + 1.55 * M   0.774
B = -4.93 + 1.50 * M   0.798
C = -7.73 + 1.10 * M   0.813
D = -0.92 + 1.02 * M   0.772
E = -0.53 + 1.25 * M   0.753
F = -1.31 + 1.43 * M   0.738
G = -4.96 + 1.55 * M   0.665
H = -8.67 + 2.00 * M   0.787
I = -63.50 + 4.75 * M   0.734
J = -66.96 + 5.63 * M   0.753
K = -4.06 + 3.35 * M   0.788

A = Face Down Left Hand   G = Face Up 1 Hand Close
B = Face Down Right Hand   H = Face Up 1 Hand Elbow
C = Side Lift 1 Hand Close   I = Face Up 2 Hand Close
D = Side Lift 1 Hand Elbow   J = Face Up 2 Hand Elbow
E = Side Lift 2 Hand Close   K = 1 Hand Stand Lift
F = Side Lift 2 Hand Elbow   M = Right Hand Pull

Table 6. Validation of Incremental Lift as Predictor
of Lifting Task Performance (Newtons)

Task	Average Error	Average Absolute Error
A	−8.55	24.72
B	7.60	25.44
C	5.48	9.67
D	3.21	13.50
E	−6.33	20.27
F	−11.40	24.50
G	0.46	21.83
H	5.97	25.75
I	−4.50	43.25
J	−4.68	46.28
K	−11.76	48.11

A = Face Down Left Hand
B = Face Down Right Hand
C = Side Lift 1 Hand Close
D = Side Lift 1 Hand Elbow
E = Side Lift 2 Hand Close
F = Side Lift 2 Hand Elbow

G = Face Up 1 Hand Close
H = Face Up 1 Hand Elbow
I = Face Up 2 Hand Close
J = Face Up 2 Hand Elbow
K = 1 Hand Stand Lift

LIFTING PHYSICAL WORK CAPACITY AS A FUNCTION OF FREQUENCY

Jeffrey E. Fernandez
Robert J. Marley
Department of Industrial Engineering

and

Nancy B. Stubbs
Department of Physical Education, Health and Recreation

The Wichita State University
Wichita, Kansas

ABSTRACT

A laboratory experiment using 15 male subjects was conducted to document lifting physical work capacity over the frequency range of 2 to 12 lifts per minute and to compare these values to bicycling physical work capacity. Results indicate that bicycling PWC is significantly higher than lifting PWC at the 0.05 level. A variation in lifting PWC as a function of frequency was observed. This could be attributed to lifting technique. Task design should, therefore, not only consider lifting PWC but also the frequency of lift.

INTRODUCTION

Manual material handling (MMH) makes up a significant percentage of work performed in today's industry despite increases in automation in many areas. Not only are MMH activities an important part of modern industry, but they also have been long recognized as the major hazard to industrial workers (NIOSH, 1981). An estimated one-fourth to one-third of all compensable, work-related injuries in U.S. industry are a result of these activities (Kroemer, 1983).

In the United States, 25 percent of all injuries are directly related to MMH tasks (NIOSH, 1981). This amounts to over 25 million lost work days and between 4 and 20 billion dollars in compensation (Goldberg, 1980; Norby, 1981; and NIOSH, 1981).

Low back pain and injury is a special concern during the performance of MMH tasks. Lifting is a particularly stressful material handling activity to the lower back (Rowe, 1969; and Ayoub, et al., 1978). Klein, Jensen, and Sanderson (1984) report that 48.1 percent of workers compensation claims initiated because of back pain or strains were a result of lifting objects. Low back injuries due to lifting can obviously result in substantial costs, both economically as well as in terms of human anguish. Therefore, the elimination, or at least a reduction in the risk of injury due to lifting tasks, is a goal of researchers and practitioners from many disciplines.

Background

The National Institute for Occupational Safety and Health describes four methods by which an individual's lifting capacity can be analyzed (NIOSH, 1981). First, the epidemiological method identifies and classifies factors which influence injury. Second, the biomechanical method determines the forces imposed upon the musculoskeletal system. This method is generally considered appropriate for predicting maximal, low frequency lifting. The third approach is the psychophysical method which estimates an individuals' lifting capacity by quantifying their subjective tolerance to the stresses of manual material handling (Ayoub, 1987).

Finally, the fourth approach is the physiological method which is concerned with the physiological stress on the body and is most often applied to repetitive lifting tasks. Oxygen consumption, heart rate, and energy expenditure are the primary physiological criteria. An important dependent variable in this method is physical work capacity (PWC). PWC is synonymous with terms such as maximal oxygen uptake (VO_2 max) or aerobic capacity and is defined as the highest volume of oxygen uptake an individual can attain during a physical activity (Astrand and Rodahl, 1977). Physical work capacity is dependent upon several factors such as age, gender, body mass, training, task, environment, testing protocol, equipment, and genetic factors (Astrand and Rodahl, 1977).

Research Problem

Task design standards using criteria from the physiological method specify that energy requirements for lifting activities should generally not exceed 33 percent of an individual's physical work capacity (Bink, 1962). Mital and Shell (1984) determined that a 29 percent limit for males and 28 percent for females are a more realistic energy demand for eight hours of work.

Petrofsky and Lind (1978), Intaranont (1983), and Fernandez (1986) observed significantly higher PWC values obtained on the bicycle ergometer, a common alternate method to lifting tasks for determining PWC. These studies indicate that current standards using the bicycle ergometer overestimate an individual's lifting capacity, based upon physiological measures.

Secondly, Fernandez (1986) documented PWC values for three frequencies of lifts per minute in freestyle lifting from floor to knuckle height. The results show that PWC increased from 2 to 6 lifts per minute but decreased from 6 to 8 lifts per minute.

Objective

The purpose of this study was to conduct a laboratory experiment to document the physical work capacity of individuals lifting in a range of 2 to 12 lifts per minute, from floor to knuckle height, and to compare these values to those attained on the bicycle ergometer.

METHOD

Subjects

Fifteen male subjects were recruited from the student population at Wichita State University. They ranged in age from 18 to 28 years of age with a mean age of 22.6 years. These subjects were financially reimbursed for participating in the experiment.

Experimental Design

Each subject performed a lifting task at 8 different frequencies of lifts per minute. These frequencies were 2, 4, 5, 6, 7, 8, 10, and 12 lifts per minute. In addition, a bicycling task on the ergometer was performed. The order of testing for the various tasks was randomized, allowing for a completely randomized block design with subjects as blocks.

Familiarization Period

All subjects participated in a familiarization period in which they practiced lifting at 4 and 6 lifts per minute to become accustomed to the task and equipment.

Procedure

Physical work capacity. Values for PWC, for all eight lifting tasks and one bicycling task, were determined by the submaximal technique described by Kamon and Ayoub (1976). This method relies upon the strong linear relationship between heart rate and oxygen consumption.

For a particular task, three progressively increasing, submaximal workloads were introduced. Heart rate (HR) and oxygen consumption (VO_2) were monitored throughout the task using the Exersentry heart rate monitor and the Beckman Metabolic Measurement Cart. The subject's steady-state HR and VO_2 were recorded at every workload. The workload intensity was variably increased and adjusted to bring about specific changes in the HR of each subject. Heart rate at steady-state of the second workload was approximately 50 percent of the estimated maximum heart rate (Max HR), defined as 220-Age, and the heart rate at steady-state of the third workload was approximately 65 percent of Max HR.

Physical work capacity was therefore calculated by simple linear regression using the 3 steady-state measures of heart rate and oxygen consumption and extrapolating to Max HR.

Lifting PWC: Subjects performed a freestyle lifting task from floor to 30 inches (approximately knuckle height). The container was 18 x 12 x 12 inches, with handles. The workload was adjusted by changing the weight in the box.

Bicycling PWC: The Monark bicycle ergometer was used for this test. Subjects pedaled at a constant rate of 65 rpm and the workload was adjusted in terms of resistance in kilograms.

RESULTS AND DISCUSSION

The data was analyzed using the PROC ANOVA procedure of the SAS mainframe statistical package (SAS Institute, 1982). Table 1 reports the means and standard deviations of PWC values for the nine different tasks.

Results from the analysis of variance procedure show a significant block effect, $F(14,112) = 15.16$, $p < 0.0001$. This indicates differences in PWC values between individuals. Furthermore, frequency of lifts also showed a significant effect, $F(8,112) = 6.27$, $p < 0.0001$.

Duncan's multiple range test was performed upon the task means. A 0.05 level of alpha was specified. Results of the Duncan test indicated three significantly distinct ranges within the different tasks. First, PWC values attained on the bicycle ergometer were significantly higher than any of the lifting tasks. This result is consistent with findings of Petrofsky and Lind, Intaranont, and Fernandez. Secondly, PWC for 5 lifts per minute through 12 lifts per minute were significantly higher than the lowest range of 2 and 4 lifts per minute.

TABLE 1

Mean and Standard Deviation for Physical Work Capacity of Various Tasks (ml/minute). N = 15

Task	Mean	S.D.
2 Lifts/min.	1869.40	582.03
4 Lifts/min.	2054.60	519.42
5 Lifts/min.	2147.27	354.32
6 Lifts/min.	2235.93	412.30
7 Lifts/min.	2183.80	518.82
8 Lifts/min.	2291.80	479.14
10 Lifts/min.	2351.87	638.53
12 Lifts/min.	2343.67	490.21
Bicycle Erg.	2651.93	758.15

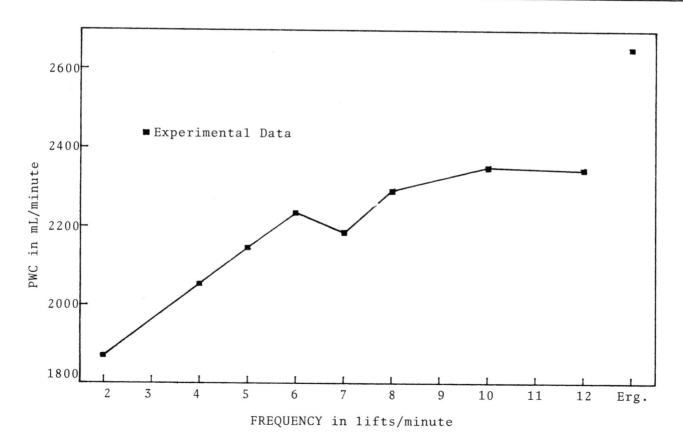

Figure 1. Physical Work Capacity as a function of Frequency.

The documentation of two distinct ranges within the spectrum of lifting frequency represents a lower rate of oxygen uptake as a function of heart rate in the 2 and 4 lifts per minute compared to the other frequencies. This is a result of the inability of these lower frequencies to stress the cardio-vascular system to the same degree as the higher frequencies.

In further examining the data, an important trend was established as shown graphically in Figure 1. It is evident that PWC increases steadily from 2 to 6 lifts per minute and then is nearly level from 8 to 12 lifts per minute. This curvilinear pattern corresponds to observed changes in lifting technique. At lower frequencies, subjects lifted with the squat style. However, at the higher frequencies, they tended to lift more in a stooping manner. This seems to be primarily a function of time constraints and a perception of fatigue at the higher frequencies. A decrease in the overall muscle mass involved during a stoop lift, compared to a squat lift, may account for a decrease in the rate of oxygen consumption as well as the estimated physical work capacity for that task.

A second trend was the decrease in PWC from 6 to 7 lifts per minute. Though not statistically significant, this drop was consistent in the data as 9 of the 15 subjects demonstrated this effect. This is very similar to the decrease from 6 to 8 lifts per minute found by Fernandez (1986). It is postulated that this reflects a transition in the lifting technique of most subjects from the squat to stoop method.

CONCLUSION

Based upon the results of this experiment, it is clear that in addition to other well documented factors that influence physical work capacity, lifting frequency and style of lift interact to further affect lifting PWC values.

This has implications for the design of manual material handling tasks in two areas. First, the capacity of a male individual performing a repetitive lifting task from floor to knuckle height should be based on a PWC value derived from lifting tasks since values from other tasks such as bicycle ergometry tend to overestimate their capacity.

Second, task design should also consider the frequency of lift. Signif-

icant differences were found between a range of 2 to 4 lifts and 5 to 12 lifts per minute. In addition, a gradual change in lifting techniques was observed. This indicates that at higher frequencies, individuals tend to stoop lift and are therefore at a higher risk of injury.

REFERENCES

Astrand, P.O., and Rodahl, K., (1977). Textbook of Work Physiology, 2nd Ed.. New York: McGraw-Hill.

Ayoub, M.M., (1987). The Problem of Manual Materials Handling. In Asfour, S.S. (Ed.), Trends in Ergonomics/Human Factors IV. New York: North-Holland.

Ayoub, M.M., Bethea, N.J., Deivanayagan, S., Asfour, S.S., Bakken, G.M., Liles, D., Mital, A., and Sherif, M., (1978). Determination and Modeling of Lifting Capacity. Final Report, HEW(NIOSH) Grant No. 5R01OH-00545-02, September.

Bink, B., (1962). The Physical Working Capacity in Relation to Working Time and Age. Ergonomics, 5(1), pp.25-28.

Fernandez, J.E., (1986). Psychophysical Lifting Capacity Over Extended Periods. Unpublished Ph.D. Dissertation, Texas Tech University, Lubbock, TX.

Goldberg, H.M., (1980). Diagnosis and Management of Low Back Pain. Occupational Health and Safety, June, pp. 14-15.

Intaranont, K., (1983). Evaluation of Anaerobic Threshold for Lifting Tasks. Unpublished Ph.D. Dissertation, Texas Tech University, Lubbock, TX.

Kamon, E. and Ayoub, M.M., (1976). Ergonomic Guide to Assessment of Physical Work Capacity. Published by American Industrial Hygiene Association. Akron, OH.

Klein, B.P., Jensen, R.C., and Sanderson, L.M., (1984). Assessment of Workers Compensation Claims for Back Strains/Pains. Journal of Occupational Medicine, 26(6), pp. 443-448.

Kroemer, K.H.E., (1983). Field Testing of Workers Involved in Materials Handling. Proceedings of U.S. Bureau of Mines Technology Transfer Seminar. Pittsburg, PA, August.

Mital, A., and Shell, R.L., (1984). A Comprehensive Metabolic Energy Model for Determining Rest Allowances for Physical Tasks. Journal of Methods Time Measurement, 11(2), pp. 2-8.

NIOSH, (1981). Work Practices Guide for Manual Lifting. NIOSH Technical Report No. 81-122. National Institute of Occupational Safety and Health, Cincinati, OH, March.

Norby, E.J., (1981). Epidemiology and Diagnosis in Low Back Injury. Occupational Health and Safety, January, pp. 38-42.

Petrofsky, J.S., and Lind, A.R., (1978). Comparison of Metabolic and Ventilatory Responses of Men to Various Lifting Tasks and Bicycling Ergometry. Journal of Applied Physiology, 45(1), pp. 60-63.

Rowe, M.L., (1969). Low Back Pain in Industry: A Position. Journal of Occupational Medicine, 11, pp. 161-169.

SAS Institue, Inc. (1982). SAS User's Guide, 1982 Ed.. Cary, NC.

Robert Marley is now at the Mechanical and Industrial Engineering Department, Montana State University, Bozeman, MT.

EFFECTS OF LIFTING IN FOUR RESTRICTED WORK POSTURES

Sean Gallagher, Richard L. Unger, and E. William Rossi
U. S. Department of the Interior - Bureau of Mines
Pittsburgh Research Center
Pittsburgh, Pennsylvania

ABSTRACT

The purpose of this study was to examine the lifting capacity of low-seam coal miners in four restricted work postures (roof heights of 36", 40", 44", and 48"), investigate the associated metabolic costs, and to examine electromyographic (EMG) data from eight trunk muscles during the lifting procedure. Subjects were thirteen underground miners accustomed to handling materials in restricted work postures. Each subject performed two twenty-minute periods of asymmetric lifting in each of four postures during the day of testing. The frequency of lifting was 10 lifts per minute. A specially designed lifting box incorporated microswitches in one handle of the box and another in the bottom of the box, in order to examine the trunk muscle function at specific points during the lfting cycle. The data collected will be used by the Bureau of Mines to make recommendations for lifting materials in low-seam coal mines.

INTRODUCTION

Statistics obtained from the Mine Safety and Health Administration's Health and Safety Analysis Center (HSAC) files for the years 1981 through 1986 have shown back injuries as being a significant and continuing problem in the underground mining industry. During this six-year period there have been a total of over 12,500 back injuries in underground coal mines. HSAC statistics also indicate that the rate of back injuries per 200,000 man-hours worked is higher the lower the seam of the coal mine (see Table 1). One reason for the increased incidence in low seams is that underground miners must perform lifting tasks in restricted working positions.

Previous Bureau research has indicated that underground miners have a reduced lifting capacity when lifting in an unrestricted kneeling position (i.e., a roof height of 48 inches), as compared to lifting in the stooped position at the same roof height (Gallagher, 1987). This research also demonstrate that myoelectric activity of the trunk muscles is significantly affected by the posture assumed when lifting (i.e., stooped or kneeling). Furthermore, the metabolic cost of lifting was found to be significantly elevated in the kneeling posture compared to stooped. The purpose of the present investigation was to examine the effects of lifting using restricted kneeling postures (i.e., roof heights of 36" and 40"), and stooped postures

TABLE 1.- Incidence of Back Injuries in Underground Coal Mines, 1981-1986. Incidence rates are expressed in terms of the number of back injuries per man-year (200,000 hours). Low-seam coal mines are defined as those with a roof height of $\leq$ 48 inches, while medium and high seam mines have a roof height > 48 inches.

	All underground coal mines	Low-seam coal mines	Medium/high-seam coal mines
1981	3.96	5.76	3.62
1982	3.74	5.08	3.51
1983	3.28	4.54	3.00
1984	3.25	4.44	2.99
1985	1.30	2.01	1.18
1986	3.13	5.67	2.71

(at roof heights of 44" and 48") on psychophysical lifting capacity, electromyography of eight trunk muscles, and the metabolic cost of performing the lifting activity.

METHOD

Subjects were 13 underground miners accustomed to handling materials in restricted work postures. Informed consent was obtained from each participant, and the subject was then prepared for the lifting tests. The eight trunk muscles investigated were identified and the skin above the muscle was prepared by shaving and cleaning the skin thoroughly with alcohol. Bipolar surface electrodes filled with an electrolyte gel were attached to the skin above the muscle (3 cm apart center to center), and a ground electrode was attached. Surface electrocardiographic electrodes were also placed for determination of heart rate during the lifting tests. Each subject performed two twenty-minute periods of lifting in each of four postures during the day of testing. In each posture, one period of lifting started with heavy box (approximately 95 pounds), and one started with a light box (approximately 25 pounds). The frequency of lifting was 10 lifts per minute.

A specially designed lifting box was used for this study. This box incorporated microswitches in one handle of the box and another in the bottom of the box (see figure 1). These switches allowed researchers to determine the following events during the lifting cycle: a) when the subject grasped the box, b) when the subject lifted the box, c) when the subject placed the box back down, and d) when the subject released his grip on the box. In addition, the collection of integrated

EMG data (at minutes 2 and 18 during the test) was triggered when the subject grasped the handle of the box. However, EMG data from all eight trunk muscles was obtained starting two seconds prior to the time the handle trigger was activated (by means of the memory buffer in the ISAAC[1] 5000 high speed data acquisition system) to assure that all EMG data associated with the lift was collected. Metabolic data was obtained during the last five minutes of every test using a Beckman Metabolic Measurement Cart I. Heart rates were taken every minute during the lifting tests. EMG data for each muscle was expressed as a percentage of the overall maximum observed for that muscle during all of the lifting tests. The MAWL data, average heart rate, oxygen consumption, ventilation volume, and respiratory exchange ratio data were analyzed using an analysis of variance (ANOVA) on repeated measures statistical package. Both mean and maximum EMG data were analyzed using a multivariate analysis of variance (MANOVA) on all muscles followed by separate ANOVAs and post hoc Duncan Range Tests performed on each separate muscle.

Lifting Box and EMG Instrumentation

EMG data during the MAWL experiments were collected using a system of surface electrodes coupled with and EMG amplifier and integrator, an analog/digital (A/D) converter, and micro-computer (figure 2). Collection of EMG data was triggered by the lifting box, which contained microswitches built into one handle and the bottom of the box to allow the marking of specific events during the course of a lift.

The lifting box was made using a welded aluminum construction and weighed approximately 19.5 lb. when empty. The box was divided into several compartments where weights could be added or removed according to the purposes of the test. Two compartments had hinged covers in which the experimenters randomly varied weight prior to the start of the test. A microswitch was built into one of the box handles, which was activated when the subject grasped the box at the start of the lift. This switch was armed by a second switch at the microcomputer to prevent false triggering due to vibration or incidental contact. When the switch was closed a +10 volt direct current (VDC) signal was sent from the lifting box to the A/D converter, triggering the collection process.

A second microswitch was attached to the bottom of the lifting box and was activated as

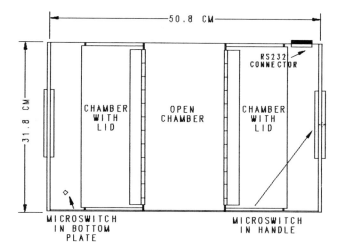

FIGURE 1. Schematic of the lifting box used in this experiment.

1 Reference to specific brands, equipment or trade names in this report is made to facilitate understanding and does not imply endorsement by the Bureau of Mines.

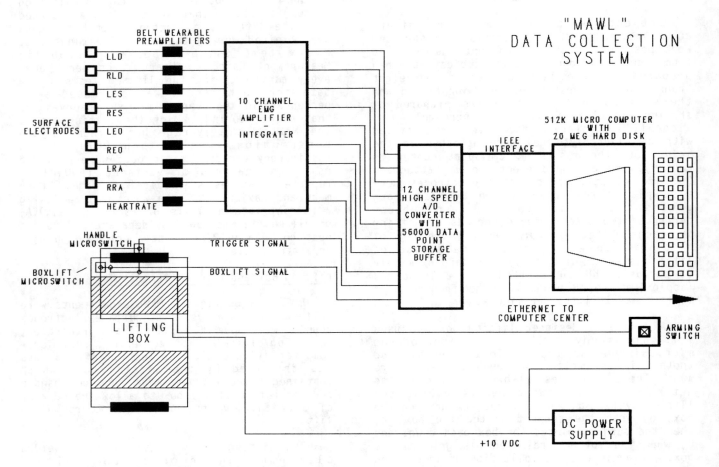

FIGURE 2. U. S. Bureau of Mines Electromyographic Data Acquisition System.

long as the box was resting on its base. When the box was lifted, the switch opened and the signal dropped from +10 VDC to zero.

A micro-computer was used to control the entire data collection process. Custom software was developed to activate the A/D converter and store the data collected. When the researcher wanted to collect EMG data the microcomputer sent the proper commands to the A/D converter, which then would collect data output from the EMG amplifier/integrator and the lifting box. The data are allowed to pass through the converter's buffer until a +10 VDC signal was recieved from the box handle microswitch, indicating that the handle has been grasped in preparation for a lift. At this point, the A/D converter stored a preset number of data points prior to the +10 VDC signal and began to store data points at the rate and duration specified by the researcher

through the software. In the present study, integrated EMG data was collected at a rate of 100 Hertz (Hz) for a seven second period of time; two seconds of data were stored prior to the lifting box handle signal being received by the A/D converter, and five seconds of data were collected after the signal was received.

The box microswitches enabled the investigators to analyze EMG data at the following points: when the box handle was grasped, when the box was lifted, when the box was set down, and when the handle was released. The EMG data described in this paper were analyzed from the time that the subject grasped the handle of the box until the handle of the box was released. After the data had been stored in the buffer of the A/D converter, the microcomputer stored the information on disk, cleared the buffer of the A/D converter, and reset the system for

TABLE 2. - Mean acceptable weights of lift and metabolic costs of lifting in the four restricted postures (N = 13)

	Kneeling		Stooped	
	36"	40"	44"	48"
MAWL (lbs)	54.2	53.5	63.5	66.0
HR (bpm)	141	133	126	126
V_E (L/min)	34.0	31.8	29.6	28.8
VO_2 (mL/kg/min)	15.8	15.3	13.7	13.8
R	0.92	0.90	0.89	0.87

another test. This data was later transferred to a mainframe computer for plotting and statistical analyses.

RESULTS

Analysis of the data indicates that restricting the posture of underground miners has a significant effect on lifting capacity ($F_{3,36} = 9.000$, $p < .001$). Table 2 lists the psychophysical lifting capacity for each of the experimental conditions along with data on the metabolic cost of performing at each roof height. The greatest amount of weight lifted by these miners was at the 48" roof height (mean = 66.0 pounds), and the least at the 40" roof height (mean = 53.5 pounds). Heart rate ($F_{3,36} = 9.543$, $p < .001$), oxygen consumption ($F_{3,33} = 8.05$, $p < .001$), and ventilation volume ($F_{3,33} = 4.880$, $p < .01$) were all significantly affected by roof height with the higher physiological values occurring in the kneeling postures. The respiratory exchange ratio ($F_{3,33} = 3.825$, $p < .05$) was also significantly affected by posture.

The results of the MANOVA for maximum EMG indicated that seam height has a significant effect on muscle activity (F = 5.115, $p < .001$). Similarly, the MANOVA for mean EMG during the lifting tasks demonstrated a significant effect due to seam height (F = 6.395, $p < .001$). Neither initial box weight (light or heavy) nor time of the data collection (minute 2 or minute 18 of the test) demonstrated a significant main effect on muscle activity in the multivariate analysis ($p > .05$). Table 3 presents a summary of significant results using the Duncan Multiple Range Test describing the effects on percentage of maximum EMG activity for the eight trunk muscles due to lifting in the four seam heights.

DISCUSSION

This study confirms the findings of an earlier investigation by the Bureau of Mines which demonstrated that underground miners have a reduced lifting capacity in the kneeling posture, while the metabolic costs of lifting in the kneeling position are greater. These findings indicate that muscular fatigue due to materials handling may occur more quickly in this posture, and that lifting heavier weights in low-seam coal mines might be better accomplished in the stooped posture (due to the higher lifting capacity in this position).

The analysis of the electromyographic data indicates that the function of the back muscles studied are quite different in the kneeling posture than when stooped. The erector spinae muscles are much more active when lifting in the kneeling posture. It seems reasonable to assume that these muscles must bear much more responsibility for the lift in this posture due to the fact that several muscle groups typically called upon for lifting are not available for use in this position. As a result, there may be an increased compressive load on the lumbar region of the spine when lifting in this position. The greater metabolic demand of lifting in this posture is likely due to an increased demand for oxygen by the back muscles.

In the stooped posture, the latissimus dorsi muscles were significantly more active during the lifting tasks than when kneeling; however, the erectores spinae were more quiescent when lifting in this posture. Many studies have shown that the erectores spinae demonstrate much less activity when the trunk is flexed, and it is assumed that the posterior group of ligaments (i.e., the posterior longitudinal, the ligamentum flavum,

TABLE 3. - Duncan range test significance for maximum EMG activity during lifting tasks. Conditions with the same letter are not significantly different at the 0.05 level. Numbers in parentheses represent the mean percentage of maximum EMG activity for the cell (N = 7). The subjects were kneeling at 36 and 40 inches, and were stooped at 44 and 48 inches.

	LLD	RLD	LES	RES	LEO	REO	LRA	RRA
36 inches	A (34.86)	A (36.63)	A (72.83)	A (77.21)	A (56.75)	A (46.91)	A (45.05)	A (33.64)
40 inches	A (26.77)	A (32.85)	A (67.01)	A (72.37)	A (56.26)	A (37.18)	A (51.99)	A (46.09)
44 inches	C (62.51)	B (60.69)	B (40.69)	B (39.39)	B (41.33)	A (48.13)	A (41.41)	A (33.37)
48 inches	B (47.98)	B (58.22)	B (30.10)	B (42.40)	A B (42.65)	A (37.90)	A (33.01)	A (35.19)

NOTE: LLD = Left Latissimus Dorsi; RLD = Right Latissimus Dorsi; LES = Left Latissimus Dorsi; RES = Right Latissimus Dorsi; LEO = Left External Oblique; REO = Right External Oblique; LRA = Left Rectus Abdominis; RRA = Right Rectus Abdominis.

the interspinous, and the supraspinous) are primarily responsible for supporting the vertebral column when the trunk is bent forward. It is likely that the subjects in the present study were dependent upon their ligaments for support during the lifting tasks, and were therefore placing those ligaments under considerable stress. Unfortunately, it is difficult to determine such stresses.

The results of the electromyography indicate that, in the stooped posture, the latissimus muscle is working at a higher percentage of its' maximum capacity than the erectores during the lift. The Bureau of Mines is currently examining the biomechanical stresses of lifting in these two postures using a model by Schultz (Schultz and Andersson, 1981; Schultz et al., 1982), in order to determine the compression and shear forces experienced by the spine when lifting in restricted working conditions.

The data presented in this paper (along with data collected in previous studies) will be used by the Bureau of Mines to make recommendations for lifting materials in the low-seam coal mining environment. The results of this study may also be valuable information for other industries where workers must handle materials in restricted work spaces.

REFERENCES

Gallagher, S. Back Strength and Lifting Capacity of Underground Miners. Paper in Proceedings: Bureau of Mines Technology Transfer Symposia, Pittsburgh, PA, July 7, 1987, St. Louis, MO, July 15, 1987, and San Francisco, CA, July 21, 1987, pp. 21-32.

Schultz, A. B., and Andersson, G. B. J. Analysis of Loads on the Lumbar Spine, Spine, v. 6, n. 1, 1981, pp. 76-82.

Schultz, A. B., Andersson, G. B. J., Haderspeck, K., Ortengren, R., Nordin, M., and Bjork, R. Analysis and Measurement of Lumbar Trunk Loads in Tasks Involving Bends and Twists, Journal of Biomechanics, v. 15, n. 9, 1982, pp. 669-675.

PHYSIOLOGICAL STRESSES ASSOCIATED WITH MANUAL HANDLING OF CONTAINERS OF VARIED SIZES AND WEIGHTS: A CASE STUDY

Ashraf M. Genaidy,[*] Jorge R. Duyos[**], and Shihab S. Asfour[**]

[*]Department of Industrial Engineering
Western Michigan University
Kalamazoo, Michigan 49008

[**]Department of Industrial Engineering
University of Miami
Coral Gables, Florida 33124

The main objective of this study was to evaluate the physiological strain imposed on individuals engaged in unloading boxes from a truck. The actual operation was performed in one of the local industries, and was videotaped and timed. This operation served as a basis for a simulation study in the authors' laboratory. Six subjects participated in this study. Since the containers utilized in the actual operation varied greatly, empirical distributions derived from the box sizes and weights handled at that industry were developed based on a sample of 466 boxes. The frequency of handling was 20 times/min. The task duration was 30 min. Heart rate and oxygen consumption were recorded every 5 min for a period of 5 min. The results showed that the task performed can be classified as very heavy work compared to industrial tasks reported in the literature. Physiological responses showed a significant increase over time.

INTRODUCTION

Despite the progress of automation, there is still a large number of industrial jobs which place a heavy physiological load upon workers performing these jobs. These heavy loads eventually lead to overstrain. Thus, by obtaining quantitative data on the physiological responses of operators as they are actually performing it a better match can be made between operator capabilities and job demands.

Several studies have been reported in the literature which investigate the effect of task parameters on the physiological responses of individuals engaged in manual materials handling (e.g.: Mital and Ayoub, 1981; Asfour et al, 1986). In most of the studies reported, the weight of load and box size were fixed within a given experimental session. The task duration was usually less than 15 min. In some industries, however, it was observed that workers were handling containers of varied sizes and weights within a given work cycle usually longer than 15 min.

This study was undertaken to evaluate the physiological strain imposed on individuals engaged in unloading boxes from a truck. The actual operation was performed in one of the local industries, and was videotaped and timed. This operation served as a basis for a simulation study in the authors' laboratory. Since the boxes utilized in the actual operation varied greatly, empirical distributions derived for the box sizea and weights handled at that industry were developed based on a sample of 466 boxes was determined. The task duration was 30 min.

METHODS AND PROCEDURES

Subjects

Six male college students volunteered to participate in this study. The physical characteristics of the subjects are given in table 1. All subjects were informed about the experimental procedures and were given an informal training prior to the data collection.

Table 1. Physical characteristics of the subjects

Variable	Mean	Range
Weight (kg)	72.91	64.09-84.09
Height (cm)	176.78	162.56-185.42
Age (years)	22.40	19.00-31.00

Description of Actual Operation

The actual operation consisted of unloading boxes from a trailer truck by using a set of rollers inside the truck and a belt conveyor outside the truck. The rollers were at a height of 125 cm from the bed of the truck and were also part of a platform which folds up into the walls of the truck from either side of the rollers.

The operation consisted of two parts. In the first part, the worker unloaded boxes by lowering them onto the rollers while he/she was standing on the platform. The height of the platform allowed the worker to reach the highest boxes. This part of the operation was continued until the worker reached the back wall and then raised the platform to the sidewalks to start the second part of the operation.

In the second part, the worker lifted boxes from the bed of the truck onto the rollers. He/she then unloaded boxes from one side, and then the other side. This part of the operation lasted approximately 40 minutes of continuous lifting, and was the focus of this study.

Description of Simulation

A physical simulation of the second phase of the operation was conducted in the authors' laboratory.

The first step in the simulation process was the determination of the box size and weight distributions. A sample of 466 boxes was taken from the actual operation. The weight of the boxes ranged from 0 to 31.82 kg. The range of weight was divided into 17 categories. Table 2 shows the weight distribution. The box size was categorized into 4 levels according to each box dimension (e.g.: length, width, and height). The distribution of the box size is given in table 3 according to each box dimension. The frequencies reported in tables 2 and 3

were used as the basis for the box size and weight distributions in the simulated study. A total of 47 boxes of different weights and dimensions were used so as to match the distributions reported.

The next step was the preparation of the lab setup as is depicted in figure 1. As shown in the figure, the

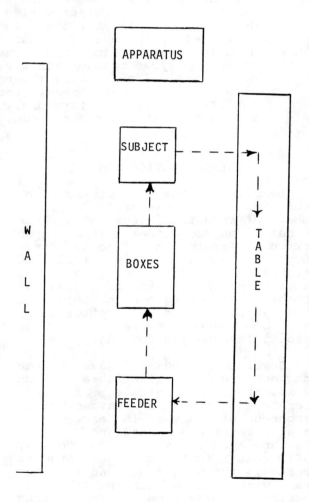

Figure 1. Lab Set-up

Table 2. Weight distribution in terms of
number of boxes and percentage

Weight (kg)	Number of Boxes	Percentage (%)
0.45	36	7.73
0.90	51	10.94
1.35	39	8.37
1.80	30	6.44
2.25	34	7.29
2.70	24	5.15
3.15	17	3.65
3.60	16	3.43
4.05	18	3.86
4.50	20	4.29
5.00-6.82	46	9.87
7.27-9.09	39	8.76
9.45-13.64	45	9.65
14.09-18.18	31	6.65
18.63-22.73	14	3.00
23.18-27.27	5	1.07
27.73-31.82	1	0.21

Table 3. Box size distribution in terms
of number of boxes (#) and percentage (%)

Size (cm)	Length		Dimension Width		Height	
	#	%	#	%	#	%
0-15	75	16.10	18	3.86	0	0.00
15-30	325	69.74	320	68.67	146	31.33
30-45	66	14.16	128	27.97	263	56.44
45-60	0	0.00	0	0.00	57	12.23

subject lifts the boxes and slides them down the table. This table simulates the rollers. The feeder, which for our simulated study served as an assistant, then removes the boxes from the table, places them on the floor and slides them back to the subject. This cycle is repeated for a period of 30 min. The wall represents the sidewall of the truck. The equipment shown in the figure was used to record the heart rate and oxygen consumption of the subject, and will be further discussed in a subsequent section.

The height of handling in the simulated study was 125 cm as in the actual operation; however, an additional height of 91 cm was also studied to test whether or not lowering the height of handling would reduce the physiological cost of the task performed.

The heart rate and oxygen consumption data were collected continuously in the following time intervals: 5 - 10 min, 15 - 20 min, and 25 - 30 min.

Experimental Design

A two within-subject factor was utilized in this study. The height range of handling and the time interval were the independent variables employed. The two height ranges were 91 cm and 125 cm. Three time intervals were used. These intervals were 5 - 10 min, 15 - 20, and 25 - 30 min.

Heart rate and oxygen consumption were the dependent variables in this study. They were computed as the average of the data collected in each of the aforementioned time intervals.

Table 4. Analysis of variance for oxygen consumption

Source	DF	Mean Square	F-Value	
Height	1	116.956	3.219	NS
Height*Subject	5	36.331		
Time	2	58.074	4.854	S
Time*Subject	10	11.964		
Height*Time	2	58.574	4.809	S
Height*Time*Subject	10	12.180		

NS - non significant at the 0.05 level
 S - significant at the 0.05 level

Table 5. Analysis of variance for heart rate

Source	DF	Mean Square	F-Value	
Height	1	506.252	1.844	NS
Height*Subject	5	274.520		
Time	2	278.775	23.872	S
Time*Subject	10	11.678		
Height*Time	2	30.335	0.764	NS
Height*Time*Subject	10	39.700		

NS - non significant at the 0.05 level
 S - significant at the 0.001 level

Equipment

A Narco-Biosystems Physiograph was used to record the heart rate of the subjects. The percentages of oxygen and carbon dioxide in expired were measured using Beckman OM-11 and SensorMedics LB-2 gas analyzers.

RESULTS AND DISCUSSION

The data collected in this study was subjected to an analysis of variance. The results of the analyses are given in tables 4 and 5 for oxygen consumption and heart rate, respectively. The main effect of height range of handling was not significant at the 5% level for both oxygen consumption and heart rate. On the other hand, the main effect of time interval was significant for both oxygen consumption and heart rate. The height range and time interval interaction was only significant for oxygen consumption.

The means and standard deviations of oxygen consumption and heart rate for both height and time interval are presented in table 6. On the average,

the 125 cm height had heart rate and oxygen consumption values that were 8 beats/min and 0.3 l/min higher than the 91 cm height range, respectively. Further, the heart rate values for the three time intervals were 136.5, 143.3, and 145.3 beats/min. Similarly, the oxygen consumption values for the three time intervals were 1.773, 1.785, and 1.831 l/min.

The results of this study clearly show that unloading a truck under the given conditions is a very demanding physical task. Astrand and Rodahl (1977) reported that a heart rate range of 130-150 beats/min or an oxygen consumption range of 1.5-2.0 liters/min can be classified as very heavy work. Wells et al (1957) pointed out that a physical task requiring a heart rate value of 140 beats/min or greater could not be continued for more than 2 hours.

Assuming that 3.0 liters/min is an average value for an individual's maximal oxygen consumption "VO2 max" as determined on a bicycle ergometer (Astrand and Rodahl, 1977), the oxygen

Table 6. Means and standard deviations (SD) of oxygen
consumption (VO2 l/min) and heart rate (HR beats/min)

Height of Handling	Time Interval (min)	VO2 Mean	SD	HR Mean	SD
Truck floor to 91	5 - 10	1.53	0.30	130.5	19.3
cm above truck floor	15 - 20	1.70	0.27	140.3	21.4
	25 - 30	1.71	0.42	142.5	25.4
Truck floor to 125	5 - 10	2.02	0.42	142.5	33.0
cm above truck floor	15 - 20	1.87	0.39	146.2	31.7
	25 - 30	1.95	0.50	148.0	26.0

consumption values reported in this study exceed 50% VO2 max. Weiser and Stamper (1978) reported that the endurance time of their subjects was on the average 59 minutes at an average workload of 62% VO2 max.

The aforementioned discussion clearly indicates that unloading a truck at a frequency of 20 times/min could not be maintained for more than 1 hour. In order to increase an individual's endurance time for such an activity, a possible engineering control would be to reduce the frequency of handling to below 10 times/min. It should be also noted that the physiological responses of individuals engaged in heavy physical tasks should be monitored continuously.

CONCLUSIONS

The findings of this study clearly indicate that containers of varied sizes and weights are being handled in industry. This could affect the physiological responses of individuals in a pattern different than those based on fixed parameters.

Further, physiological responses showed significant increase during the course of work. Therefore, it is recommended that future research in the field of manual materials handling should investigate the effect of varied task paramaters and working time on the physiological reponses of individuals.

REFERENCES

Asfour, S. S., Ayoub, M. M., Genaidy, A. M. (1986). A data base of physiological responses to manual lowering. In W. Karwowski (Ed.), Trends in ergonomics/human factors III (pp. 811-818). Amsterdam: Elsevier Science Publishers.

Astrand, P. O., and Rodahl, K. (1986). Textbook of work physiology. New York: Mc-Graw Hill.

Mital, A., and Ayoub, M. M. (1981). Effect of task variables and their interactions in lifting and lowering tasks. American Industrial Hygiene Association Journal, 42, 134-142.

Weiser, P.C. & Stamper, D.A. (1978). Psychophysiological interactions leading to increased effort, leg fatigue, and respiratory distress during prolonged, strenuous bicycle riding. In G. Borg (Ed.), Physical work and effort (pp. 401-416). London: Taylor & Francis.

Wells, J.G., Balke, B., & Van Fossan, D.D. (1957). A suggested standardization of work classification. Journal of Applied Physiology, 10, 51-55.

MANUAL MATERIALS HANDLING IN UNUSUAL POSTURES: CARRYING OF LOADS

M. M. Ayoub and J. L. Smith
Texas Tech University
Department of Industrial Engineering
Lubbock, Texas 79409

ABSTRACT

Carrying capacity of individuals was established for four height constrained postures: unconstrained, upright walking, walking in a semi-stooped posture (at 80% of standing height), walking in a full-stooped posture (at 60% of standing height), and crawling (at 40% of standing height). A dynamic strength test, the isoinertial elbow height lift was used to develop prediction models for each of the carrying activities. Carrying capacity decreased as ceiling height decreased. On the average, females handled slightly over 50% of what the males handled under the given task conditions.

INTRODUCTION

The purpose of this study was to determine the carrying capacity of aircraft maintenance technicians. This study is a portion of a larger research and development effort.

This study was conducted with the following objectives:

1. To determine the carrying capacity in simulated maintenance activities using unusual postures.

2. To perform a series of anthropometric and strength measurements on the sample population.

3. To determine the relationship between performance on simulated carrying activities, and scores on four strength tests.

SIMULATED CARRYING TASKS

The simulated carrying tasks were defined by the following postures, which were determined by ceiling constraints corresponding to various percentages of stature:

1. *Two-hand carrying by stand-walk* (normal upright standing height without ceiling constraint). Box held with both hands and lifted from the floor, carried in front of the body for 10 feet using normal upright carrying posture, then lowered to the floor.

2. *Two-hand carrying by semi-stoop* (80% of stature). Box held with both hands, lifted from the floor, carried in front of the body for 10 feet using normal upright carrying posture of slightly flexed neck and trunk, then lowered to the floor.

3. *Two-hand carrying by full-stoop* (60% of stature). Box held with both hands, lifted from the floor, carried in front of the body for 10 feet, and then lowered to the floor. The carrying consisted of flexed neck and very flexed trunk and knees. Two forms of carrying emerged:
   1) load held near chest/resting on thighs
   2) load held in front of knees

4. *Two-hand carrying by crawl* (40% of stature). Box placed in front of body in crawling position; container with handles grasped by both hands and lifted, carried forward, and lowered with both hands; then crawl to container with knees (hands remain in contact with box handles throughout movement).

5. *One-hand carrying by crawl* (40% of stature). Tool-box container placed on preferred side of body; container lifted, carried forward, and lowered with preferred hand (while other hand supports body); then crawl to container with knees (preferred hand remains in contact with tool-box handle throughout movement)

A 24" x 12" x 6" container with handles was used for the carrying task at 40% of stature. A container of the same size, but without handles, was used for all other two-hand carrying tasks.

For the one-hand carrying task male subjects used a 15" x 10.5" x 10.75" [outside dimensions: length x width x depth] tool box container and females used a similar container of dimensions 15" x 8" x 8". The tool box containers had one middle handle. The 5 lb incremental unit for the tool box containers was two 1" diameter steel bars, 11.25" in length. Rectangular styrofoam inserts were used at the inside ends of the tool boxes to minimize weight movement.

When performing the carrying tasks, the subjects were instructed to lift the container, carry it 10 feet, and then lower it onto the floor. Subjects used freestyle lifting technique, but they were restricted to forward movement and duckwalking was prohibited. Subjects were not allowed to drag or slide the box on the floor.

After each successful carry, the experimenter asked if the subject was able to safely carry more weight. An increment of five pounds was added to, or deleted from the container, as appropriate. This was repeated until the subject judged that he/she could not safely carry more weight, or until the experimenters deemed the trial invalid based on defined criteria. These criteria were: sliding/dragging the container along the floor, dropping the container more than 3 inches from the floor, difficulty in lifting/carrying/lowering the container, and physical overexertion (as displayed by straining and/or verbal and facial expression).

Initial weight of the load to be carried was determined based on the subject's incremental elbow lift (IEL) performance. The following formulae which relate IEL to initial weight of load were determined by inspection of pilot data:

Posture	Ceiling Height (% of stature)	Initial Weight Males	Females
Stand	100	IEL-20 lbs	IEL-10 lbs
Semi-stoop	80	IEL-20 lbs	IEL-10 lbs
Full-stoop	60	IEL-50 lbs	IEL-20 lbs
Crawl (2 hand)	40	30% of IEL	50% of IEL
Crawl (1 hand)	40	30% of IEL	50% of IEL

A 10 minute break was given between each task. The order in which the tasks were executed was determined randomly.

STRENGTH TESTS

To determine if there was a relationship between the scores on a series of strength tests and the performance on the carrying tasks, the following strength tests were administered to each subject:

1. Isoinertial incremental six foot lift.
2. Isoinertial incremental elbow level lift
3. Isometric 38 cm vertical pull
4. 70 pound hold endurance (males) or 40 pound hold endurance (females)

Isoinertial incremental six foot lift- The equipment used in this test was the Incremental Weight Lifting machine. The subject was instructed to:

a) Face the Incremental Weight Lifting machine
b) Place his/her feet approximately shoulder width apart
c) Grasp the bar with an overhand grip
d) Keep his/her back straight
e) Lift the weight to a height of six feet
f) Lift with a smooth motion without lifting his/her heals from the ground

The starting weight was 40 pounds. This weight was increased in ten pound increments until the subject was unable to lift the weight to a six foot height. The last weight successfully lifted to six feet was recorded as the test score.

Isoinertial incremental elbow height lift- The equipment used in this test was the Incremental Weight Lifting machine. The subject's elbow height was found and marked on the Incremental Weight Lifting machine. The subject was instructed to:
a) Face the Incremental Weight Lifting machine
b) Place his/her feet approximately shoulder width apart
c) Grasp the bar with an overhand grip
d) Keep his/her back straight
e) Lift the weight to his/her elbow height
f) Lift with a smooth motion without lifting his/her heals from the ground

The starting weight was the subject's maximum weight lifted to six feet during the isoinertial incremental six foot lift test. This weight was increased in ten pound increments by the researcher until the subject was unable to lift the weight to his/her elbow height. The last weight successfully lifted to elbow height was recorded as the test score.

Isometric 38 cm vertical pull- The equipment used in this test was a double handled bar. The bar was connected to a calibrated load cell. The load cell was connected to a length of chain which brought the total length of the apparatus to 38 cm. The chain was attached to a platform on which the subject stood. The subject was instructed to:

a) Grasp the bar with alternating over/under hand grip
b) Place his/her feet approximately shoulder width apart
c) Bend his/her knees until he/she could attain a position with his/her back as perpendicular to the floor as possible
d) Keep his/her back straight
e) Pull vertically with an increasing force until the maximum force is reached

The subject was asked to maintain his/her maximum force for approximately three seconds. The researcher observed the load cell display and read and recorded the maximum force obtained. This test was repeated a minimum of three times and the average value was recorded as the test score.

70/40 Pound Endurance Hold- The equipment used in this test was the Incremental Weight Lifting machine and a stop watch which recorded time in 1/100s of a minute. A 70 pound weight was used as the held weight for males and a 40 pound weight was used for female subjects. The subject's elbow height was found and marked on the Incremental Weight Lifting machine using a magnetic marker. The subject was instructed to:
a) Face the Incremental Weight Lifting machine
b) Place his/her feet approximately shoulder width apart
c) Grasp the bar with an underhand grip
d) Bring the bar to elbow height and hold it there for as long as he/she could
e) The subject was instructed to not lock his/her elbows into his/her stomach, but could into his/her side

The experimenter recorded the time that the subject could hold the given weight at his/her elbow height. The time was used as the score for this test.

SUBJECTS

Forty subjects from the student body of Texas Tech University participated in the project. Twenty subjects were male and twenty subjects were female. All subjects underwent a physical examination prior to participating in the study to ensure that the subjects were in good health and fit to perform lifting/carrying/lowering activities. All subjects satisfied the Air Force height and weight selection criteria, and were between 18 and 30 years of age. The participants were selected to represent a cross section of the population in terms of this height and weight criteria.

Anthropometric Characteristics

The following anthropometric measurements were taken on each subject:

(1) Stature
(2) Body Weight
(3) Sitting height
(4) Functional reach
(5) Cervicale height (R, L)
(6) Suprasternale height (R, L)
(7) Xiphoid height
(8) Umbilical height
(9) Crotch height
(10) Eye height (R, L)
(11) Mastoid height (R, L)
(12) Acromion (R, L)
(13) Iliac crest (R, L)
(14) Anterior superior iliac spine (R, L)
(15) Trochanter height (R, L)
(16) Patella height (R, L)
(17) Tibiale height (R, L)
(18) Lateral malleolus height (R, L)

These anthropometric data were collected using standard anthropometric calipers and data collection protocol (NASA, 1978).

EXPERIMENTAL PROCEDURE

The experiment was conducted in two hour sessions with subjects permitted to attend one session per day. The initial session for each subject consisted of a familiarization period to familiarize each subject with the various carrying tasks and the experimental setup. At the same session, the anthropometric data were measured and the strength data were collected.

Following the first sessions, carrying data were collected for the subjects. The sequence of the carrying tasks was randomly determined for the subjects. Each carrying task was run until the maximum weight was determined (either determined by the subject or stopped by the attending experimenters) prior to the start of the next task. Subjects were allowed no rest between trials of the same carrying task, however, there was a minimum of 10 minutes rest between two different tasks.

Ten male and ten female subjects were randomly selected to repeat the experiment and strength testing in order to determine the reliability of the experimental methodology.

RESULTS

A total of 5 carrying tasks were performed by 40 subjects (20 male, 20 female). The descriptive statistical analysis on the carrying data, anthropometric data, and the strength data were based on the combined male and female data sets. The regression models were generated based on the combined data set. The pair-wise t test was conducted on the test and retest data of carrying tasks and strength tests for both 10 male and 10 female subjects.

Descriptive Statistics

Carrying Tasks. Figure 1 presents graphically the average maximum acceptable weight of carry for each of the tasks for combined, male, and female subjects. The general tendencies, as shown in the plots are, as the ceiling height decreases the maximum acceptable weight of the load carried decreases.

Figure 1. Plot of Average Maximum Acceptable Weight of Carry for Combined, Male, and Female Subjects While Subjected to Four Carrying Tasks

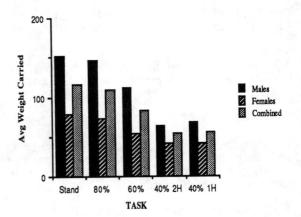

The data presented in Figure 1 point out significant differences in carrying task performance between male and female subjects. Table 1 presents, for each task, the average weight of carry, comparing the five task categories [Two hand carry with unrestricted ceiling (Stand); Two hand carry with ceiling at 80% of standing height (80%); Two hand carry with ceiling at 60% of standing height (60%); Two hand carry with ceiling at 40% of standing height (40% 2H); and One hand carry at 40% of standing height (40% 1H)] of male data, female data and a combined male and female data set. The standing, unconstricted carrying tasks consistently resulted in the highest values (116 lb.). There was a tendency for the weight to decline for bent-back and bent-knee two-handed carrying (110.2 lb. for 80% and 83.5 lb. for 60%). The carrying while crawling tasks resulted in the smallest values (52.9 lb. for 40% 2H and 54.6 lb. for 40% 1H). However, there was no significant difference of weight of carry between tasks 40% 2H and 40% 1H. The reason behind the different weight of carry for the five tasks is explained largely by the biomechanical / postural stresses associated with each task category. The Stand task, without ceiling constraints, allowed the subject to keep the back straight and utilize the body more effectively to help support the container and resulted in the higher carrying values. As the ceiling constraint was applied, the degree of freedom of movement was reduced and the center of gravity of the container was moved further away from the body. As a result, an additional effort was required to balance the container as well as the body while carrying. It was also noticed that the speed of carrying was significantly reduced. This was particularly true for those tasks where the ceiling was at 40% of subject's standing height.

Table 1: Descriptive Statistics of Weight of Carry of All Tasks

Combined Data Set (N=40)					
	Stand	80%	60%	40%2H	40%1H
Ave	116.0	109.9	83.6	52.9	54.6
Std Dev	44.1	42.2	36.7	13.7	15.7
min	55.0	55.0	35.0	30.0	25.0
max	210.0	205.0	165.0	85.0	87.0
CV	38.1	38.4	43.9	25.9	28.7
Male Data Set (N=20)					
	Stand	80%	60%	40%2H	40%1H
Ave	153.0	146.5	113.3	64.3	67.6
Std Dev	30.3	26.5	27.5	8.8	9.6
min	105.0	100.0	65.0	50.0	42.0
max	210.0	205.0	165.0	85.0	87.0
CV	19.8	18.1	24.3	13.7	14.2
Female Data Set (N=20)					
	Stand	80%	60%	40%2H	40%1H
Ave	79.0	73.3	54.0	41.5	41.5
Std Dev	14.1	11.2	12.6	5.9	7.3
min	55.0	55.0	35.0	30.0	25.0
max	105.0	100.0	80.0	50.0	50.0
CV	17.9	15.2	23.4	14.1	17.5

Table 2. Descriptive Statistics of Strength Test Data

Combined Data Set (N=40)				
	6'Lift	70#EH	Inc Elb	38cm L
Ave	83.0	0.54	117.3	270.5
Std Dev	38.2	0.30	48.8	91.4
min	30.0	0.13	50.0	128.0
max	170.0	1.50	200.0	443.0
CV	46.1	55.07	41.6	33.8
Male Data Set (N=20)				
	6'Lift	70#EH	Inc Elb	38cm L
Ave	116.5	0.59	162.0	341.9
Std Dev	23.5	0.24	22.8	69.7
min	80.0	0.15	110.0	248.0
max	170.0	1.01	200.0	443.0
CV	20.1	40.81	14.1	20.4
Female Data Set (N=20)				
	6'Lift	70#EH	Inc Elb	38cm L
Ave	49.5	0.50	72.5	199.2
Std Dev	9.4	0.35	12.5	39.3
min	30.0	0.13	50.0	128.0
max	70.0	1.50	90.0	305.0
CV	19.1	70.05	17.3	19.7

Strength Tests. Table 2 presents descriptive statistics for the four strength tests for the male and female subjects separately and combined. Female average strengths ranged from 30% (6 foot incremental lift) to 58% (38 cm isometric lift) that of male average strengths.

Correlational Analysis

Prior to the development of models whereby carrying task performance could be predicted based on one or more strength test scores and/or anthropometric characteristics were determined, correlation analysis were performed to determine the relationship between the strength score and task performance. The purposes of the correlation analysis was to identify the promising independent variables for entry into prediction models.

Correlations Between Carrying Data and Strength Test Performance. Table 3 presents the correlations between the various strength tests and five carrying task performance measures. For all the carrying tasks, strong, positive, and significant correlations exist between weight of carry and six foot lift, incremental elbow lift, and 38 cm lift. Correlations for the elbow height holding task were not considered due to the fact that the weights held were different (70 pounds for males and 40 pounds for females).

Based on the results of the correlation analysis, it was decided that development of stepwise regression models using strength test performance: six foot lift, incremental the incremental elbow lift has the highest correlations with the carrying tasks.

Table 3. Correlation Between Strength Test and Weight of Carry

Correlation Matrix for Variables: X_1 ... X_8

	6'Lift	Inc EL	38cm L	Stand	80%	60%	40%2H	40%1H
6'Lift	1							
Inc EL	.938	1						
38cm L	.877	.847	1					
Stand	.874	.926	.842	1				
80%	.895	.937	.865	.983	1			
60%	.862	.927	.814	.961	.955	1		
40%2H	.797	.876	.729	.883	.9	.894	1	
40%1H	.866	.912	.782	.897	.917	.905	.909	1

Regression Analysis

Models predicting weight of carry (combined sex data set) based on strength test performance were developed using step-wise regression analysis. A step-wise regression program from SAS (Version 5.16) was used in the development of regression equations. Maximum improvement in R-square was the criterion for the selection of variables to enter the model. The significant level (SL) was set at 0.01 and the selected candidate variables were the incremental elbow lift, the incremental six foot lift, and the 38 cm lift.

Table 4 presents the summary of step-wise regression analysis on the carrying tasks: Stand, 80%, 60%, 40% 2H, and 40% 1H. The models' predictors were selected based on the justifications that the incremental elbow lift scores, the six feet lift scores, and the 38 cm lift scores are significantly correlated with the carrying tasks performance.

Table 4. Regression Models Predicting Stand Carrying Capacity as a Function of Strength Tests Scores

	Model				R^2
Stand Carry	=	17.835	+	0.837(IEL)	.86
80% Carry	=	15.018	+	0.809(IEL)	.88
60% Carry	=	1.928	+	0.697(IEL)	.86
40% Carry (2H)	=	24.11	+	0.245(IEL)	.77
40% Carry (1H)	=	20.24	+	0.293(IEL)	.83

Validation of Experimental Data

Twenty subjects (10 males, and 10 females) were randomly selected out of the 40 subjects to repeat the strength tests and the carrying tasks. The pair-wise t test was performed on the test and retest data (separated by male and female). The test results indicated that there are no significant differences between the pairs of data (test vs retest) in either the male data set or the female data set.

CONCLUSIONS

As a result of simulating a series of carrying activities, the following points need to be emphasized:

(1) The data presented in Figure 1 points out significant differences in carrying task performance between male and female subjects of the experimental sample. Across all tasks, for males the average weight carried to completion was approximately 109 lb., whereas for females the average weight carried to completion was approximately 58 lb.

(2) Comparing the five task categories (Stand, 80%, 60%, 40% 2H, and 40% 1H), the 40% 2H and 40% 1H tasks consistently resulted in the lowest values of weight carried. The Stand task consistently resulted in higher weight of carry values than other tasks with ceiling restrictions. The reason is explained largely by the biomechanical/postural differences associated with each task category.

(3) Differences in carrying task performance can be largely explained in terms of postures adopted by the subjects. The importance of posture in carrying task performance has broad implications for job and workplace design.

(4) For all the carrying tasks, strong and significant positive correlations existed between weight of carry

and the incremental elbow lift test score. The relationship is not surprising, since intuitively, carrying tasks should be a function of upper body strength and hence the high correlation with IEL strength score.

(5) Models predicting weight of carry based on the incremental elbow lift score, the 6 foot incremental lift score, and the 38 cm lift score were developed using step-wise regression analysis. In general, the prediction models produced R^2 value ranging from .77 to .89.

(6) The pair-wise t tests on the test and retest data of both the strength tests and carrying tasks for both the male and female data sets showed no significant differences.

REFERENCES

1. Ayoub, M. M., Denardo, J. D., Smith, J. L., Bethea, N. J., Lambert, B. K., Alley, L. R., and Duran, B. S., (1982), *Establishing Physical Criteria for Assigning Personnel to Air Force Jobs- Final Report*, Air Force Office of Scientific Research, Contract No. F49620-79-C-0006.

2. Ayoub, M. M., Smith, J. L., Selan, J. L., and Fernandez, J. E., (1985a), *Manual Materials Handling in Unusual Positions- Phase I*, Final Report prepared for the University of Dayton Research Institute.

3. Ayoub, M. M., Smith, J. L., Selan, J. L., Chen, H. C., Fernandez, J. E., Lee, Y. H. and Kim, H. K., (1985b), *Manual Materials Handling in Unusual Positions- Phase II*, Final Report prepared for the University of Dayton Research Institute.

4. Ayoub, M. M., Smith, J. L., Selan, J. L., Chen, H. C., Lee, Y. H. and Kim, H. K., (1987), *Manual Materials Handling in Unusual Positions- Phase III*, Final Report prepared for the University of Dayton Research Institute.

5. NASA, (1978), *Anthropometric Source Book*, NASA Reference Publication 1024.

THE EFFECTS OF ASYMMETRY, LOAD LEVEL, START POSITION AND LOAD VELOCITY ON LUMBAR MOTION

Gary A. Mirka
Biodynamics Lab
Department of Industrial and Systems Engineering
The Ohio State University
Columbus, Ohio 43210

ABSTRACT

In order to completely understand and model the dynamic lumbar spine one must understand not only the forces acting about the spine but also the motions of the lumbar region as well. This study was an investigation in to the effects of certain lifting parameters on the motions of the lumbar spine. The effects of asymmetry, load velocity, start position and load weight on the range of motion, angular velocity and angular acceleration of the lumbar spine were investigated in three planes: coronal, sagittal and transverse. The results indicate trade offs between the coronal and sagittal planes as a result of changes in asymmetry and changes in the lifting style as a result of increasing lift velocity. The results also indicate that the location in time of the peak accelerations are varied as a result of different lifting conditions.

INTRODUCTION

The cost of low back injuries is a major problem facing industry today. The research that has been done to combat this problem can be divided into two factions. The first group of researchers approach the problem by building whole body models and studying capacity for lifting. (Chaffin 1969, Schultz and Andersson 1981, Kroemer 1983, Frievalds et al. 1984, Aghazedeh and Ayoub 1985, Mital et al. 1986, Pytel and Kamon 1981) The other group of researchers concentrate their efforts on the intervertebral joint set apart from the rest of the body and study the forces and force producers that act around the joint. (Gracovetsky and Farfan 1986, Anderson et al 1985, McGill and Norman 1986, Schultz and Andersson 1981, Marras et al. 1987) The goal of this research was to bridge the gap between these two approaches by studying the motions of the lumbar spine as the subject does a whole body lift while the starting position (straight vs. bent leg), asymmetry (sagittally symmetric vs. 90 degree turn from sagittaly symmetric), load velocity (37.5 cm/sec vs 75 cm/sec) and weight (180N vs. 270N vs. 360N) were varied. In this way the effect of certain whole body parameters on the motions of the lumbar spine are documented.

METHODS

Subjects

Sixteen male volunteers with no history of back pain were tested in this study. They had an average age of 27.1 years (range 19-35 years), an average height of 185 cm (std. dev. 9.6 cm.) and an average weight of 80.5 kg (std. dev. 14 kg.)

Equipment

The equipment used for this study consisted of a Kin/Com isokinetic dynamometer, a lumbar motion monitor (LMM) and a Compaq portable micro-computer connected to a Lab Master A/D board. The Kin/Com was altered in much the same way as Aghazadeh and Ayoub (1985) altered the Cybex. A large wooden wheel was attached to the rotating axis of the Kin/Com. Wound around this wheel was a cable that was attached through a series of pulleys to a load cell strain gauge that was located in the handle arrangement (see figure 1).

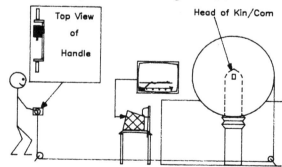

Figure 1. Experimental Apparatus

The LMM is a piece of equipment developed and built in the Biodynamics Lab that monitors the position of the lumbar spine in three dimensions. The LMM is essentially an exoskeleton that follows the movements of the spine and uses potentiometers to reflect the instantaneous position. The data from the LMM was sent through a cable to a Lab Master A/D board where it was digitized and sent to the Compaq computer for storage and later computations.

Independent Variables

The independent variables in this study were 1) the starting position 2) the asymmetry of the lift, 3) the vertical linear velocity of the handle being lifted and 4) the weight being lifted. The subjects were asked to lift the weight from the ankle to the level of the shoulder.

The starting position variable had two levels: straight leg and bent leg. The asymmetry variable also had two levels: 0 degrees (sagittally symmetric) and 90 degrees rotated from sagittally symmetric to the left. Based on previous literature two levels of vertical load velocity were used: 37.5 cm/sec and 75 cm/sec. In this study the subject was required to regulate the external force and keep this force to one of three specified levels: 180N, 270N and 360N. The force originated at the cable/load cell interface and was regulated using a CRT screen as a feedback system wherein the subject monitored his force output and its relation to the dotted line marker that was placed at the specified force level. In order for the trial to be accepted the subject had to keep his force within 40N of the specified level.

Dependent Variables

There were a total of 12 dependent variables that were investigated in this study. These variables of spinal motion included range of motion, average angular velocity, peak angular velocity and the peak angular acceleration of the lumbar spine. Each of these variables was calculated from the data for each of the three planes of motion: sagittal, lateral (coronal) and twisting (transverse).

The range of motion of the lifting trial in all three planes was defined as the angular range traversed during the concentric portion of the lift. The average angular velocity was calculated over the same range but was the first derivative of the position data. The maximum velocity and maximum accelerations were, on the other hand, instantaneous measures of motion. These were the peak values that occurred in the concentric range of motion.

Procedure

The subject was allowed a short training period to get a feel for the amount of vertical force they have to exert to maintain the constant force through the range.

The order of presentation of conditions was counterbalanced between subjects. Each subject was asked to lift the handle arrangement under every set of conditions. If their performance did not meet the requirement

of maintaining the target value plus or minus 40 N they were asked to repeat the trial. They were also asked to repeat the trial if they failed to use the correct lifting posture throughout the lift. The data was collected only during the concentric lift. The lumbar motion monitor data and three channels of Kin/Com data were collected through an A/D board and stored on a personal computer.

Data Conditioning

The conditioning of the experimental data consisted of running it through two computer programs that change the raw voltage values, that came from the potentiometers of the LMM, into the dependent measures of interest. The first program input the raw voltage values of the potentiometers and converted those into angular position readings for each of the three planes of motion. The position data was then smoothed to reduce the effects of the digital data collection system. This program then calculated, in each plane of the body, the instantaneous velocity and acceleration at each point in time using the position data. The second computer program took this data and calculated the values of the dependent variables.

ANALYSIS APPROACH

A statistical analysis was performed to determine which lifting parameters affected the lifting motions. This analysis was performed in steps. The first step was a MANOVA that grouped all three planes of motion and all dependent measures together. From this MANOVA the independent variables that affected the lifting motions as a whole were found. A second MANOVA was performed that still grouped the planes of motion together but analyzed each dependent measure separately. Finally individual ANOVA's were performed on each dependent measure in each plane of motion to determine how each reacted individually.

RESULTS

The first MANOVA was an attempt to find those independent variables that affected the lifting performance regardless of the plane of the body or dependent measure. This MANOVA indicated that there were three factors that had an effect on the lifting performance. These factors were: the vertical velocity of the load, the asymmetry of the lifting posture, and the interaction of these two factors.

When the data was analyzed in the second MANOVA the planes of motion remained grouped together but now the dependent measures were analyzed separately. When the range of motion was the measure of interest it was concluded that load velocity and asymmetry were the only significant factors. The second dependent measure, average angular velocity, reacted significantly to four factors: vertical velocity of load, starting position, asymmetry of lift and the interaction between the velocity and the asymmetry. The maximum angular velocity and the maximum angular acceleration of the spine had three significant factors: vertical velocity of load, asymmetry of lifting stance and the interaction of these two factors.

The final analysis consisted of breaking this data down still further into each plane of motion. A summary of significant variables and interactions is presented in table 1.

Table 1. Summary Table of Main Effects, Two-Factor Interactions and Three-Factor Interactions

	V	S	F	A	VS	VF	VA	SF	SA	FA	VSF	VSA	VFA	SFA
LATERAL PLANE ROM	X		X	X						X		X		
AVEL	X		X	X	X		X					X		
MVEL	X		X	X									X	
MACC	X		X	X										
SAGITTAL PLANE ROM	X		X				X							
AVEL	X		X				X	X					X	
MVEL	X		X					X						
MACC	X													
TWISTING PLANE ROM				X										
AVEL				X										
MVEL				X				X						
MACC				X				X			X			

X's denote significant effects at p<.05

V—LOAD VELOCITY
S—STARTING POSITION
F—LOAD FORCE
A—ASYMMETRY
ROM—RANGE OF MOTION
AVEL—AVERAGE VELOCITY
MVEL—MAXIMUM VELOCITY
MACC—MAXIMUM ACCELERATION

The results indicate that there are some general trends that are of importance. It was reported that on the average an increase in asymmetry (going from 0 to 90 degrees) caused several changes. First, there was an increase in the range of motion in the lateral plane (+4.91 degrees) and twisting plane (+11.64 degrees), while there was a decrease in the range of motion in the sagittal plane (-4.07 degrees). The average angular velocity of the spine also increased with increasing asymmetry in the lateral plane (+2.71 degrees) and the twisting plane (+5.5 degrees), while a decrease was found in the sagittal plane (-2.1 degrees).

A trend that occurred as a result of increasing load vertical velocity (going from 37.5 to 75 cm/sec) was a decrease in the range of motion in the sagittal plane (- 5.07 degrees) and an decrease in the lateral plane (-1.08 degrees).

DISCUSSION

The documentation of trunk motion is important because any motions of this part of the body greatly influence the strain in the muscles of the L5/S1 joint which can lead to high spinal loadings and low back injuries. The mass of the body above the L5/S1 joint is very large. Any accelerations and of a mass this large requires a great amount of force to be exerted by the muscles of the trunk. In addition, the center of gravity of the trunk has a great mechanical advantage over the low back musculature in forward bending postures. This accentuates the problems encountered in accelerations and decelerations. Once an understanding of the motions of the body are well documented, dynamic models containing this information can be built so that these models will be more accurate in their predictive ability.

A better understanding of the internal implications of the lifting motion parameters taken in this study would be helpful in eliminating unsafe working practices. For example, if in the future it is shown that the greatest risk for injury is the peak instantaneous velocity in the lateral plane, work places should be designed in such a way to reduce this peak velocity by altering those variables that contribute to high peak velocities. From the information gathered in this study, it is possible to isolate the two and three factor interactions as well as the main effects that should be closely monitored in order to reduce risk of injury.

Effects of Asymmetry

There are several portions of this study that have indicated a series of trade-offs that seem to be occurring between the planes of motion. One example is the range of motion trade-off. That is, as the asymmetry of the lift increases, the range of motion in the sagittal plane decreases. The subject still had to bend down the same vertical distance under both levels of asymmetry but under the asymmetric condition a large portion of this vertical range was taken by the lateral bend of the torso.

The trade-off for range of motion between the planes of the body appears to be almost a one to one degree trade-off. For example, under sagittally symmetric conditions the range of motion in the sagittal plane was 43.75 degrees while the range in the lateral plane was 3.91 degrees. Under the asymmetric condition the sagittal range of motion fell 5.06 degrees to 38.69. The range of motion in the lateral plane rose 4.90 to 8.81 degrees.

Effects of Load Velocity

In general, the range of motion is a function of the load velocity for all planes of motion. The amount of motion was greatest at the slowest load velocities. This can be explained by a change in the technique that the subject used at different velocity levels. At the high load velocity the subject used more of his upper extremities to exert the designated force because at these high velocities he did not have time to use the large trunk muscles to maintain   and control the required force level. When he uses his upper extremities the trunk motion is minimized because the subject gets to the vertical posture and does not move from that position thus allowing the extremities to do the work.

At the low load velocity it was observed that the subject used a shift in his body's center of mass forward and a hyper-extension of the trunk beyond vertical to maintain the desired level of force. This "optimization" of style could cause the greater range of motion under the low velocity for each of the three planes of motion.

Peak Time of Instantaneous Measures

It is of importance to those who wish to model the internal forces to know when the instantaneous measures reach their peaks. This time lapse information is important because as a person is performing a lift the anatomical positions and moment arms are constantly changing. This instantaneous data can be of great importance to those researchers who are creating models that calculate loadings on the spine and how they change over time (Reilly and Marras 1988). The dependent measures of this experiment that are instantaneous are the maximum velocity and maximum acceleration. It was found that depending on the asymmetry of the lift the location of these peaks changes.

In general during a sagittally symmetric exertion the location of the maximum acceleration and maximum velocity are reached at approximately the same time with the maximum acceleration occurring slightly before the maximum velocity. The location in time of when these peaks occurs seems to be dependent on the vertical load velocity of the lift. For the faster lifts the peaks occur at the beginning of the lift while for the slower lifts the peaks are reached as the subject approaches verticality.

An explanation for this phenomenon may lead us back to the hypothesis of different type of lifts under different lift velocities. When a person is performing an upper extremity lift the majority of the angular motion of the trunk occurs early in the lift and the extremities perform the remainder of the lift.

The peak angular velocities would therefore be located early in the lift for the high velocity conditions. Under the low velocity conditions the trunk is performing most of the lift and it spreads its performance out over the whole range of motion.

Under sagittally symmetric conditions the location in time of the peak velocity and peak acceleration changed with the velocity of the lift but they are usually found together in time. This is not the case for the asymmetric conditions. Under asymmetric conditions the maximum acceleration is typically reached in the beginning of the lift while maximum velocity is reached near the end of the exertion, as the subject is approaching the vertical position. It is interesting to note that these trends seem to occur regardless of the plane of motion.

The location in time of these measures can dramatically affect the loading on the spine. Under the high load velocity conditions since these peaks are reached when the person is still bent over the moment arm of the body's center of gravity is its greatest just as the angular accelerations are reaching their peaks. This combination of accelerations and moment arm will cause the internal structures to exert much more force and therefore the internal load encountered by the spine will increase dramatically.

The results of asymmetric lifts showed that the peak accelerations were reached early in the lift. Asymmetrical lifts of any sort cause shear forces in the spine because the muscles on one side of the body are exerting more force that those on the other side. When accelerations and increased moment arm are added, the shear force will also increase dramatically.

One of the assumptions of the NIOSH Lifting Guide is that the person doing the lifting is doing a slow controlled lift. This means that the load being lifted has a constant vertical velocity. From the results of this experiment it can be concluded that the load velocity is not always a good predictor of the angular velocity of the L5/S1 intervertebral joint. The velocity of the load can be extremely smooth and the angular velocity of the lumbar spine can either be smooth, great in the beginning and small at the end or small in the beginning and great at the end of the lift. The location in time of the angular accelerations of the joint can vary even more. By assuming that the angular velocity at a joint is constant when it may not be, leaves gross underestimates of the compressive forces at the vertebral level. The more accurately the motions of the lumbar spine are measured and used in lumbar models the more accurate our estimates of compression will be.

CONCLUSIONS

Describing the motions of the intervertebral joint as a function of the external linear motion of the load is a very complicated matter. There appears to be a transfer function going from this linear motion to the angular motion that contains as its inputs the asymmetry of the lift and the velocity of the lift and various interactions. There are a few generalizations that can be made about this transfer function. Increased asymmetry brings about a decrease in average velocity in the sagittal plane and an increase in average velocity in twisting and lateral planes. Increasing the lift velocity brings about greater angular velocity in all planes of motion.

Any work that is done in the area of low-back research should have as its goal a greater understanding of how the back functions and thereby a greater understanding of the mechanism of low-back injury. The nature of this region divides itself very easily into two types of analysis: single joint analysis and whole body analysis. Each of these methods gives us valuable information in relation to how the low back functions but each by itself cannot completely describe the functions of the back. Knowledge of both areas and how they interact will lead to a greater understanding of the mechanism of low-back injury.

The important issues raised in this study point the direction for future research. These issues included the time dependent measures and the interactions of the lifting parameters. In this study the peak accelerations occured when the subject was at the greatest iomechanical disadvantage whether it be bent over or at high asymmetry. If future studies could find combinations of lifting parameters that caused the peak accelerations from the point of greatest biomechanical disadvantage to a point in time when the muscles do not have to work quite so hard to perform the task. The interactions of lifting parameters varied greatly from dependent measure to dependent measure. It is therefore of great importance to find which dependent measure has the greatest detrimental impact on the lumbar region. Once the internal impact of the range of motion, velocities and accelerations are known, the value of experiments like this one will aid designers to reduce the rate of low-back injuries in industry.

LIST OF REFERENCES

Aghazadeh, F. and Ayoub, M. M. "A Comparison of Dynamic- and Static-Strength Models for Prediction of Lifting Capacity." Ergonomics, 1985, 28, 1409-1417.

Anderson, C. K., Chaffin, D. B., Herrin, G. D. and Matthews, L. S. "A Biomechanical Model of the Lumbrosacral Joint During Lifting Activities. " Journal of Biomechanics, 1985, 18, 571-584.

Chaffin, D. "Computerized Biomechanical Models- Development and Use in Studying Gross Body Actions." Journal of Biomechanics, 1969, 2, 429-441.

Freivalds, A., Chaffin, D., Garg, A., Lee, K. "A Dynamic Evaluation of Lifting Maximum Acceptable Loads." Journal of Biomechanics, 1984, 17, 251-262.

Gracovetsky, S. and Farfan, H. "The Optimum Spine." Spine, 1986, 11, 543-574.

Kroemer, K. H. E. "An Isoinertial Technique to Assess Individual Lifting Capacity." Human Factors, 1983, 25, 493-506.

Marras, W. S., Rangarajulu, S. L. and Wongsam, P. E. "Trunk Force Development During Static and Dynamic Lifts" Human Factors, 1987, 29, 19-29.

McGill, S. M. and Norman, R. W. "Partitioning of the L4-L5 Dynamic Moment Into Disc, Ligamentous, and Muscular Components During Lifting." Spine, 1986, 11, 666-677.

Mital, A., Aghazadeh, F. and Karwowski W. "Relative Importance of Isometric and Isokinetic Lifting Strength in Estimating Maximum Lifting Capabilities." Journal of Safety Research, 1986, 17, 65-71.

Pytel, J. and Kamon, E. "Dynamic Strength Test as a Predictor for Maximal and Acceptable Lifting." Ergonomics, 1981, 24, 663-672.

Schultz, A. B. and Andersson, G. B. J. "Analysis of Loads on the Lumbar Spine." Spine, 1981, 6, 76-82.

Reilly, C. and Marras, W. "SIMULIFT: A Simulation Model of Human Trunk Motion." In Press, Spine, 1988.

SAFETY OF BACK!   WHAT IS THE MARGIN?

Shrawan Kumar and Anil Mital
University of Alberta, Canada and
University of Cincinnati, USA

ABSTRACT

Despite many efforts to control low-back pain problem it still continues to be a concern. Since the causation of low-back pain is under multifactorial control, it will always be the factor most vulnerable at a given time which will determine safety. The present study is an integrative inferential synthesis of the published work to discern the margin of safety. An attempt has been made to conclude, on the basis of objective evidence, an all encompassing criteria to ensure system safety. Though psychophysical approach integrates biomechanical and physiological variables the role of sensory conditioning needs to be addressed.

INTRODUCTION

The economic burden on the modern society due to low back pain has been emphasised by many authors (Andersson 1981, Magora 1970, Kelsey and White 1980, Holbrook et al 1984, NSC 1978 and others). Variety of approaches have been taken in an effort to control the problem. This has included many epidemiological studies (Chaffin and Park 1973, Chaffin et al 1977, Ayoub et al 1978 and others), biomechanical studies (Chaffin 1975, Evans and Lissner 1959, Sonoda 1962 and others), physiological studies (Brown 1972, Garg and Saxena 1979, Kumar 1984 and others), and psychophysical studies (Snook 1978, Ayoub 1978, Mital 1985 and others). At different times based on the information uncovered different recommendations have been made. These have taken place in form of load limits for handling based on epidemiological information. (ILO & others), the method of lifting (bend knees and straight back), a flexible limit based on the metabolic cost (5.2 Kcal - NIOSH 1981), the self selection of the magnitude of the load based on the task characteristics using psychophysical methodology (Snook 1978, Ayoub 1978, Mital 1985, and others). NIOSH (1981) incorporated all four of these criteria in its Work Practices Guide of Manual Lifting for a floor to knuckle lift in a sagittal plane. However, injury statistics negate any significant impact on global picture of low-back pain profile in modern society. The question then arises, is there a margin of safety for human back?

Perhaps due to the diversity of approaches and a lack of unifying theme an all encompassing control strategy has eluded us. Since the back is under multi- factorial control, it is logical to surmise that the variable stressed beyond its threshold may contribute pre- cipitously to the low back injury. In order to explore the extent of the margin of safety the current study was conducted. Here an inferential synthesis of published work was carried out with a view to integrate concepts and criteria of epidemiology biomechanics (mechanical properties, strength, biomechanical modeling), physiological

kinesiology (range of motion, electromyography, intra-abdominal pressure), physiological cost, and psychophysics. Most industrial and occupational tasks tax most of the variables most of the time. It is possible to overstress one variable to failure. In normal working environment all variables are simultaneously involved, may be to different extent. Therefore, it is also conceivable that one variable may indicate the safety buffer zone for some or most relevant variables. Discovery of such a coupled relationship may significantly assist the control strategy by allowing a criteria integrated approach of margin of safety to be incorporated in task design and management of health and safety.

Epidemiology

Due to lack of clarity in the cause and effect relationship between identifiable and quantifiable risk factors and the low-back pain, the epi- demiological picture of this disorder is somewhat fuzzy. However, a variety of risk factors have been identified. These fall under three categories a) psycho- logical b) genetic, and c) mechanical (Kumar 1989). While little can be done to manipulate genetic and psychological factors, the mechanical factors lend themselves for design efforts. Using epidemiology as information base and design as a potential strategy, Snook (1978) proclaimed that up to one third of the back injuries can be prevented. In a survey carried out by Liberty Mutual Insurance Company he found that up to one-fourth of the jobs required manual handling of a magnitude acceptable to less than 75% of the workers. These jobs were responsible for half the back injuries reported, indicating three fold increase in susceptibility when the tasks require such exertion. This implies that two-thirds of all back injuries associated with manual materials handling can be prevented if the job is designed to fit at least 75% of the work force. One third of the injuries will occur despite the design effort. Counting the injuries which were not associated with heavy manual handling Snook (1978) calculated an over all possible reduction in low back injuries by 33% for industrial workers if an appropriate design strategy were to be instituted. Though such claims may be hard to conclusively substantiate experimentally yet it indicates approximately 83% collective margin of safety for the work force.

Biomechanics

Since pain is the physiological mechanism of ensuring system safety, and every back injury has a mechanical derangement, disruption or abnormality associated with such pain episodes; the biomechanical aspects become pivotal in such considerations. Conversely if biomechanical safety and normality can be maintained for the spinal segments and elements, the problem of low-back pain can be largely contolled. Several workers have looked into the material properties of the spinal structures. Based on the combined data of Evans and Lissner (1959) and Sonoda (1962) a consensus on the mean ultimate compressive strength of the lumbar spinal units was reached. For people under 40 years of age 6700 N and for people 60 years or over 3400 N was agreed. NIOSH (1981) incorporated these values in its "Work Practices Guide for Manual Lifting" designating them as maximum permissible limit and action limit respectively. As generic norms these values may serve a useful purpose. However, the concept of safety margin can only be addressed if individuals maximum voluntary contraction and thereby generated compressive forces were to be compared with the ultimate compressive strength of his spinal units. Clearly such an experiment is impossible for obvious reasons. Therefore, a comparison of the voluntarily generaged maximal compressive forces of an experimental sample with the ultimate compressive strength of the spinal units of an age, sex and body weight matched sample will shed light on the issue.

Kumar et al (1988) studies isometric and isokinetic stoop lifting strength capabilities at 20 cm, 60 cm and 100 cm per second velocities of ten male volunteers. They reported mean peak strength of 726 N, 672 N, 639 N and 597 N respectively for those four conditions. Under these experimental conditions the mean peak compressive forces generated were 6,933 N, 6,329 N, 6,017 N, and 5,613 N. A compression of 6,933 N exceeded the NIOSH Maximum permissible limit of 6,700 N which represents the mean ultimate compressive strength of lumbar spinal units. Anderson and Schultz (1979) reported a maximal voluntary isometric stoop lifting effort of 900 N from a sample of five male subjects. This was calculated to have generaated a compression of 7,863 N on the lumbar spine. In other studies also such efforts are commonly reported.

Hulton and Adams (1982) tested the ultimate compressive strength of 16 lumbar spinal units from 8 male subjects whose age and body weight matched the experimental samples of Andersson

and Schultz (1979) and Kumar et al (1988) (Table 1).

Table 1. Demographic details of the experimental samples of Kumar et al, Andersson and Schultz and Hutton and Adams.

	Kumar et al	Andersson & Schultz	Hutton & Adams
1. Sex	M	M	M
2. Mean Age (range)	26.6(23-34)	30(22-43)	33(22-46)
3. Mean body weight (range)	76.5(65-97)	74(66-84)	76(65-86)
4. Number of Subjects	10	5	8

The ultimate compressive strength of their sample was 10,219 N. Since the ages and body weights of the subjects in the three studies were matched the results may be extrapolated from one study to the other. The significance of the body weight lies in the stimulus it provides the skeletal system. The mechanical milieu plays an important role in bone growth, density and hypertrophy. Though the mean age of the subjects in Kumar et al's study was slightly younger, they came from a similar age range. Further, an identical body weight across the three studies will ensure a similar tissue strength characteristics. A comparison between the ultimate compressive strength and the magnitude of the compression voluntarily generated reveals a 32% and 23% difference considering Kumar et al's (1988) and Andersson Schultz (1979) study respectively (Table 2). This difference is suggested to represent the margin of safety.

The validity of Hutton and Adams (1982) data is further enhanced for such a comparison due to their experimental protocol involving testing of the spinal units in flexed position, being the natural position in which lifting is performed.

The structural failure of spinal units can precipitate either in acute or chronic conditions. Most of the studies have concentrated on determination of peak forces in activities of known high stresses. However, biological tissues are like other physical materials with a finite life, and similarly subject to wear and tear. They are capable of self repair but also subject to mechanical deformation upon load application. All biological tissues are viscoelastic in nature and prolonged load may result in permanent deformation. Repeated load application may also result in cumulative fatigue reducing their stress bearing capacity. Such changes may

Table 2. Maximum voluntary compression, ultimate compressive
strength and the margin of safety.

Study	Maximum Voluntary Contraction MVC	Calculated lumbo-sacral compression	Margin of safety
	N	N	%
Kumar et al (1988)			
Isometric	726	6,933	32%
Isokinetic			
20 cm.s^{-1}	672	6,329	38%
60 cm.s^{-1}	639	6,017	41%
100 cm.s^{-1}	597	5,613	45%
Andersson & Schultz (1979)			
Isometric	900	7,863	23%

reduce the threshold stress at which the tissues fail. One may, therefore, examine the fatigue failure charactersistics of the spinal units.

Brinckmann et al (1987) and (1988) investigated the fatigue failure of the lumbar spine. In their experimental protocol, loads of between 20-30%, 30-40%, 40-50%, 50-60%, 60-70% and 75% of the estimated ultimate compressive strength (UCS) of spinal units were applied at a frequency of .25 Hz. They found that both, the magnitude of the load and the number of the cycles, affect the failure. At lower loads high repetition and at higher load lower repetition produced fatigue fractures. When their specimens were loaded between 50-60% of the UCS 92% of the specimens suffered fatigue fracture after 5000 cycles. A 91% fatigue fracture was reported after 500 cycles when the load was increased by additional 10%. At a load of 75% of UCS the fatigue fractures were precipitated in ten cycles. In an epidemiological study Kumar (1989) also reported a statistically significant difference between the cumulative load exposure to a group of institutional aides suffering low-back pain when compared to sex, age, body weight and height matched fellow workers with no back pain. Long range low grade loading of the spine will be difficult to control and measure. However looking at the results of Brinckmann et al (1987, 1988) it would appear that physiological limitations strongly favour biological safety. To begin with, the compression generated by the maximum voluntary contraction ranges between 68 to 77% of the ultimate compressive strength allowing 23 to 32% of margin of safety. When one considers that the MVC can be sustained only for a few seconds and it decays exponentially with the duration of the hold (Romert) it is obvious that such compressions can not be self generated. Also the MVC cannot be repeatedly generated without long rest pauses. Rapid repeated trials degenerated quickly with drastic reduction in magnitude thereby preventing total (load x cycle) exposure from rising. In addition glycogen depletion and lactate accumulation render human body physically unable to accumulate large number of load cycles due to

general exhaustion and muscle fatigue. Therefore, it is stated that the loading frequency used by these authors are physiologically unattainable even in a highly paced industrial environment. It must also be borne in mind that such rapid cyclic loading does not allow much needed recovery time to the viscoclastic biological tissues. This in turn will progressively accentuate the deformation rendering the tissues more vulnerable to suffer injury. Even when performing submaximally to determine maximum acceptable load of the lift for males, Snook's (1970) figures show that it dropped to 30-40% for floor to knuckle and shoulder to arm reach heights; to around 50% for knuckle to shoulder height lift when the frequency was increased from one lift per hour to 12 lifts per minute. Thus, an initial difference between the UCS and the compression at MVC (23-32%) when combined with the physiological limitation of inability of repetition will only increase the margin of the safety over and above 30% range stated before.

PHYSIOLOGICAL KINESIOLOGY

In addition to failure in compression mode, the tissues can be strained beyond their physiological limit and precipitate injury. To investigate this aspect Adams and Hutton (1986) compared the maximal in vivo range of flexion with that of osteo-ligamentous preparations. The active range of flexion was reported by them to be 10° short of the osteoligamentous preparation. Such a difference between extreme forward flexion and elastic limit of osteoligamentous preparation ensures the safety from possible strain injuries by preventing excessive deformation and generation of high tensile stresses. In their experiment Adams and Hutton (1986) also report that in a typical motion segment at 2° reduction in flexion at its elastic limit reduces the resistance to bending moment by 50% and hence a 50% reduction in the tensile stresses of the intervertebral ligaments and annulus. At the limit of the active flexion the osteoligamentous prepration provides half of the resistance to bending moment exerted by the upper body in forward bending (Adams et al

1980). Considering most activities of daily living or occupational activities it is obvious that only modest ranges of motion are commonly used. Thus such an interplay between posture and material properties ensures at least a safety margin of 50% in force enduring capacity. A given degree of muscle contraction is evoked for postural stability and readiness to move to the next phase of activity. Any sudden force may, however, tend to overcome the viscoelastic resistance of the muscle due to passive stretching. Such forces may result in sprains and strains as minor injuries before structured damage can occur.

Kumar and Davis (1983) studied the electromyographic activity of the erectores spinae and the intra-abdominal pressure during a static weight hold and a dynamic lift of the same weight. Both, the EMG and intra-abdominal pressure values for static posture activities were only a third to half of those obtained for handling the same weight in lifting activities. They suggested that an in- creased biomechanical demand in a dynamic lift was due to overcoming inertia and the need for postural stability. Freivalds et al (1984) also found that the dynamic efforts increased the load by as much as 40%. Park (1973) reported that acceleration during dynamic lifting increased the biomechanical stress by 15 to 20%. Since Kumar and Davis (1983) found that the physiological variables for static loading evoked only a third to half of the dynamic activity it is suggested that the excess response was to meet the need of dynamic activity. Since the acceleration and inertia effects have been shown to increase the load between 20 to 40% (Park 1973, Frievalds et al 1984), the increased response may well contribute from 30 to 45% toward the safety of the execution of the task. Marras et al (1987) also reported that under sudden loading conditions of the trunk the peak muscle responses increased by 35% in expected and 50% in unexpected conditions. This heightened response over the static values clearly indicate the intrinsic mechanism guiding the response overshoot to ensure system safety.

PSYCHOPHYSICS, METABOLIC COST AND BIOMECHANICS

The safe load for lifting has generally been determined on the basis of consideration of only one of the following criteria: a) biomechanical stress b) physiological cost, and c) psychophysical assessment. In Work Practices Guide for Manual Lifting, for the first time, all these criteria were considered together along with the epidemiological factors (NIOSH 1981). However, a complete integration and a clear and exhaustive interrelationship has not yet emerged. Combining the data developed by Snook 1978 and Ayoub 1978 and using psychophysical methodology maximal acceptable load for lifting to different vertical heights, at three horizontal distances away from the body, and six different lifting frequency was tabulated (NIOSH 1981). With increase in any of the three task variables the maximum acceptable load declined. However, quantitatively none of the other two criteria could be compared from those studies. In a simultaneous consideration of psychophysical and physiological criteria for determination of safe level of load for lifting Mital (1985) reported that the former always yielded a value less than or equal to obtained by the latter across low as well as high frequencies. Mital and Kromodihandjo (1986) reported a high correlation between the psychophysically determined maximal acceptable load for lifting and the back compressive forces. Using the data of Hutton and Adams (1982) they extrapolated by means of a significant regression equation the ultimate compressive strength of the lumbar spine of the subjects. Comparing the ultimate compressive strength with the compressive load developed during maximum acceptable weight revealed a difference of 30 to 50%. This difference is considered to be the margin of safety of the human back. Thus, a self selection of load enhanced and ensured the margin of safety in excess of 30%.

Conclusion

The maximum acceptable weight for lifting and maximum voluntary contraction have been calculated to generate a compression of .5 to .7 ultimate compressive strength of lumbar spinal units. The fatigue failure of the lumbar spinal units can occur by rapid application of maximal voluntary contraction. However, it is physiologically impossible, thereby protecting biomechanical safety of the system. A psychophysical self selection of work load for lifting activity has also been found to favour the system safety by arriving at a load equal to or lower than what may be arrived at by using the physiological criteria. It, therefore, appears that psychophysical critera may have an integrative role. However, the sensitivity of the psychophysical methodology over a very long term due to sensory conditioning needs to be investigated. It is conceivable that this methodology may not allow the detention of the threshold level of cumulative load where vulnerability is heightened as shown by Kumar (1989).

REFERENCES

Adams, M.A., Hutton, W.C., and Stott, J.R.R., (1980). The resistance to flexion of the lumbar inter-vertebral joint. Spine 5(3), 245-253.

Adams, M.A., and Hutton, W.C., (1986). Has the lumbar spine a margin of safety in forward bending. Clinical Biomechanics, 1, 3-6.

Andersson, G., (1979). Low back pain in industry: epidemiological aspects. Scand. J. Rehabil. Med. 11(4), 163-168.

Andersson, G.B.J. and Schultaz, A.B., (1979). Transmission of moments across the elbow joint and the lumbar spine. Journal of Biomechanics, 12, 747-755.

Ayoub, M.M., Dryden, R.D., McDanial, J.W., Knipfer, R.E. and Aghazadeh, F., (1978). Modeling of Lifting Capacity as a Function of Operator and Task Variables. In: Safety in Manual Materials Handling. C.G. Drury, Ed. DHEW (NIOSH) Publication No. 78-185, pp 120-130.

Brinckmann, P., Johannelweling, N., Hilweg, D. and Biggemann, M., (1987). Fatigue fracture of human lumbar vertebrae. Clinical Biomechanics, 2, 94-97.

Brinckmann, P., Biggemann, M., and Hilweg, D., (1988). Fatigue fractures of human lumbar vertebrae. Clinical Biomechanics Supplement No. 1.

Brown, J.R., (1972). Manual lifting and related fields. An annotated bibliography. Labour Safety Council of Ontario.

Chaffin, D.B., (1975). On the validity of biomechanical models of the low-back for weight lifting analysis. The American Society of Mechanical Engineers, 1-13.

Chaffin, D.B. and Park, K.S., (1973). A Longitudinal Study of Low Back Pain as Associated with Occupational Lifting Factors. Am. Ind. Hyg. Assoc. J., 34, 513-525.

Chaffin, D.B., Herrin, G.D., Keyserling, W.M. and Garg, A. (1977). A method for evaluating the biomechanical stresses resulting from manual materials handling jobs. Am. Ind. Hyg. Assoc. J. 38, 662-675.

Evans, F.G., and Lissner, H.R. (1959). Biomechanical Studies on the Lumbar Spine and Pelvis. The Journal of Bone and Joint Surgery, 41A(2), 278-290.

Freivalds, A., Chaffin, D.B., Garg, A. and Lee, K.S., (1984). A dynamic biomechanical evaluation of liting maximum acceptable loads. Journal of Biomechanics, 17, 251-262.

Garg, A., and Saxena, U., (1979). Effects of lifting frequency and technique on physical fatigue with special reference to psychophysical methodology and metabolic rate. Am. Ind. Hyg. Assoc. J. 40, 894-903.

Holbrook, T.L., Grazies, K., Kelsey, J. and Stauffer, R.N., (1984). The frequency of occurence, impact and cost of selected musculoskeletal conditions in the United States. American Academy of Orthopaedic Surgeons.

Hutton, W.C., and Adams, M.A., (1982). Can the lumbar spine be crushed in heavy lifting? Spine, 7(6), 586-590.

Kelsey, J.L., and White, A.A., (1980). Epidemiology and impact of low-back pain. Spine 5, 133-142.

Kumar, S., (1984). The physiological cost of three different methods of lifting in sagittal and lateral planes. Ergonomics 27,425-433.

Kumar, S., (1989). Cumulative stress as a risk factor for back pain. Manuscript submitted to spine.

Kumar, S., and Davis, P.R., (1983), Spinal loading in static and dynamic postures: EMG and intra-abdominal pressure study. Ergonomics, 26, 913-922.

Kumar, S., Chaffin, D.B., and Redfern, M., (1988). Isometric and isokinetic back and arm lifting strengths: device and measurement. Journal of Biomechanics, 21, 35-44.

Magora, A. (1970). Investigation of the relation between low back pain and occupation. I age sex community and other factors. Industr. Med. Surg. 39, 465-471.

Marras, W.S., Rangarajulu, S.L., and Lavender, S.A., (1987). Trunk loading and expectation. Ergonomics, 30, 551-562.

Mital, A., (1985). A comparison between psychophysical, and physiological approaches across low and high frequency ranges. Journal of Human Ergology, 14, 59-64.

Mital, A., and Kromodihardjo, S., (1986). Kinetic analysis of manual lifting activities: Part II - Biomechanical analysis of task variables.

NIOSH, (1981). Work practices Guide for Manual Lifting. Department of Health and Human Services.

National Safety Council, (1978). Accident Facts.

Park, K., (1973). A computerised simulation model of postures during manual materials handling. Ph.D. Dissertation, University of Michigan.

Sonoda, T., (1962). Studies on the Compression, Tension, and Torsion Strength of the Human Vertebral Column. J. Kyoto Prefect Med. Univ., 71, 659-702.

Snook, S., (1978). The design of manual handling tasks. Ergonomics, 21, 963-985.

Snook, S.H., Irvine, C.H., and Bass, S.F., (1970). Maximum Weights and Workloads Acceptable to Male Industrial Workers. Am. Ind. Hyg. Assoc. J., 31, 579-586.

THREE DIMENSIONAL MEASURES OF TRUNK MOTION COMPONENTS DURING MANUAL MATERIALS HANDLING IN INDUSTRY

W.S. Marras, L.R. Sudhakar, and S.A. Lavender
The Ohio State University
Department of Industrial and Systems Engineering
Biodynamics Laboratory
1971 Neil Ave.
Columbus, Ohio 43210

The objective of this study was to monitor and document the three dimensional spine motion components experienced during the performance of industrial work that is associated with various risks of low back disorder (LBD). An industrial study was performed that examined on-the-job trunk motions of 64 workers from 13 different industries. Trunk range of motion, velocity and acceleration were documented. Worker anthropometry, health history, external load moments, job satisfaction and risk (identified from OSHA 200 logs) were also recorded for the various jobs. The results identified and quantified those trunk motion characteristics as well as other workplace variables that were associated with high risk jobs. A regression model of job related LBD risk was also created based upon this information. The relationship between these motions and biomechanical loading of the spine as well as means to reduce the risk of LBD in the work place (based upon this study) are discussed.

INTRODUCTION

Low back disorders (LBD) have been identified as one of the most costly and widespread problems facing society today. It has been well established that the risk of suffering a LBD is closely associated with the work requirements of industrial employees. Many biomechanical evaluations have attempted to understand the risk associated with these jobs by evaluating the static, two dimensional sagittally symmetric moments imposed about the spine when performing a lifting task. However, a review of typical manual materials handling (MMH) situations in industry shows that almost all lifting tasks involve three dimensional dynamic motions of the trunk. Furthermore, no studies could be found in the literature that examine the three dimensional dynamic motions required by the trunk during the performance of industrial MMH tasks. It is important to understand these motion components since previous laboratory investigations based upon dynamic analyses have shown that the loading of the spine is greatly influenced by dynamic trunk motion (Freivalds et al., 1984; McGill and Norman 1985; Reilly and Marras 1989).

Current laboratory studies are exploring, theoretically and under highly controlled experimental conditions, how the internal loads within the trunk react to dynamic motion of the spine during lifting. However, epidemiologic industrial based studies are needed to determine the influence of the dynamic workplace trunk motion upon the risk of suffering a LBD due to the occupational task requirements.

Thus, the objective of this study was to monitor and document the three dimensional motion components that are experienced by the spine during the performance of industrial MMH work associated with various risks of LBD.

METHOD

Subjects

In this experiment 64 male and female MMH workers from various industries throughout the Midwest volunteered as subjects. Thirty eight of these subjects were males and 29 were females. The mean age of this population was 36.6 years with ages ranging from 19 to 63 years. The mean height of this population was 172.2 cm (SD=10.8) and the mean weight of this population was 74.8 kg (SD=15.2). None of the subjects were experiencing a LBD at the time of the experiment although 60.5 percent of the males and 50.0 percent of the females had a history of LBD.

Experimental Design

An industrial study was performed to evaluate the trunk motion and workplace components associated with the risk of LBD in industry. The study consisted of several steps. First, medical records and OSHA 200 logs were examined in various companies so that an estimate of normalized risk was established for each job. Risk was defined in three ways consisting of 1) normalized incident rate of LBD (over 200,000 hours of exposure), 2) days lost (over 200,000 hours of exposure), and 3) normalized turnover rate (% per year) for each job.

Jobs associated with high and low risk of LBD were identified and used as the independent variables in this study. High and low risk jobs were defined by identifying the upper and lower one-third of each risk distribution (incidence, days lost, or turnover) respectively. Low risk jobs based upon incidence rate were defined as those repetitive MMH tasks with incident rates less than or equal to 3 incidents per 200,000 hours of exposure. High risk jobs were those that had incidence rates of above 8 incidences per 200,000 hours of exposure. Low risk jobs based upon days lost were considered those jobs involving less than or equal to 4 days lost. High risk jobs were identified as those jobs involving more than 56 lost days. Low risk jobs based upon turnover rate were those jobs involving less than or equal to 10 % turnover per year. High risk jobs involved jobs associated with greater than 56 % turnover per year. In this experiment, 64 subjects sampled while performing 58 different industrial jobs in 13 different industries.

Second, each worker was fitted with a lumbar motion monitor (LMM) and was told to perform their jobs as they normally would. All subjects performed primarily repetitive manual materials handling tasks. Trunk motion characteristics (range of motion, velocity, acceleration and repetition) in all three planes of the body (sagittal, coronal and transverse) were collected so that tasks that were representative of the work cycle were recorded. Measures of worker anthropometry, worker satisfaction, health history, job history and factors associated with external load moments were also collected during this time. In all over 114 directly measured and derived workplace variables were considered for each job investigated.

Third, profiles of the trunk motion characteristics were determined for each job and this information along with risk and job history measures were entered into a data base. These trunk motion characteristics and workplace variables were considered the dependent variables in this study. This data base was then evaluated to determine the contribution to risk of each of the dependent measures.

Apparatus

A three dimensional LMM has been developed in our laboratory that is capable of assessing the instantaneous position of the lumbar spine in three dimensional space. Calibration experiments have shown that this device is accurate to within one and a half degrees. The LMM is essentially an exoskeleton of the spine that has been instrumented with sensors. The LMM is shown in Figure 1. The outputs of these sensors are multiplexed and transmitted via telemetry to a demultiplexer and A/D board in a portable computer. Each channel of data was sampled at a rate of 60 Hz. The computer calculates the instantaneous position, velocity and acceleration of the lumbar spine throughout a MMH exertion and evaluates the various motion component profiles in each plane of the body. A schematic representation of the data collection procedure is shown in Figure 2.

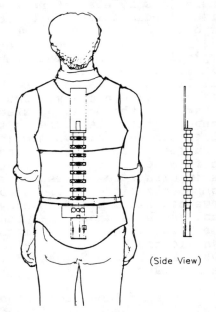

(Side View)

Figure 1: Lumbar Motion Monitor

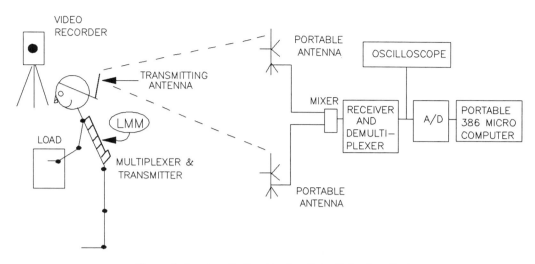

Figure 2: Lumbar Motion Monitor Data Collection System

Data Analysis

The data were first evaluated using analysis of variance procedures (ANOVA) to determine which of the dependent variables reacted in a statistically significant manner to each risk category. This analysis indicated which variables could be used to characterize high versus low risk jobs. Of the 114 variables collected, 95 were evaluated statistically. The remaining variables were variables that could not be evaluated quantitatively (i.e. job number, etc.).

Correlation analyses were also performed on the 95 dependent variables. Correlation matrices were developed that were used to determine which of the variables were correlated with each of the three risk indices and the strength of the correlation.

Finally, those variables showing statistically significant correlations ($p \leq .10$), and the variables showing statistically significant changes ($p \leq .10$) in the ANOVA procedures were entered into a stepwise regression analysis. This analysis would allow for the prediction of risk levels based on the back motion characteristics, the workplace variables, and anthropometry. This analysis procedure structure is shown graphically in Figure 3.

RESULTS

The results of these analyses are far too extensive to mention explicitly in this paper. However, since external

factors, such as load moment have been well investigated in previous literature, this description will focus upon back motion components. The ANOVA evaluation allowed us to describe those workplace variables that were associated with high and low risk jobs. Generally, increased range of motion and increased velocities in the transverse plane, in addition to the increased rate of motion changes in the sagittal and transverse planes characterized the high risk incident rates. Increases in the coupled range of motion, and increases in the magnitudes of the loads handled external to the body were factors which tended to increase lost time. Finally, high turnover rates can best be described by increased twisting velocity, increased rate of motion changes in all planes, increased numbers of interaction between trunk motions and external workplace factors.

Generally, the correlation analysis showed that workplace factors, lateral motion and twisting motion, and changes in the sagittal motion were correlated with higher LBD incidence rates. Lost days were correlated with

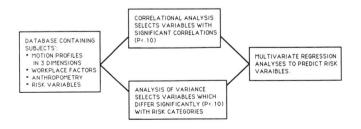

FIGURE 3: DIAGRAM OF DATA ANALYSIS PROCEDURE

workplace factors, peak accelerations in the coronal plane, and velocity by load interactions in the transverse and coronal planes. When correlations with turnover rates were examined, significant correlations were found with workplace factors as well as age, motion in the transverse plane, changes in motion in sagittal planes.

The results of the stepwise regression models for each risk indicator are summarized in Table 1. This table indicates the six workplace variables' contribution to each of the three risk indicators. The top row shows the dependent measure in the analysis and the variance accounted for by each regression model. The left hand column of this table grossly classifies the workplace variables into back motion planes, coupling of back motions, workplace external factors, anthropometric and job related factors. Each of the remaining column shows the combination of variables that best describe the various risk indices and their relative contribution (in percent). Finally, the last row of the table shows the amount of regression model variability that is explained by each model (R^2).

DISCUSSION

These results represent preliminary findings of an ongoing study. As more industrial data are collected these analyses will be updated and the statistical power associated with the analyses will increase. However, this analysis has indicated that this study represents a promising approach to studying industrial LBD.

Several new developments have resulted from this research effort. First, this study demonstrates that we are now able to quantitatively describe the three dimensional motion activities of the lumbar spine during work. Furthermore, it is also possible to investigate these motions as a function of workplace factors. Second, we have been able to document the relative contributions of workplace variables, and specifically, trunk motion parameters to LBD risk, defined in various ways. Third, this study has made it possible to quantitatively describe the characteristics of workplace factors and trunk motion variables associated with high and low risk jobs.

Table 1. Summary of Regression Model Analysis

VARIABLES	INCIDENT RATE ($R^2 = 0.75$)		LOST TIME ($R^2 = 0.42$)		TURNOVER RATE ($R^2 = 0.78$)	
BACK MOTION	PARAMETERS	% WT	PARAMETERS	% WT	PARAMETERS	% WT
CORONAL	MEAN AVG VEL	9.68	PEAK MAX VEL	24.76	PEAK RATE OF CHANGE IN MOTION	13.23
	MEAN RANGE	11.96	PEAK VEL * PEAK ACCEL	33.50		
	LIFT RATE * WBOX * RANGE	11.23				
SAGITTAL	PEAK ACCEL	10.80			(PEAK RATE OF CHANGE IN MOTION)^2	14.46
	MEAN RATE OF CHANGE IN MOTION	8.87				
TRANSVERSE					PEAK AVG VEL	15.82
					PEAK RATE OF CHANGE IN MOTION	6.99
					(MEAN RATE OF CHANGE IN MOTION)^2	8.60
COUPLING	LAT. RANGE * SAG. RANGE * TWIST RANGE	7.38				
WORKPLACE FACTORS	WEIGHT LIFTED PER HOUR	13.06	PEAK WBOX	41.75	PEAK WBOX	10.15
	MEAN START HEIGHT	10.42			MEAN WBOX	14.73
					WORK DURING LIFT DOWN	6.60
ANTHRO-POMETRY					AGE	12.90
JOB RELATED	SATISFACTION	16.61				

The analyses have emphasized two main points related to MMH. First, trunk motion is a factor that is associated with a greater risk of LBD. Second, much of LBD risk is associated with three dimensional trunk motion. These findings have been suspected by researchers for some time, but now we have been able to document this relationship in a non- epidemiologic industrial study as well as document the exact motions that increase LBD risk.

These findings also have several ergonomic and biomechanic implications. First, ergonomic interventions need to consider the effects of motion and asymmetric lifting. This analysis has demonstrated that motion is interactive with position. For example, while there is some contribution to risk from motions in the sagittal plane, once motion occurs in the coronal and transverse planes LBD risk, as determined by incidence and turnover rates, increases substantially. Laboratory studies (Marras and Mirka, 1989a; 1989b) have shown that asymmetric trunk motion reduces trunk strength as motion increases as well as increases the activity of the trunk loading musculature. Thus, models and lifting guides need to be expanded to evaluate MMH activities that are associated with asymmetric motion. Current guides that are applied (incorrectly) to realistic asymmetric motion conditions could greatly misinterpret the load on the spine as well as the strength capabilities of the worker.

Finally, these results may also be supported by biomechanical implications. Studies have shown that disc tolerance decreases as the trunk becomes more asymmetric (Troup and Edwards, 1985). Thus, not only is strength reduced when workers lift asymmetrically which would increase the risk of an overexertion injury, but the motion would tend to increase disc loading in positions where the disks are weakest.

CONCLUSIONS

This study has made it possible to identify those trunk motions and conditions that are associated with greater risks of LBD. The significance of these findings is that this information could be used to help interpret the biomechanical significance of these motions and conditions. This information could be used to assist the ergonomist in minimizing the risk of injury due to the design of the MMH task as well as help us understand the nature and conditions associated with LBD injuries.

REFERENCES

Freivalds, A., Chaffin, D.B., Garg A., and Lee, A., 1984, A dynamic Biomechanical evaluation of lifting maximum acceptable loads. J. Biomech. 17(4), 251-262.

Marras, W.S. and Mirka G.A., (1989a), "Trunk Responses to Asymmertric Acceleration", Journal of Orthopaedic Research, (in review).

Marras, W.S. and Mirka G.A., (1989b), "Trunk Strength During Asymmetric Trunk Motion", Human Factors, (in review).

McGill S.M. and Norman R.W., 1985, Dynamically and statically determined low back moments during lifting, J. Biomech. 18(12), 877-885.

Reilly C.H. and Marras W.S., 1989, Simulift: A simulation model of human trunk motion, Spine, 14(1), 5-11.

Troup, J.D.G. and Edwards, F.C. 1985, Manual Handling and Lifting. London: Her Majesty's Stationery Office.

ACKNOWLEDGEMENT

Financial support for this project was provided by the Industrial Commission of Ohio, Division of Safety and Hygiene.

BIOMECHANICAL MODELING OF ASYMMETRIC LIFTING
TASKS IN CONSTRAINED LIFTING POSTURES

Sean Gallagher, Christopher A. Hamrick, and Arnold C. Love
U.S. Bureau of Mines
P.O. Box 18070
Pittsburgh, PA 15236

Twelve subjects participated in an investigation of the biomechanical stresses of asymmetric lifting in stooped and kneeling postures. Three factors were manipulated in this study: Posture (stooped or kneeling), height of lift (35 or 70 cm), and weight of lift (15, 20, or 25 kg). Subjects were required to lift or lower a box every 10 seconds for a period of 2 minutes. Electromyography (EMG) of eight trunk muscles was collected during a lift in this period. The EMG data, normalized to maximum extension and flexion exertions in each posture, were input to a biomechanical model and used to predict compression and shear forces at the L_3 level of the lumbar spine. Results from the EMG-driven biomechanical model indicated that compression was greater when lifting to a higher shelf ($p < 0.001$), and indicated a significant interaction between posture and the weight of the lifting box ($p < 0.01$). Peak lateral shear was not significantly affected by any main effects or interactions ($p < 0.05$). Anterior shear was increased with increasing height of lift ($p < 0.001$), and also by the posture x weight interaction ($p < 0.01$). A multivariate analysis of variance (MANOVA) indicated a complex relationship for recruitment of the eight trunk muscles, with the triple interaction being significant ($p < 0.001$). The results of this investigation will be used to evaluate safe loads for lifting in these restricted postures.

INTRODUCTION

The height of an underground coal mine is generally determined by the thickness of the coal seam. In several cases, the coal seam of a mine may be less than 48" high. Such mines are often called "low-seam" coal mines. Workers in these mines often have to lift heavy materials in restricted postures (usually stooped or kneeling). There is reason to believe these postures result in significant compressive and shear loading on the lumbar spine.

Previous Bureau of Mines research has described the psychophysical lifting capacity, metabolic demands, and electromyography (EMG) of trunk muscles when performing tasks in restricted postures (Gallagher, et al., 1988; Gallagher and Unger, in press). Other investigators have researched the intraabdominal pressure associated with lifting in these postures (Davis and Troup, 1966; Ridd, 1981; Sims and Graveling, 1988). However, little research has been performed estimating the internal forces on the lumbar spine due to trunk muscle contraction in these postures. Marras and Sommerich (1990) recently described a biomechanical model driven by trunk muscle EMG that allows estimates of forces acting on the lumbar spine. A paper describing the results of this model's estimates of forces during symmetric lifting in restricted postures can be found elsewhere (Gallagher, et al., 1990). The purpose of the current investigation was to estimate the internal forces acting on the

lumbar spine during asymmetric lifting tasks in stooped and kneeling postures.

METHOD

Subjects

Twelve healthy male subjects ($M = 35.7$ years of age $\pm$ 6.8 S.D.) volunteered to participate in a study examining the biomechanics of asymmetric lifting in restricted postures. Nine of the subjects were experienced underground miners, while three of the subjects were volunteers from the U.S. Bureau of Mines. All participants operated under terms of informed consent.

Experimental procedure

Three independent variables were manipulated in this experiment -- posture (P) for the lift (stooped or kneeling), height (H) to which the box was lifted (35 or 70 cm), and weight (W) of the lifting box (15, 20, or 25 kg). A within subjects repeated measures design was employed. EMGs of eight trunk muscles (l. and r. erectores spinae, latissimus dorsi, external oblique, and rectus abdominis) were collected during lifting tasks, and were later digitized and input to a dynamic biomechanical model (Marras and Sommerich, 1990). The model output included estimates of compression, anterior-posterior shear, and right lateral shear (at the L_3 level of the lumbar spine), as well as torques about the X, Y, and Z axes, using the coordinate system described by Schultz and Andersson (1981). In addition, the model produced estimates of muscle forces for the eight trunk muscles. The assumption was made that the maximum stress that could be exerted by a muscle was 50 N/cm^2 (Reid and Costigan, 1987). Estimates of the cross-sectional areas of trunk muscles were obtained through anthropometric measurements, as described by Schultz, *et al.* (1982). All dependent variables except muscle forces were analyzed using 2 x 2 x 3 (P x H x W) analyses of variance (ANOVA). The data on muscle forces were subjected to a 2 x 2 x 3 (P x H x W) multivariate analysis of variance (MANOVA). Critical alpha levels were .05 for all statistical tests.

The lifting tasks were performed under a 1.2 m roof that restricted the subject's posture. An aluminum lifting box (50.8 x 33.0 x 17.8 cm) with two covered compartments was used to perform the lifting tasks. The subject performed a series of twelve asymmetric lifting and lowering tasks in a counterbalanced order. The subject was required to lift or lower the box to the appropriate height every 10 seconds for a period of 2 minutes. EMG data were collected during the third lift of this sequence.

RESULTS

Table 1 summarizes significant main effects and interactions for the dependent variables (outputs from the biomechanical model) in this investigation. Table 2 contains average model estimates of compression, shear, and torque for each of the experimental conditions.

Table 1.-- Summary of significant main effects and interactions for all dependent variables.

	Pos	Ht	Wt	PxH	PxW	HxW	PxHxW
Peak Compression	*	***	***		**		
Peak Lat. Shear	.067		.091		.080		
Peak Ant. Shear	.100	***	***		**	.075	
Peak X Torque	*	**	***				
Peak Y Torque							
Peak Z Torque	**			***	*		
Peak Mus. Force	***	***	***	**	***	***	***
Ave. Compression	**	***	***		*		
Ave. Lat. Shear							
Ave. Ant. Shear		***	***		*		
Ave. X Torque	***	**	***				
Ave. Y Torque	*	.090					
Ave. Z Torque	*	*		**	*		
Ave. Mus. Force	**	**	***	**	***	***	***

* $p < 0.050$
** $p < 0.010$
*** $p < 0.001$

Ergonomics and Musculoskeletal Disorders

Table 2.-- Model estimates of compression, shear, and torque for all experimental conditions. Values represent the mean for all 12 subjects. Numbers in parentheses represent the standard deviation.

	Compression	A-P Shear	R-L Shear	X Torque	Y Torque	Z Torque
Kneeling 35 cm, 15 kg	1599.6 (383.4)	187.3 (86.0)	21.9 (16.0)	56.5 (18.5)	22.0 (11.2)	5.9 (5.2)
Kneeling 35 cm, 20 kg	1788.8 (345.7)	207.5 (88.0)	20.0 (16.7)	62.9 (20.0)	24.7 (11.2)	6.2 (4.3)
Kneeling 35 cm, 25 kg	2114.2 (374.2)	275.2 (144.8)	22.3 (16.4)	75.6 (21.1)	25.4 (12.9)	9.4 (4.8)
Kneeling 70 cm, 15 kg	1942.1 (378.6)	244.5 (114.2)	24.0 (17.2)	69.4 (18.0)	22.3 (10.1)	8.0 (5.5)
Kneeling 70 cm, 20 kg	2188.2 (322.8)	262.5 (98.3)	24.4 (16.0)	78.2 (17.0)	24.9 (11.1)	8.9 (4.9)
Kneeling 70 cm, 25 kg	2560.2 (343.5)	360.1 (150.9)	23.6 (18.5)	89.4 (17.7)	22.1 (9.8)	11.1 (6.3)
Stooped 35 cm, 15 kg	1375.0 (450.9)	159.8 (75.8)	24.8 (17.1)	44.8 (23.0)	13.6 (9.3)	6.6 (4.3)
Stooped 35 cm, 20 kg	1521.6 (502.0)	185.9 (75.0)	27.1 (14.4)	49.2 (25.2)	13.5 (8.8)	9.4 (5.3)
Stooped 35 cm, 25 kg	1719.3 (433.3)	194.5 (66.4)	23.4 (11.5)	58.2 (21.0)	15.9 (9.7)	9.4 (4.8)
Stooped 70 cm, 15 kg	1593.5 (507.6)	208.5 (63.7)	27.7 (16.6)	49.7 (22.9)	15.5 (10.4)	12.8 (9.0)
Stooped 70 cm, 20 kg	1810.1 (549.6)	244.3 (99.6)	25.7 (13.9)	57.3 (27.3)	17.5 (12.7)	13.0 (7.7)
Stooped 70 cm, 25 kg	1974.3 (600.5)	279.6 (97.0)	31.4 (21.2)	64.3 (31.6)	20.2 (10.8)	17.1 (9.0)

Compression

As can be seen in this Table 1, there was a significant PxW interaction on compression. Table 2 shows that compression increased at a higher rate for kneeling compared to stooping; however, kneeling compression was always higher than that observed when stooped. The height of lift had the expected effect on compression (i.e., lifting to a higher shelf increased compression).

Shear Forces

Lateral shear was not significantly affected by any independent variables. However, peak anterior shear was significantly affected by lifting height, as well as by the PxW interaction. Examination of Table 2 shows that anterior shear was increased with increasing height of lift, while anterior shear increased at a higher rate in the kneeling posture, compared to stooped.

Muscle Forces

The MANOVA on muscle forces for the eight trunk muscles studied indicated a complicated recruitment pattern due to the changing conditions. The P x H x W interaction achieved significance at the 0.001 level. Figure 1 illustrates the average over the 12 subjects of peak EMG activity during lifts for the eight trunk muscles studied for all experimental conditions. These EMG data were normalized to maximum exertions for each of the subjects, using procedures described by Marras (1987).

DISCUSSION

Compared to a previously reported study on symmetric lifting in restricted postures (Gallagher, et al., 1990), many of the internal responses to asymmetric lifting were affected by interactions of the independent variables. For example, both compression and anterior shear were significantly affected by the PxW interaction. This phenomenon was not observed in the symmetric lifting study. Several similarities between the two studies can be noted, however. For example, in both studies compression was consistently higher in the kneeling posture when compared to the stooped posture. This is primarily the result of increased erectores spinae activity when lifting in the kneeling position.

In the study examining symmetric lifting (Gallagher, *et al.,* 1990), peak lateral shear was found to be significantly greater in the stooped posture. In contrast, the present study showed a trend (non-significant) towards greater peak lateral shear in the kneeling posture. This reflects a difference in recruitment of the latissimus dorsi muscles in these two postures according to the symmetry of the lifting task. As with the previous study, anterior shear was generally greater in the stooped posture, indicating greater activity of the abdominal obliques in this position.

Another similarity of the two studies was the complex recruitment pattern of the trunk musculature for the treatments studied. However, certain observations can be made with regard to trunk muscle activity. For example, the activity of the latissimus dorsi were higher in the stooped posture than when kneeling. However, increased erectores spinae activity was consistently demonstrated in the kneeling posture. The rectus abdominis were slightly more active in the stooped posture.

While the current model appears to give a good picture of the muscular loading on the spine, it is worth noting that this model is as yet unable to address other biomechanical factors that may be responsible for production of low back pain. For instance, in the stooped posture investigated here, a large portion of the restorative moment to maintain the position of the trunk is provided through the muscle-sparing action of the posterior ligaments (Gracovetsky and Farfan, 1986). The strain on these ligaments is apt to be quite considerable, but unfortunately remains difficult to quantify, and is not addressed by the current model.

The data reported in this study will assist in the development of recommendations for manual lifting tasks in the underground mining environment. The model estimates of compression will be compared to known compression tolerance limits of the spine, and recommended load limits will be designed so that these tolerance limits are not exceeded.

REFERENCES

Davis, P.R., and Troup, J.D.G., 1966. Effects on the trunk of erecting pit props at different working heights. *Ergonomics*, 9,6, 475-484.

Gallagher, S., Hamrick, C.H., and Love, A.C., 1990. Biomechanical Modeling of Symmetric Lifting Tasks in Constrained Lifting Postures, Paper in *Advances in Industrial Ergonomics and Safety II* (B. Das, ed.), London, Taylor and Francis, pp. 559-566.

Gallagher, S., Marras, W.S., and Bobick, T.G., 1988. Lifting in stooped and kneeling postures: effects on lifting capacity, metabolic costs, and electromyography of eight trunk muscles. *International Journal of Industrial Ergonomics*, 3,1, 65-76.

Gallagher, S., and Unger, R.L., in press. Lifting in four restricted lifting conditions: Psychophysical, physiological, and biomechanical effects of lifting in stooped and kneeling postures. *Applied Ergonomics*.

Gracovetsky, S., and Farfan, H.F., 1986. The optimum spine. *Spine*, 11, 6, pp. 543-573.

Marras, W.S., 1987. Preparation, recording and analysis of the EMG signal. Paper in *Trends in Ergonomics/Human Factors IV* (S.S. Asfour, ed.), Amsterdam, North-Holland, pp. 701-708.

Marras, W.S., and Sommerich, C.M., 1990. A Three-Dimensional Motion Model of Loads on the Lumbar Spine, Part 1: Model Structure, unpublished manuscript.

Reid, J.G., and Costigan, P.A., 1987. Trunk Muscle Balance and Muscular Force, *Spine*, v. 12, pp. 783-786.

Ridd, J.E., 1985. Spatial restraints and intra-abdominal pressure. *Ergonomics*, 28, 1, 149-166.

Schultz, A.B., and Andersson, G.B.J., 1981. Analysis of Loads on the Lumbar Spine, *Spine*, v. 6, pp. 76-82.

Schultz, A. B., Andersson, G. B. J., Haderspeck, K., Ortengren, R., Nordin, M., and Björk, R., 1982. Analysis and Measurement of Lumbar Trunk Loads in Tasks Involving Bends and Twists. *Journal of Biomechanics*, 15, 669-675.

Sims, M.T., and Graveling, R.A., 1988. Manual handling of supplies in free and restricted headroom. *Applied Ergonomics*, 19,4, 289-292.

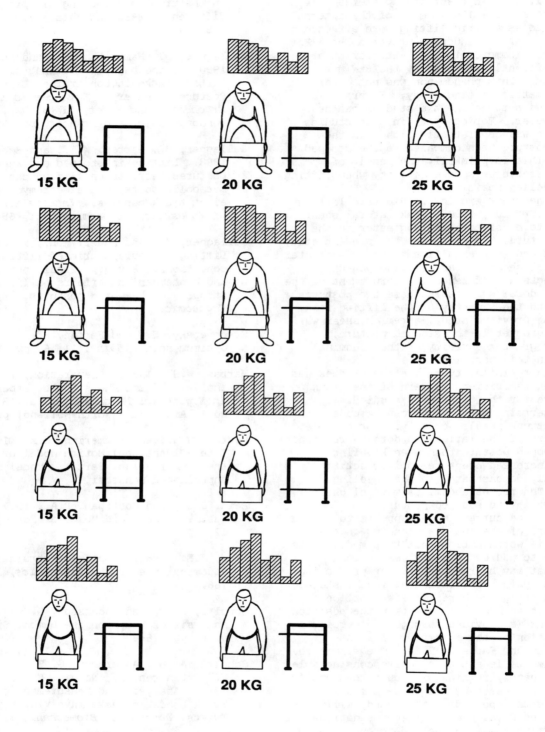

Figure 1. Normalized peak EMG for each experimental condition. From left to right, muscles represented in histograms are l. and r. latissimus dorsi, l. and r. erectores spinae, l. and r. external obliques, and l. and r. rectus abdominis.

SYMMETRIC AND ASYMMETRIC STOOP-LIFTING STRENGTH

Shrawan Kumar
Department of Physical Therapy
University of Alberta
Edmonton, Alberta, Canada

ABSTRACT

The strength capability of workers is an essential design criterion. Though true job simulated strength is most appropriate, availability of strength values in many standardized postures will permit selection of most suitable or a series of suitable strength values. Six normal young male subjects performed a maximal lifting effort in a stooping posture in isometric (with hip angles 60° and 90° flexion) and isokinetic modes (50 cm per second linear velocity). The isokinetic lift was performed from floor to knuckle height. Each lift was performed at half, three-quarters, and full reaches by each subject in sagittal, 30° lateral and 60° lateral planes. The strength was measured using the Dynamic Strength Tester (DST) described by Kumar et al (1988). The strength was inversely related to reach distance (p<0.0001). The largest declines in strength were between half and three-quarters reach distances (45% to 50%). From three-quarters reach to full reach, the drop was between 10% to 30%. With increasing asymmetry, the strength declined significantly (p<0.0001). The extent of this decline decreased at longer reaches. Based on these findings, a case is built for a biomechanical standardization of strength measurements.

INTRODUCTION

One of the essential demands of the industrial work is strength exertion. NIOSH (1981) stated that more than half of all workers have to exert significant strength on job. Despite a significant drive to automate all economically viable operations, manual material handling continues to be a prevalent industrial activity. In fact, Asfahl (1984) claims that for every ton of product, 50 to 180 tons of materials are manually handled. NIOSH (1981) reported that overexertion is the major cause of prevalent industrial back disorders. Statistics Canada (1988) reported that the maximum number of injuries in industry occurred to the back (27%). In terms of physical characteristics, most injuries involved sprains and strains (48%). Of all the events which directly resulted into a work injury, the most common was overexertion (Statistics Canada, 1988). The proportion of injury was highest in product fabricating, assembling and repairing occupations followed by service and construction trades for a total of 36% (Statistics Canada, 1988). Job strength rating (JSR: the ratio of strength required on the job to the strength of the individual) has been shown to be a good predictor of the trends of the low back injuries (Chaffin, 1974). He showed that as the JSR increased from .5 to 1.0 there was a progressive increase in the incidence rate of the low-back pain. However, most of the studies carried out on the strength capability have been restricted to trunk (Langarana and Lee, 1984; Smith et al, 1985). Studies of the functional activities have been generally made in the isometric mode in a self selected optimally comfortable posture (Chaffin et al, 1978). However, it must be emphasized that the vast majority of industrial activities are performed in dynamic mode.

Furthermore, frequently most of the job parameters are fixed and not under operator's control. Under such conditions, the operator has to fit in the job. Thus, dynamic activity with varying mechanical disadvantage and asymmetry of the posture is rule rather than exception. Kumar et al (1988) reported a significant and systematic difference in strength capability in dynamic activities as compared to static ones. A significant and progressive decline was found by these authors with increasing velocity of activity. Kumar (1987a and 1987b) reported a significant difference between static and dynamic strength capability for arm-lift. Further, static as well as dynamic strength capabilities were inversely related to the horizontal reach distance at which the strength was tested. In a round table panel discussion, Kumar (1990) also reported a significant effect of asymmetry on arm-lift strength. However, such a combination of physical factors have not been reported for stoop-lift in strength literature. The present study attempts to fill this void.

MATERIALS AND METHODS

Subjects and Tasks

Six normal male subjects without any history of low-back pain in the last 12 months volunteered for the study. These subjects were also checked for musculoskeletal and cardiovascular disorders. The demographic details of some of the characteristics of the sample follows (Table 1).

The recruited subjects were given the full details of the experiments and trained at submaximal level for the task. Such trained and prepared subjects performed 28 different stoop-

Table 1. Anthropometric characteristics of the experimental sample

Stats	Age yrs	Weight kg	Height cm	Knuckle Height cm	Reach cm
Mean	26.7	73.5	175.2	77.1	62.2
SD	11.2	7.9	5.9	4.3	3.2

lifting tasks. Of these, 18 were in isometric mode, 9 in isokinetic mode and one was in standard stoop posture as defined by Chaffin et al (1978). For standard posture stoop-lift strength measurements, the subjects were asked to acquire an optimal stooping posture which gave them an ability to exert their maximal lifting strength. Besides standard posture, the subjects performed stoop-lifting efforts in isometric mode with hip angle 60° and 90°. These activities were performed in sagittal, 30° left lateral and 60° left lateral planes at half reach, three-quarters reach and full reach distances. The full reach distance was measured as the distance between the shoulder joint center and the center of the grip. When the subjects assumed the required posture for the pre-determined randomized test condition, the instrumented handle bar was handed over to the subject. The subjects were instructed to gradually build their lifting strength against the stationary handle, without jerking, to their maximum within the first couple of seconds and keep it at that level for a period of another 3 seconds. The isokinetic lifts were performed at a 60 cm per second linear velocity of handle displacement. Like isometric lifts, these were also performed at half reach, three-quarters reach and full reach in sagittal, 30° left lateral and 60° left lateral plane. The starting position for the isokinetic lifts was floor level and the termination position was knuckle height. During these lifts, subjects were asked to keep their knees straight and bend their backs as deep as was required to reach the initial position. The subjects were also instructed not to lift with a jerk, instead to build to their maximal force within the first 10 cm from the starting position. The 27 experimental conditions are shown in Figure 1.

Measuring Devices

The measuring devices consisted of a SM 1000 load cell mounted under the handle bar of the strength tester. The latter was a modified version of the device described by Kumar et al (1988). For isometric testing, the latter allowed appropriate placement of the handle bar by cranking a winch on a continuous chain and locking it in position. For isokinetic strength testing, the device allowed the movement of the handle bar at a selectable constant velocity. This was achieved by coupling the handle bar through a flexible metal strap and one-way clutch to a shaft rotating at a fixed preset

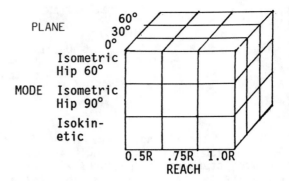

Figure 1. The experimental conditions

speed. The clutch coupled the mechanical resistance of the motor driven system until the threshold speed was reached. This allowed a resistance free movement of the handle bar below the preset speed. When the threshold speed was reached, the clutch engaged the constant speed shaft and controlled the speed with a very high load resistance. A displacement measuring potentiometer was coupled with the flexible metal strap to measure the linear displacement of the handle bar. The outputs of the load cell and the displacement measuring potentiometer were fed to a microcomputer (IBM XT) through a Data Translation A to D board. The system was thoroughly calibrated prior to the experiments.

Experimental Design and Data Analysis

The 28 experimental conditions were treated as blocks. The sequence of these blocks was randomized. Each block was repeated three times, with the average of the three readings taken to represent the block. The peak and average strength of the 28 conditions were numerically and graphically compared. They were also subjected to an analysis of variance to see the effect of mode, plane and reach of the experimental conditions. A post hoc analysis was also carried out.

RESULTS

The peak and average stoop-lifting strengths of the experimental sample at half, three-quarters and full reach distances in sagittal, 30° lateral and 60° lateral planes are presented in Tables 2 and 3. The average strength was invariably significantly lower than the peak strength for all experimental conditions. The strengths recorded for both isometric modes were greater than that of the isokinetic mode. However, the difference between the two isometric modes was small with a consistent trend for higher strength to occur with greater trunk flexion. However, these differences were not statistically significant.

The analysis of variance for the peak strengths revealed the following (Table 4).

Table 2. Mean peak stoop-lifting strength

Mode	Stat	Sagittal Plane			30° Lateral Plane			60° Lateral Plane		
		Reach			Reach			Reach		
		Half	T-Q	Full	Half	T-Q	Full	Half	T-Q	Full
Isometric	Mean	638	329	201	502	308	196	432	217	145
H = 60	SD	169	67	92	143	93	25	101	62	27
Isometric	Mean	675	356	198	567	311	189	379	223	139
H = 90	SD	229	99	52	172	112	74	181	72	45
Isokinetic	Mean	533	285	227	500	287	202	323	215	156
	SD	209	86	43	179	111	75	133	85	52

Table 3. Mean average stoop-lifting strength

Mode	Stat	Sagittal Plane			30° Lateral Plane			60° Lateral Plane		
		Reach			Reach			Reach		
		Half	T-Q	Full	Half	T-Q	Full	Half	T-Q	Full
Met	Mean	512	263	159	384	236	146	322	172	113
60	SD	185	80	38	143	76	52	128	43	35
Met	Mean	540	278	156	444	236	142	296	172	108
90	SD	165	65	43	139	195	48	35	56	42
Kin	Mean	342	180	128	312	177	128	193	127	102
	SD	123	49	42	129	165	35	87	42	23

Table 4. Analysis of variance for peak strength

Part of Model	F-ratio	DF	Probability
Mode	0.39	2	NS
Plane	72.80	2	.000
Reach	62.15	2	.000
Mode*Plane	3.25	4	.018
Mode*Reach	3.28	4	.017
Plane*Reach	16.58	4	.000
Mode*Plane*Reach	2.25	8	.028

While there was no significant difference between the three different modes of lifting (isometric with hip at 60°, isometric with hip at 90° and isokinetic), the plane in which the lift was performed and the reach distance at which the lift was performed were highly significant (p < 0.0001). The strength was inversely related to the reach distance at which the effort was applied. The drop was greatest between half reach and three-quarters reach, ranging between 47% and 49% in sagittal plane among the three modes of lifting. In 30° lateral plane, this drop ranged between 39% and 46%. However, in 60° lateral plane, it was more spread. The drop in isokinetic mode was only 34%, whereas in isometric mode with hip at 90°, demonstrated a drop of 42%. The drop in strength value from three-quarters reach to full reach was more pronounced in isometric modes than in isokinetic, 20-25% as opposed to 11-17%. Statistically, the post hoc analysis revealed that other than the difference between the three-quarters and full reaches in isokinetic mode and isometric modes in 60° lateral plane were significant with one exception. This was between the half and three-quarters reaches in 60° lateral plane during isokinetic activity. The effect of plane on the strength production ability was also significantly affected by the asymmetry of the posture (p<0.0001). The drop in peak strength within the modes of lifting was greatest between sagittal and 60° lateral planes. The differences between sagittal plane and 30° lateral planes were not as pronounced (Table 5).

The post hoc analysis revealed that at half reach the peak strength in 60° lateral plane was invariably different from those in sagittal plane and 30° lateral plane. However, there was no significant difference between sagittal and 30° lateral plane. The 60° lateral plane was also significantly different from sagittal plane at three-quarters reach in isometric lift with hip at 90°.

The analysis of variance of average strength during stoop lifting was similar to that of peak strength with one exception (Table

Table 5. The decline in peak strength in 30^0 and 60^0 lateral planes from sagittal plane values (expressed as percent of sagittal plane corresponding activity)

Mode of Activity	30^0 Lateral Plane			60^0 Lateral Plane		
	Reach			Reach		
	Half	Three-Quarters	Full	Half	Three-Quarters	Full
Isometric Hip 60^0	22	6.4	2.5	33	34	28
Isometric Hip 90^0	16	13	4.5	44	37.5	30
Isokinetic	6	0	11	40	25	32

6). The mode of lifting (isokinetic, isometric 60° and isometric 90°) was also a significant factor in determining its magnitude. However, the effect of mode was only significant at half reach. In each of the three planes at this reach, the isokinetic mode was different from the isometric modes. The differences did not reach significance at any other reaches. Further, as the reach distance increased the magnitude of the difference declined. The statistically significant differences as revealed by the post hoc analysis for average strength was identical to those of peak strength. The magnitude differences between different reach distances had a trend similar to those seen with peak strengths. The asymmetry

Table 6. Analysis of variance for average strength

Part of Model	F-ratio	DF	Probability
Mode	6.19	2	.005
Plane	73.36	2	.000
Reach	59.19	2	.000
Mode*Plane	4.18	4	.004
Mode*Reach	8.54	4	.000
Plane*Reach	19.27	4	.000
Mode*Plane*Reach	2.23	8	.030

of the posture of exertion had a similar effect on average strength as on peak strength. In fact, all statistically significant differences for peak were also significant for average strength. In addition, a significant difference was revealed between sagittal plane and 60° lateral plane for isometric lift with hip at 60° at three-quarters reach. The trend of the magnitude difference between different level of asymmetry for average strength was similar to that of the peak strength.

DISCUSSION

The value of strength as a design variable and a screening variable is obvious in face of the fact that most of the work related injuries (up to 47%) are overexertion injuries (NIOSH, 1981; Statistics Canada, 1988). Perhaps, it is due to this consideration that numerous studies have been reported in the literature. Kroemer (1986) stated that such studies have been conducted in the fields of physiology, physical education, psychology, ergonomics, human factors and others. However, he also pointed out there is such a diversity of information that at times they appear contradictory. Drury (1986), however, stated that the biomechanical factors have been carefully controlled and standardized in these studies. Comparison of many studies clearly reveals that, in fact, a biomechanical standardization is lacking. Though, the strength, in addition to biomechanical factors, is also affected by motivation and perception (Drury, 1986). With a standardization of interface between the operator and the task, one can attempt to normalize the perceptual aspects. Given the nature of the repetitious industrial tasks, the motivational factor is expected to even out over a longer period. It is the biomechanical factor which, therefore, is the most consequential. If a standardization of biomechanical factors in terms of mechanical disadvantage, postural asymmetry and the velocity of lift is achieved in strength testing, it may be of significant value. Much the way MTM is used, it may be possible to derive the relevant strength values, broken according to the task characteristics, from pre-determined and pre-recorded reference values. With this consideration, on an assumption of a bilateral symmetry, the total workspace was divided into two identical halves. One half was investigated for two isometric postures and one isokinetic exertion through the range. The velocity of isokinetic exertion was chosen at 50 cm per second to be closer to the speed of execution of many industrial activites.

Stoop lift was investigated because it continues to be favoured in industrial activities over squat lift, perhaps due to lesser physiological cost (Kumar, 1984 and 1988; Garg & Sexena, 1979; Brown, 1972; and others). Further, the postures of shallow and deep stoop was investigated for isometric lift due to significant sensitivity of strength to posture (Kumar & Garand, 1988; Mital & Das, 1988; Chaffin, 1975; and others). Postural asymmetries and mechanical disadvantage being the most frequently changing variables in industrial setting are most relevant for investigation.

As has been shown before (Kumar and Chaffin, 1985; Kumar et al, 1988), the isometric strengths were higher than those during isokinetic trials. The significant need of postural stability during constantly changing posture in isokinetic activity does not permit as high a torque production. A consistently lower value of strength with a shallow stoop, though not statistically significant, is a clear example of subtle effects of posture on strength. It is possible that in deeper stoop more powerful gluteal muscles were more active compared to a shallower posture. Though the effect of mode of lifting was not a statistically significant factor in determining the peak strength, it was so for average strength. It would appear, therefore, that the average strength may be a more sensitive variable. Furthermore, it is also thought that the average strength may be a better indicator of peoples' ability for industrial activities. The average strength is less affected by the sudden, short-lived motivational spurts. Instead, it represents less variable more objective interaction between muscular strength and the biomechanical parameters of the task.

The mechanical disadvantage of the task, as reflected in the reach distance of the lift, was the factor which affected the strength value the most, causing it to decline up to 70% from half reach to full reach. A significant decline in arm-lift strength with increasing mechanical disadvantage has been reported before (Kumar, 1987a and 1987b). These results clearly indicate that people are strongest closer to the body. Regardless of the condition, the highest drop in strength occurred between half and full reach. Generally, the drop between three-quarters reach and full reach was half or less of what was found between half and full reach. The postural asymmetry was a factor which also significantly influenced the strength value. The greater the asymmetry the lesser the strength (Table 5). The industrial activities involve changing mechanical disadvantage and postural asymmetries. In light of the information obtained in this study, it is suggested that isometric strength measured in an optimal posture (mechanical disadvantage and asymmetry) does not faithfully represent a person's ability to generate strength. The concept of biomechanically standardized strength tables for industrial application based on industrial population is advocated.

REFERENCES

Asfahl, C.R. (1984), Engelwood Cliffs, N.J. Prentice Hall

Brown, J.R. (1972), Ontario: Ministry of Labour

Chaffin, D.B. (1974), J Occup Med; 16:248-254

Chaffin, D.B. (1975), Am Indust Hyg Assoc J; 36:505-511

Chaffin, D.B., Herrin, G.D., Keyserling, M.S. (1978), J Occup Med 20:403-408

Drury, C. (1986), Proceedings of the Human Factors Society - 30th Annual Meeting pp 968-969

Garg, A. and Saxena, U. (1979), Am Ind Hyg Assoc J; 40:894-903

Kroemer, K.H.E. (1986), Proceedings of the Human Factors Society - 30th Annual Meeting. pp. 977-981

Kumar, S. (1984), Ergonomics 27(4), 425-433

Kumar, S. (1987), Trends in Ergonomics/Human Factors IV. Editor S.S. Asfour, Publisher North-Holland, pp. 675-683

Kumar, S. (1987), In Musculoskeletal Disorders at Work, Ed. P. Buckle, Pub. Taylor and Francis, London, pp. 37-42

Kumar, S. (1988), International Journal of Industrial Ergonomics, 2:273-284

Kumar, S. (1990), International Conference of Institute of Industrial Engineers. San Francisco

Kumar, S., Chaffin, D.B. (1985), Paper Presented at American Society of Biomechanics Con. 1985

Kumar, S., Chaffin, D.B., and Redfern, M. (1988), J. Biomech, 21:35-44

Kumar, S., and Garand, D. (1988), Proceedings of the 21st Annual Conference of Human Factors Association of Canada, pp. 183-186

Langarana, N., Lee, C.K. (1983), Spine 9:171-175

Mital, A. and Das, B. (1987), Clinical Biomechanics, 2:97-106

National Institute for Occupational Safety and Health (1981), NIOSH Rpt No. 81-122, Cin. Ohio

Smith, S.S., Mayer, T.G., Gatchel, R.J., Becker, T.J. (1985), Spine 21:757-764

Statics Canada (1988), Ottawa

The Effect of Rigid Container Shape on Maximum Acceptable Weight of Lift

Lee T. Ostrom[1], James L. Smith, and M.M. Ayoub

Texas Tech University, Lubbock, TX

[1] Presently with EG&G, Idaho, Idaho Falls, ID

In the chemical, pharmaceutical, and paint manufacturing industries raw and finished materials are frequently shipped in cylindrical rigid containers instead of boxes. Psychophysical manual materials handling lifting capacity studies to date have not considered whether rigid container shape has an effect on maximum acceptable weight of lift. Two psychophysical studies were designed and conducted in order to answer this question.

During the first experiment ten subjects lifted boxes with three different heights, at three different frequencies, and through three different lift ranges in order to establish a data base in which to compare the results of the second experiment. During the second experiment the same group of subjects lifted cylinders with three different heights at the same frequencies of lift and through the same lift ranges as the first experiment. The subjects adjusted the weight of the box until they felt they could safely lift that weight for eight hours. This weight was the maximum acceptable weight of lift (MAWL). The experimental design for both experiments was a 3 factor randomized complete block design with blocking on subjects. The dependent variables were oxygen consumption (VO_2), MAWL, and Borg's rating of perceived exertion.

The results indicated that when one designs a lifting task for a rigid container the shape of the rigid container, whether a cylinder or a box, does not need to be taken into account and that current psychophysical lifting capacity prediction models can be used to design cylinder lifting tasks.

INTRODUCTION

In the chemical, pharmaceutical, and paint manufacturing industries raw and finished materials are frequently shipped in cylindrical rigid containers instead of boxes. Psychophysical manual materials handling lifting capacity studies to date have not considered whether rigid container shape has an effect on maximum acceptable weight of lift. Two psychophysical studies were designed and conducted in order to help answer this question.

METHOD

Prior to the begining of the experiments, each subject participated in a two week training program which was designed to eliminate the effect of learning on the data. The training program was based on the program devised by Asfour (1980) and was designed to increase physical work capacity (PWC), muscular strength and to improve muscular endurance.

The first of the two experiments in this study involved having five male and five female Texas Tech University students lift boxes, with sagittal and frontal plane dimensions of

35.64cm, with three different heights (30.48, 60.96, and 91.44cm), through three different lift ranges (floor to knuckle, knuckle to shoulder, and floor to shoulder height), at three different frequencies of lift (6, 9, and 12 lifts per minute). The experimental design of the first experiment was a 3 factor randomized complete block design with blocking on subjects. The dependent variables in this experiment were oxygen consumption (VO_2), Borg's rating of perceived exertion (RPE) (Borg, 1974) and maximum acceptable weight of lift (MAWL). This experiment served as a data base in which the results of the second experiment were compared. The second experiment involved having the same group of subjects lift cylinders, with diameters of 35.46cm, with three different heights (30.48, 60.96, and 91.44cm) through the same lift ranges and at the same frequencies of lift as the first experiment. The experimental design of the second experiment was also a 3 factor randomized complete block design with blocking on subjects and the dependent variables for the second experiment were the same as the first experiment.

In both experiments the lift range, container, and frequency of lift were selected at random for each experimental session. The starting weight of the container was set randomly at either relatively heavy or relatively light. The subjects were allowed to adjust the weight of load of the container to the maximum they felt they could lift for eight hours without strain, discomfort, without becoming tired, weakened, overheated or out of breath. They could adjust the weight of the container as many times as they wanted. The weight adjustment period lasted for 20 minutes (Ayoub et al., 1978), however, the subjects made most of the weight adjustment in the first 10 to 15 minutes of this period. The final weight at the end of this period was the maximum acceptable weight of lift (MAWL). The subjects then rested for three minutes while the apparatus was prepared for collecting VO_2 data. After the rest period, the subjects lifted his or her maximum acceptable weight of lift for four minutes. During the last three minutes the subjects' VO_2 was measured by means of the Beckman Metabolic Cart. Immediately after the four minute weight lifting session the subjects were asked their perception of exertion on Borg's RPE scale which was on a 35.46cm by 60.96cm card and posted adjacent to the weight lifting platform.

The apparatus used in these experiments were:

Beckman Metabolic Cart

A 35.46cm x 60.96cm card with Borg's RPE scale printed on it in 2.54cm numbers and letters

Three boxes with sagittal and frontal plane dimensions of 35.46cm and heights of 30.48, 60.96, and 91.44cm

Three cylinders with diameters of 35.36cm and heights of 30.48, 60.96, and 91.44cm

A lifting platform with a wide range of adjustments

Various sizes of lead and steel weights

A tape recorder and recordings with tones at certain intervals to time the three frequencies of lift.

RESULTS

The data collected in both experiments were analyzed for normality by using the SAS procedure PROC UNIVARIATE (SAS, 1985). The hypotheses that the MAWLs, VO_2s, and Borg's RPEs collected in this study were normally distributed could not be rejected at the $\alpha = 0.01$ level. Due to a subject not completing the second experiment, two of the 270 data points were missing. SAS missing data point option was used to calculate data points for those which were missing (SAS, 1985).

ANOVA tables were developed for both data sets by means of PROC ANOVA (SAS, 1985). The ANOVA tables for the two data sets were examined for differences in statistical significance for all the combinations of the independent variables for each dependent variable. Table 1 shows a summary of the ANOVA tables.

A student t-test was used to determine whether there were statistically significant differences between the means of the MAWL, VO_2, and Borg's RPE data for the two experimental data sets. Table 1 shows these comparisons. There were no statistically significant differences in the overall means of the MAWL and Borg's RPE data. There was a statistically significant difference in the overall means of the VO_2 data. The mean of the VO_2 data was lower for the second

TABLE 1

SUMMARY OF ANOVA TABLES

Source	Experiment One			Experiment Two		
	MAWL	VO2	Borg's RPE	MAWL	VO2	Borg's RPE
Subjects	Y	Y	Y	Y	Y	Y
Fr	Y	Y	Y	Y	Y	Y
BH	Y	N	N	Y	N	N
BH * Fr	N	N	N	N	N	N
LR	Y	Y	Y	Y	Y	N
LR * Fr	Y	N	N	Y	Y	N
BH * LR	N	N	N	N	N	N
BH * LR * Fr	N	N	N	N	N	N

Note: 1.   Y indicates statistical significance at the $\alpha = 0.05$ level, N indicates that it was not statistically significant at the $\alpha = 0.05$ level.

2.   Fr = frequency of lift; BH = box height; LR = lift range.

experiment. This indicated that either lifting cylinders is less physiologically demanding, so the VO2s were lower for lifting cylinders, or the subjects became more skilled at lifting during the first experiment. As more skilled lifters the subjects may have found postures that were less stressful during the cylinder lifting experiment. Lifting postures were observed to be different for the cylinder lifting experiment. When lifting the cylinders, the subjects could not utilize the obviously missing corner for gripping and had to position one hand along the curved surface of the container.

One possibility for the decrease in oxygen consumption in the second experiment which could not be analyzed from the experimental design utilized in the study was container volume. The cylinders used in experiment two had 78.5% of the volume of the boxes used in experiment one.

DISCUSSION

The results indicated that when one designs a lifting task for a rigid container, the shape of the rigid container, whether a cylinder or a box, does not need to be taken into account and that current psychophysical lifting capacity prediction models can be used to design cylinder lifting tasks.

REFERENCES

Asfour, S.S., "Energy Cost Prediction Models for Manual Lifting and Lowering Tasks," Texas Tech University: Ph.D. Dissertation, Lubbock, TX, 1980.

Ayoub, M.M., N.J. Bethea, S. Deivanayagam, S.S. Asfour, G.M. Bakken, D. Liles, A. Mital, and M. Sherif, Determination and Modeling of Lifting Capacity, Final Report to NIOSH. Institute of Biotechnology, Texas Tech University, Lubbock, TX 1978.

Borg, G., "Psychological Aspects of Physical Activities", in L.A. Larson (ed.), Fitness, Health, and Work Capacity, p.141, Macmillan, New York, 1974.

SAS Institute, Inc., SAS User's Guide: Statistics, Version 5 Edition, Cary, N.C., 1985.

TABLE 2

SUMMARY OF t-TESTS

Maximum Acceptable Weight of Lift
(kilograms)

Experiment	N	Mean	S.D.	t Critical	t-Test Ho: Di = 0	Power of Test
One	270	25.39	9.76			
Two	268	24.47	10.03	1.96	Accept	0.212

Oxygen Consumption
(milliliters of oxygen consummed per minute)

Experiment	N	Mean	S.D.	t Critical	t-Test Ho: Di = 0	Power of Test
One	270	1056.6	409.5			
Two	268	933.7	368.4	1.96	Reject	

Borg's Rating of Perceived Exertion
(relative scale)

Experiment	N	Mean	S.D.	t Critical	t-Test Ho: Di = 0	Power of Test
One	270	13.01	1.45			
Two	268	13.12	1.56	1.96	Accept	0.210

PHYSIOLOGICAL MODELS AND GUIDELINES FOR THE DESIGN OF HIGH FREQUENCY SHOULDER LIFTING TASKS

S.S. ASFOUR[*], M. AKCIN[*], M. TRITAR[*] and A.M. GENAIDY[**]

[*]Department of Industrial Engineering
University of Miami
Coral Gables, Fl, 33124, U.S.A.

[**]Department of Mechanical, Industrial and Nuclear Engineering
University of Cincinnati, OH, 45221, U.S.A.

The main objective of this study is to investigate the effect of weight of load, frequency of handling, and age of industrial workers on endurance time and physiological responses for prolonged lifting tasks from shoulder to reach height, to Sixteen male industrial workers participated in this study. The independent variables were the frequency of handling, the weight of load, and the age of workers. The dependent variables were endurance time, oxygen consumption, minute ventilation, heart rate and the subjective responses of the subjects (ratings of perceived exertion). The results indicated that (1) endurance time decreased with increasing frequency or load, (2) Endurance time, heart rate, oxygen consumption and minute ventilation were significantly affected by frequency and load. (3) Age did not have a significant effect on endurance time, heart rate and oxygen consumption but it was significant for minute ventilation.

INTRODUCTION

Manual material handling (MMH) is the principal source of compensable work injuries in the United States, and four out of five of the injuries that are due to MMH affect the lower back (Federal Register, 1986). From an epidemiological perspective, the NIOSH "Work Practice Guide for Manual Lifting" (1981) cites studies revealing that frequency rates (number of injuries per man-hour) and severity rates (number of hours lost due to injury per man-hour on job) of musculoskeletal injuries increase significantly when: (1) heavy objects are lifted, (2) the object is bulky, (3) the object is lifted from floor, and (4) objects are frequently lifted. Therefore, the reduction of risk of such injuries has been of major concern to practitioners of occupational safety and health, federal agencies, insurance companies, and researchers. Ergonomics job design (EJD) is one of the engineering controls used for preventing MMH injuries. The goal EJD of is to maintain job requirements within the capabilities of the workforce. Traditionally, EJD is evaluated for MMH tasks through the development of a job stress index which is defined as the ratio of job demands and workers capacity. In manual materials handling, job demands are often expressed as the weight handled at a certain frequency over a given range, with containers of certain size. The worker capacity can be expressed in terms of his/her material handling capability based on physiological, biomechanical, or psychophysical criteria. In the literature, three job stress indices based on either metabolic, isometric strength, and psychophysical acceptable weight of lift were developed. These indices were referred to as degree of loading "DL" (Bonjer, 1962), lift strength ratio "LSR" (Chaffin et al. 1977), and job severity index "JSI" (Ayoub et al. 1978).

OBJECTIVES

In many industries workers are asked to continuously handle relatively light to moderate loads at high frequencies of lift for work shifts lasting 1 to 8 hours. Based on observations a typical example of such an activity is lifting an average load of 8 kg at an average frequency of 17 times/min. Based on NIOSH recommendations, one can only determine the amount of load for either 1 or 8 hour durations. Therefore depending only on these recommendations would result in either underestimating or overestimating the capabilities of workers. Thus the main objective of this study is to fill the gaps found in the MMH literature for continuous high frequency lifting tasks for shoulder to reach height.

METHODS AND PROCEDURES

To achieve the objective of this study, sixteen male industrial workers from local industries were asked to participate in this study. Prior to the start of this study a qualified physician examined all the subjects. Subjects were excluded from participating in this study if they had musculoskeletal injuries, heart problems, high blood pressure, or other medical restrictions. All the subjects were informed about the experimental procedures. The informed consent form and the general information packet were signed by each subject.

An experiment was conducted to determine the endurance limits and the physiologic fatigue limits of the subjects while performing the lifting tasks.

Independent variables

The independent variables investigated in this study were the frequency of lift, the weight of load and the age of workers. The literature reviewed indicates that an increase in either frequency or the load increases the physiological responses of the individual performing the lifting task. The frequency levels studied were 8, 12 and 16 lifts/min. The levels of loads studied were 5, 10 and 15 kg. These three levels of frequency and three levels of loads were selected to cover a wide range of loads and frequencies encountered in industry. Subjects of two age groups were tested in this study, namely 20-29 and 30-39 years of age.

Dependent Variables

The dependent variables investigated in this study were the endurance time, oxygen consumption, minute ventilation and heart rate.

Endurance time. The endurance time was defined in this study as the maximum length of time during which an individual is capable of lifting a given load at a given frequency continuously. The upper limit for endurance time was set at 8 hours.

Oxygen Consumption, Minute Ventilation and Heart Rate. In this study oxygen consumption, and minute ventilation were measured while the subject was performing the task using the microcomputer based physiologic monitor program described by Asfour et al. (1987). The heart rate was measured while the subject was performing the task using the microcomputer based cardiac monitor program described by Genaidy et al. (1987).

Experimental Design

A mixed design was used in this study. The weight of load, frequency of lift and age group of subjects were the independent variables used in the experiment. The weight of load and frequency of lift were treated as the within-subject factors. Subjects were nested within each age group. Each subject performed a total of 9 treatments of frequency-weight combinations.

Experimental procedure

Sixteen paid industrial workers were utilized in this study. Subjects were asked to avoid participation in any strenuous activity for at least 48 hours prior to experimental sessions and to get their normal amount of sleep. Each subject performed only 3 experimental sessions per week. The upper limit for endurance time was set at 8 hours. Each subject was given 10 minutes of rest every 50 minutes of work and 1 hour for lunch after the fourth hour of work. At the end of every 50 minutes of work and upon termination of the experimental session the rating of perceived exertion (leg, back, arm, hand, shoulder, and overall body) and fatigue scales for the overall body were recorded. Each subject was asked to lift a compact box (38 x 38 x 25 cm) from shoulder height (127cm) to reach height (178 cm) in an environmentally controlled laboratory. The box was lowered automatically using a mechanical device.

Average Physiological Response. The average subjective response is defined as the mean of the observations taken at every five minute intervals to a time of ET, where ET is the endurance time. Mathematically,

$$X_{pr} = (X_5 + X_{10} + X_{15} + ... + X_{ET})/N$$

where,

X_{pr} = Average physiological response
X_5 = Observation taken at min 5
X_{10} = Observation taken at min 10
X_{15} = Observation taken at min 15
X_{ET} = Observation taken at a time equal to endurance time
N = Number of observations

Average Subjective Response. The average subjective response is defined as the mean of the observations taken at the end of each 50-minute work period to a time of ET, where ET is the endurance time. Mathematically,

$$X_{sr} = (X_{50} + X_{110} + X_{170} + ... + X_{ET})/N$$

where,

X_{sr} = Average subjective response,
X_{50} = Observation taken at min 50
X_{110} = Observation taken at min 110
X_{170} = Observation taken at min 170
X_{ET} = Observation taken at a time equal to endurance time
N = Number of observations

Table 1. The Analysis of Variance Results for the Physiological and Subjective Responses

	ET	HR	VO$_2$	VE	LRPE	ARPE	SRPE	BRPE	ORPE
LD	**	**	**	**	**	--	**	**	**
FR	**	**	*	**	--	--	**	**	**
AG	--	--	--	**	--	--	--	--	--
LD*FR	--	**	--	--	--	--	--	--	--
AG*LD	--	--	--	--	--	--	--	--	*
AG*FR	--	--	--	--	--	--	--	--	--
AG*LD*FR	--	--	--	--	--	--	--	--	--

** significant at 1% level; * significant at 5% level; -- not significant

LD	load	**ET**	endurance time
FR	frequency	**HR**	heart rate
AG	age	**VO$_2$**	oxygen consumption
VE	minute ventilation	**LRPE**	leg RPE
ARPE	arm RPE	**SRPE**	shoulder RPE
BRPE	back RPE	**ORPE**	overall RPE

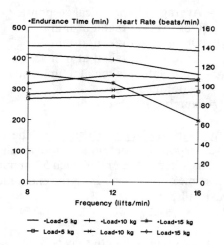

Fig. 2 Mean Endurance Time and Heart Rate vs Frequency for Different Loads for Knuckle to Shoulder Height.

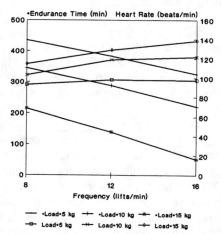

Fig. 1 Mean Endurance Time and Heart Rate vs Frequency for Different Loads for Shoulder to Reach Height.

RESULTS AND DISCUSSION

As demonstrated in Table 1., the response variable AGE was found to be significant only for the minute ventilation. Also all interactions with age were not significant. The main effects of LD and FR were statistically significant for ET, HR, VO_2, VE at 0.05 level, but no significance was discovered for any two way interaction except the LD*FR interaction for the heart rate. Endurance time and heart rate are plotted vs frequency of lift in Fig. 1 (detailed results are shown in Akcin 1991). The results demonstrate very clearly that ET, and HR are significantly affected by the frequency of lift. An increase in the frequency of lift or the load is followed by a decrease in the ET and a significant increase in the HR. These results are in agreement with the findings of Jomoah (1990). Whereas he obtained much higher endurance times with somewhat lower recorded physiological responses as shown in Fig. 2. ET values reported by Jomoah (1990) for loads of 5, 10 and 15 Kgs at 16 lifts/min were 427, 350 and 199 minutes respectively, for knuckle to shoulder lifts. The corresponding HR values were 94, 106 and 107 beats/min. In the present study the ET values for the same loads were 320, 220 and 53 minutes respectively. For shoulder to reach lifts the corresponding HR values were 97, 120 and 137 beats/min. This clearly indicates that ET was much lower for shoulder to reach lifts than for knuckle to shoulder lifts while the HR values were somewhat higher. It seems that subjects muscular fatigue rather than physiological fatigue was the cause of the lower ET values. These results ere not in agreement with the previous studies (Mital et al., 1984). Mital et al reported that the physiological responses were less for shoulder to reach heights.

CONCLUSIONS

The main conclusions of this study are: (1) Endurance time decreases with the increase in load or frequency of lift; (2) heart rate, minute ventilation and oxygen consumption increases with the increase in the load and/or frequency of lift; (3) ET values recorded in the present study, for shoulder to reach lifts are much lower than those reported by Jomoah (1990) for knuckle to shoulder lifts.

ACKNOWLEDGEMENTS

This study was supported by the National Institute for Occupational Safety and Health Through Grant No. 1R01OH02591-02.

REFERENCES

Akcin, M., 1991. Physiological Models and Guidelines for the Design of Prolonged Arm Lifting Tasks, Unpublished Thesis, Department of Industrial Engineering, University of Miami, Coral Gables, Florida, USA.

Asfour, S. S., Genaidy, A. M., and Khalil, T. M., 1987. An On Line Microcomputer Based Metabolic Monitoring System, The International Journal of Industrial Ergonomics, vol. 1, 169-177.

Bonjer, F. H., 1962. Actual Energy Expenditure in Relation to Physical Working Capacity, Ergonomics, vol.5 29-31.

Chaffin, D. B., Herrin, G. D., Keyserling, W. M., and Foulke, J. A., 1977. Preemployment Strength Testing, DHEW (NIOSH), Publication no., 77-163.

Genaidy, A. M., Asfour, S. S. and Khalil, T. M., 1987a. An on line Microcomputer Based Cardiac Monitoring System. The International Journal of Industrial Ergonomics, vol.1, 273-283.

Jomoah, I., 1990. Physiological Models and Guidelines for the Design of High Frequency Ar Lifting Tasks, Advances in Industrial Ergonomics and Safety Research, III.

Mital, A. Shell, R. L., Mital, C., Sanghavi, N. and Ramanan, S., 1984. Acceptable Weights of Lifting for Extended Workshifts, NIOSH.

NIOSH, 1981. Work Practice for Manual Lifting, DHHS (NIOSH), Publication No. 81-122, Cincinnati, OH.

DETERMINING PERMISSIBLE LIFTING LOADS: AN APPROACH

M. M. Ayoub
Texas Tech University
Lubbock, Texas, USA 79409

Three approaches have been used to determine permissible loads for lifting. These approaches are (1) the biomechanical approach which utilizes the compressive forces on the spine as the criterion, (2) the physiological approach which utilizes the metabolic energy expenditure requirements as the criterion, and (3) the psychophysical approach which utilizes the perceived exertion by the subjects as the criterion for determining the maximum weight to be lifted.

Simplified prediction equations for estimating the oxygen consumption rate and the maximum compressive force on the spine based on a dynamic study for three different lifting heights were developed. These models were a function of frequency, weight of the load, body weight, and their interactions.

Based on these models and the psychophysical maximum acceptable weight of lift data, it was found that biomechanics, physiology, and psychophysics became limiting at different frequencies: biomechanics generally, but not always, is limiting at low frequencies; psychophysics is limiting at intermediate and at times high frequencies; and physiology is usually limiting at high frequencies.

INTRODUCTION

For nearly four decades, manual materials handling (MMH) has been a major topic of interest to professionals from a number of disciplines, including engineering, ergonomics, physical therapy and rehabilitation, orthopedic surgery, work physiology and biomechanics. The primary reason for this interest is the devastating cost and human suffering caused by the severity of materials handling related injuries. Prevention and control of such injuries is a global concern, shared by many researchers and organizations.

Back injury occurs with alarming frequency. In a comprehensive survey of 7,729 men at Western Electric from 1950-1957, only 32 percent had normal backs (Becker, 1961). Troup (1965) stated that in the United Kingdom, about 19 percent of all reported accidents affect the spine and trunk. About 40 percent of back injuries are from lifting and 33 percent are from twisting movements of the spine. More recent studies show that back injuries in industry are still a major source of lost time and compensation claims. Caillet (1981) estimated that 70 million Americans have suffered back injuries, and that this number will increase by 7 million annually. According to Khalil, Asfour, and Moty (1984) low back pain is the second largest pain problem, headaches being the first.

Within the area of ergonomics, three approaches have been used extensively in the past to determine the safe or permissible load for occasional and repetitive lifting. These approaches are:

(1) The biomechanical approach utilizes the compressive forces on the spine as a criterion for determining the maximum weight to be lifted, given certain job conditions;

(2) The physiological approach utilizes the energy expenditure requirements as the criterion for determining the maximum weight to be lifted given certain job conditions;

(3) The psychophysical approach utilizes the perceived exertion by the subjects, who decides on the maximum weight which he can lift given certain job conditions.

Job conditions may include such variables as a) frequency of lift, b) container or load size, c) range of lift, which is defined by the starting and ending points of lift, d) containers with and without handles, e) presence or absence of trunk twisting.

There are also some differences between (1) biomechanical and psychophysical stresses and (2) physiological and psychophysical stresses. For infrequent lifting, the maximum acceptable weights of the load based on biomechanical stress criteria are generally higher than those based on the psychophysical fatigue criteria. For repetitive lifting the maximum acceptable weights based on psychophysical fatigue criteria are lower at low lifting frequencies and higher at high frequencies than those based on physiological fatigue criteria.

METHODS

Using laboratory experiments, prediction equations for estimating the oxygen consumption rate and the maximum compressive force on the spine based on a dynamic study for the three different lifting heights were developed. The oxygen consumption rate and the maximum compressive force on L5/S1 could be expressed as a function of task variables, frequency and the weight of the load, a personal variable - body weight, and their interactions. The frequency effect and its interactions with body weight and the weight of the load were not significant for the maximum compressive force on L5/S1 disc. These prediction equations for estimating the oxygen consumption rate and the maximum compressive force on the spine are shown in Table 1.

RESULTS AND DISCUSSION

Using these developed equations and the current NIOSH biomechanical and physiological MPL (Maximum Permissible Limit) limits (650 Kg and 5 Kcal/min), the biomechanical criterion is the limiting criterion up to a frequency of 5.5 lifts/min for the range of FK (floor to knuckle height) lifting. The physiological criterion becomes limiting after this frequency. This transition from biomechanically limiting criterion to psychophysically limiting criterion occurs at a frequency of 3.5 lifts/min for the range of FS (floor to shoulder height) lifting. However, for the range of KS (knuckle to shoulder height) lifting, the biomechanical criterion is limiting up to the frequency of 2.5 lifts/min; the psychophysical criterion becomes limiting above this frequency.

With the current NIOSH biomechanical and physiological AL (Action Limit) limits (350 Kg and 3.5 Kcal/min), the biomechanical approach does not appear to be realistic for the range of FK and FS lifting. However, for the range of KS, the biomechanical AL criterion (350 Kg) is limiting up to 9 lifts/min. Either the psychophysical criterion or the physiological criterion becomes the limiting criterion after this frequency. These transition frequencies are shown in Figures 1, 2, and 3 for the range of lifting height FK, FS, and KS, respectively.

RANGE	VO_2 and COMP Models
F-K	$VO_2 = 3.8941*B + 0.5211*F*B + 3.0072*L + 3.0577*F*L$ $COMP = 4.6153*B + 12.2923*L$
F-S	$VO_2 = 5.7152*B + 0.3087*F*B + 0.1976*L + 5.2392*F*L$ $COMP = 4.8600*B + 10.6654*L$
K-S	$VO_2 = 3.7471*B + 0.1896*F*B + 3.7880*L + 3.1313*F*L$ $COMP = 2.8119*B + 14.6556*L$
	VO_2 = Oxygen Consumption Rate (ml/min) COMP = Maximum Compressive Force (Kg) B = Body Weight (Kg) F = Frequency of Lift (lifts/min) L = Weight of Load (Kg)

Table 1.   Prediction equations for estimating the oxygen consumption rate and the maximum compressive force on L5/S1.

The average dynamic maximum compressive force ranged from 4 to 40% higher than the static maximum compressive force for the same task for this study. Based on this finding, for illustrative purposes, if the biomechanical AL and MPL criteria were raised to 500 and 800 Kg, respectively, then the raised biomechanical MPL criterion would be limiting up to 3 lifts/min for the range of FK. The psychophysical approach would be limiting between 3 and 5.5 lifts/min; the physiological approach would be limiting above 5.5 lifts/min for the range of FK. The raised biomechanical AL criterion would be limiting up to 5 lifts/min, and the physiological approach would be limiting after 5 lifts/min. However, the maximum frequency suggested by this physiological approach is 8.5 lifts/min (Figure 1).

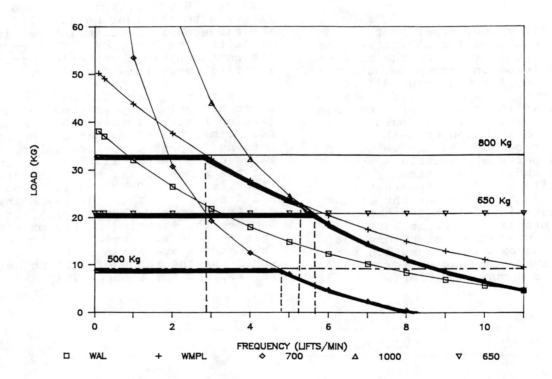

WAL = Weight of load (Kg) from the psychophysical model for AL criterion
            (99% male population).
WMPL = Weight of load (Kg) from the psychophysical model for MPL criterion
            (25% male population).
700 = Weight of load (Kg) from the physiological model for AL criterion
            (700 ml-O_2/min).
1000 = Weight of load (Kg) from the physiological model for MPL criterion
            (1000 ml-O_2/min).
650 = Weight of load (Kg) from the biomechanical model for MPL criterion
            (650 Kg Comp Force).
500 = Weight of load (Kg) from the biomechanical model for a raised AL
            criterion (500 Kg Comp Force).
800 = Weight of load (Kg) from the biomechanical model for a raised MPL
            criterion (800 Kg Comp Force).

Figure 1. Comprehensive model for maximum weight of load for FK lifting.

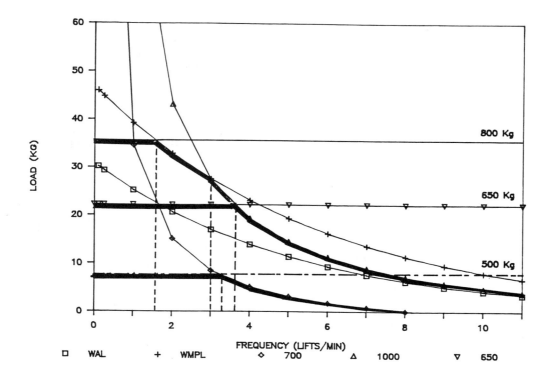

Figure 2.   Comprehensive model for maximum weight of load for FS lifting.

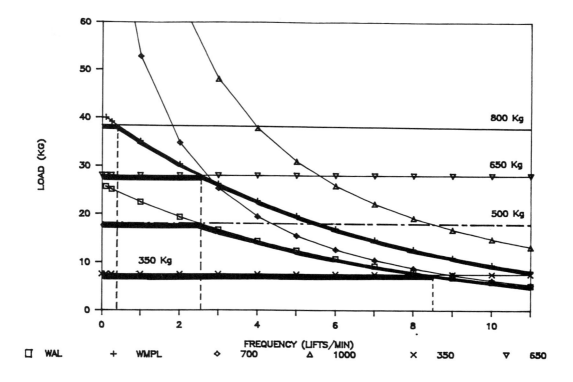

Figure 3.   Comprehensive model for maximum weight of load for KS lifting.

The raised biomechanical MPL criterion would be limiting up to 1.5 lifts/min, and the psychophysical approach would be limiting between 1.5 and 3 lifts/min for the range of FS. The physiological approach would be limiting above the 3 lifts/min for the range of FS. The raised biomechanical AL criterion would be limiting up to 3 lifts/min, and the physiological approach would be limiting after 3 lifts/min. However, the maximum frequency suggested by the physiological approach would be 8 lifts/min. At this frequency, the body movement alone would require a metabolic energy rate of 3.5 Kcal/min for the range of FS (Figure 2).

The raised biomechanical MPL criterion would be limiting up to 0.5 lifts/min after which the psychophysical approach would be limiting for the range of KS. The raised biomechanical AL criterion would be limiting up to 2.5 lifts/min, after which the psychophysical approach would be limiting. Beyond 9 lifts/min, the psychophysical and physiological approaches were in agreement (Figure 3).

SUMMARY

The physiological approach is limiting at the frequencies higher than 5 lifts/min for the range of FK, whether the biomechanical criteria are based on the elevated AL and MPL or the original NIOSH MPL criteria. The physiological approach is limiting at the frequencies higher than 3 lifts/min for the range of FS, whether the biomechanical criteria are based on the elevated AL and MPL or the original NIOSH MPL criteria (Kim, 1990).

Based on the results of this study, it appears that the NIOSH AL and MPL criteria should be re-examined.

REFERENCES

Becker, W. F. (1961). Prevention of Low Back Disability, *Journal of Occupational Medicine*, (3), 329-335.

Caillet, R. (1981). *Low Back Pain Syndrome*, F. A. Francis Co., Philadelphia.

Khalil, T. M., Asfour, S. S. and Moty, E. A. (1984). Case studies in low back pain, *Proceedings of the 28th Annual Meeting of the Human Factors Society*, 465-470.

Kim, H. K. (1990). Development of a Model for Combined Ergonomic Approaches in Manual Materials Handling Tasks. Dissertation, Department of Industrial Engineering, Texas Tech University, Lubbock, Texas, USA.

NIOSH Technical Report. (1981). *Work Practices Guide for Manual Lifting*, U. S. Department of Health & Human Services, Cincinnati, Ohio, USA.

Troup, J. D. G. (1965). Relation of lumbar spine disorders to heavy manual work and lifting, *The Lancet*, 857-861.

THE EFFECT OF CONFINED SPACE ON THE MAXIMUM ACCEPTABLE WEIGHT OF LIFT

Waldemar Karwowski and Hashim Alsabi
Center for Industrial Ergonomics
University of Louisville
Louisville, Kentucky 40292, USA

The main objectives of this study were to analyze the effect of space confinement on the maximum acceptable weight of lift (MAWL), and determine the minimum acceptable space for lifting. Two different laboratory experiments were conducted. The first experiment was aimed to determine the effect of floor space confinement with adequate headroom, if any, on the maximum acceptable weight of lift, while the second experiment was designed to determine the minimum space that would be considered by the subjects as acceptable for lifting tasks. The results showed that although within the space restriction imposed in the first experiment the effect of space confinement on the selected maximum acceptable weight of lift was not statistically significant, subjects lifted on the average about 9% more weight as the width and length of the working space were increased by 10 and 20 cm, respectively. Subjects also reported significantly higher levels of spacial comfort under increased floor space conditions. On the average, subjects selected a floor area of 66 x 90 cm as the minimum acceptable space they would prefer for lifting a compact load between the floor and two tables on the sides.

INTRODUCTION

When working space is too small for the given task, a decrement in performance and an increased operator stress can be expected. Therefore, appropriate workspace clearances and access dimensions are very useful, and should be used for ergonomic design purposes (Nobe, 1982). The spacial aspects of lifting tasks have been identified as one of the environmentally-related risk factors in manual handling of loads (NIOSH, 1981; Ridd, 1985). While restricted space affects human performance on tasks that require body movements, many of the industrial MMH tasks are performed in confined spaces (Drury, 1985).

Although an adequate amount of space is needed to allow for free body movements and to facilitate muscular recovery from physical stress (Drury, 1985), only limited scientific data is available on the effect of space confinement on manual handling capacity. For example, Mital (1986) studied the effect of restricted space on manual carrying capabilities, and found that both males and females carried about 13% less load through a 56 cm inside passage as compared to an open space environment. Since loading/unloading in service industry and many maintenance activities are often performed in restricted spaces, more research is warranted to understand the effect of space confinement on human lifting abilities.

OBJECTIVES

Several environmental variables are important factors that may modify the human capability in manual load handling. Currently, with the exception for load carrying, no data is available on the effect of space restriction on manual lifting capacity. According to Drury (1985), there are at least three different levels of space restriction that need to be considered from the ergonomic design point of view. These levels are as follows:

Range 1. Task spaces so small that the operator will not fit into the space and/or the task cannot be accomplished successfully.

Range 2. Task spaces in which the task can be accomplished but where the addition of more space will increase task performance (speed, accuracy) or decrease operator stress.

Range 3. Task spaces so large that further increases in space produce no increase in performance or decrease in stress.

The main objectives of this study were: 1) to analyze the effect of space confinement on the maximum acceptable weight of lift (MAWL), and 2) to determine the minimum acceptable floor space for manual lifting tasks.

METHODS AND PROCEDURES

Eight male college students were selected for the study. Age and physical characteristics of the subjects are shown in Table 1. The isometric strength measurements (arm, shoulder and back) were made according to the procedures described by Chaffin (1975), while the isokinetic strength tests were done following the procedures outlined by Pytel and Kamon (1981). Before beginning the experiment, each subject was required to fill out a personal consent form and personal data form. Two different laboratory experiments were conducted. The first experiment was aimed to determine the effect of space confinement, if any, on the maximum acceptable weight of lift, while the second experiment was designed to determine the minimum space that would be considered by the subjects as acceptable for lifting tasks.

Table 1. Age, anthropometry [cm] and strength [lbs] characteristics of the subjects [N = 10].

Variable	Mean	S.D.	Range
Age [years]	22.75	3.80	17-28
Body weight [lbs]	164.08	15.79	149-193.5
Stature [cm]	174.80	5.11	170.5-183.7
Acromial height	144.56	4.68	139.5-153.9
Arm length	73.26	4.63	67.8-81.9
Shoulder breadth	44.63	1.59	42.8-47.5
Static arm strength	76.75	11.11	58-91.5
Static back strength	201.33	39.81	124-267
Static composite strength	263	48.11	201.5-329
Static leg strength	292.25	73.99	154-375
Dynamic arm strength	51.38	7.33	41-62
Dynamic shoulder strength	39.25	3.34	34.5-44.5
Dynamic back strength	92.63	18.29	69-124.5
Dynamic lift strength	111.06	31.19	58-148

Table 2. Definition of space confinement levels for experiment 1.

Level of confinement	Width[1]	Length
Level 1	Shoulder breadth +10 cm	83 cm
Level 2	Shoulder breadth +15 cm	93 cm
Level 3	Shoulder breadth +20 cm	103 cm

[1] Width was defined as the distance between two tables.

The subjects were instructed to adjust the weight lifted to the maximum acceptable level according to Snook's (1978) psychophysical methodology. The subject was asked to perform four lifts per minute for thirty minutes. He was to lift the box from the floor and place it on one rack, and then to lift it again and place it on the second rack, then back to the floor.

The box used was 35 x 35 x 34 centimeters and had a false bottom to minimize visual cues. Two racks with overall dimensions of 63.5 x 82.5 x 111.0 centimeters were used. The racks each had a shelf placed 79.0 cm from the floor. Irregular shaped pieces of lead were provided for use as weights.

Upon completion of the experiment, the subject was asked to rate his perceived exertion level in the arms, low-back and overall. The subject was also asked to rate his spacial comfort on a scale from 1 to 7, with 1 being "extremely cramped", and 7 being "extremely comfortable".

In order to determine the differences in lifting performance, a lifting test in open space was also performed. In this test the subject was positioned between the two racks. The distance between the two racks was set to be double the subjects' shoulder breadth. The subject was asked to lift the weight box four lifts per minute for thirty minutes. The subject was also asked to do the ratings as described previously.

EXPERIMENT 1

The effect of space confinement on the maximum acceptable weight of lift

For this experiment, a special purpose room was constructed inside the laboratory to simulate total space confinement. Two racks were placed inside the room, and three different wall positions were constructed to achieve different degrees of confinement. The distance from the wall to the first confined position was set at 82.5 centimeters, with an increment of 10 centimeters each for the next two positions. The distance between the racks was set to be the subjects' shoulder breadth plus 10 centimeters for the first degree of confinement, and an increment of 5 centimeters each for the next two levels of confinement (see Table 2).

Pheasant (1986) proposed that shoulder breadth be used as an indicator of the clearance at shoulder level. Shoulder breadth (or bideltoid) was defined as the maximum horizontal breadth across the shoulders measured to the protrusions of the deltoid muscles. The 95th percentile value of shoulder breadth for the US population is 51.50 cm.

EXPERIMENT 2

Determination of the minimum acceptable space for manual lifting tasks

For the purpose of the second experiment, a movable wall was constructed, and two racks were used. The racks were placed in the enclosed area against room dividers. The experiment had two phases. In phase one, the subject was set up in an area that was called "minimum starting condition". The distance between the racks was set to be the subjects' shoulder breadth plus an increment of ten centimeters. The distance to the wall constructed above was set to be 82.5 centimeters.

Table 3. Definition of initial floor space dimensions for the starting condition in experiment 2.

Starting condition	Width[1]	Length
Small (minimum)	Shoulder breadth+10 cm	82.5 cm
Large (maximum)	[Shoulder breadth] x 2 +20 cm	same as width

[1] Width was defined as the distance between two tables.

The weight of the box was set to MAWL selected by the subject in an open space. The subject was asked to adjust the space to the minimum level that he was comfortable with. This was achieved by instructing the subject to request the change in the position of either the facing wall (length) or the rack (width) or both by using the linguistic variables "little", "some", and "a lot".

In the second phase, the distance between the racks and the distance to the facing wall was set to double the subjects' shoulder breadth plus an increment of twenty centimeters. This was called the "maximum starting condition". The subject was asked to make inward movements of the facing wall and the movable rack. The weight box used in this phase was the weight selected in the open space using the psychophysical methodology first proposed by Snook (1978). The final dimensions of the work area were calculated after the completion of the experiment.

RESULTS AND DISCUSSION

Results of experiment #1

The response variables included the maximum acceptable weight selected (MAWL), perceived exertion, and perceived spacial comfort. The results of analysis of variance for randomized complete block design are illustrated in Tables 4-7.

The results showed that although within the space restriction imposed in the first experiment the effect of space confinement on the selected maximum acceptable weight of lift was not statistically significant, subjects lifted on the average about 9% more weight as the width and length of the working space was increased by 10 and 20 cm, respectively. Subjects also reported significantly higher levels of spacial comfort under the increased floor space conditions.

Table 4. Anova table for the maximum acceptable weight of lift in confined space.

Source	d.f.	Type I SS	F	P
Subject	7	1256.36	14.46	0.0001
Work space	2	38.43	1.55	NS
Error	14	173.73		
Corrected total	23	1468.52		

NS - not significant at $p \leq 0.05$.

Table 5. Main effects for the maximum acceptable weight of lift [lbs] in confined space.

Space confinement	Mean*	S.D.	Range	N
Level 1**	a 32.76	5.66	26.50-44.80	8
Level 2	a 34.93	9.62	26.10-53.20	8
Level 3	a 35.76	8.93	26.10-49.00	8

N - number of observations
* - different letters indicate differences between means at $p \leq 0.05$
** - amount of space available increases from level 1 to level 3

Table 6. Anova table for spacial comfort in confined space.

Source	D.F.	Type I SS	F	p
Subject	7	8.29	4.06	0.02
Confined space	2	10.58	18.14	0.0001
Model	9	18.87		
Error	14	4.08		
Corrected total	23	22.95		

R-square = 0.82

Table 7. Main effects for spacial comfort in confined space.

Space confinement	Mean*	S.D.	Range	N
Level 1	a 3.50	.92	2.00-5.00	8
Level 2	b 4.25	.95	3.00-5.00	8
Level 3	c 5.12	.64	4.00-6.00	8

* Different letters indicate significant differences between means at $p \leq 0.05$.

Results of experiment #2

The starting condition had a significant effect on the selected minimum acceptable space for lifting tasks (see Tables 8-10). Subjects tended to select the space with significantly greater ($p<0.05$) width and length when starting with the minimum space allowance, and extending the available space than when starting with the maximum space condition. It should be noted, however, that under both starting conditions, subjects were able to either restrict or enlarge the preferred space. For comparison, the recommended clearances for aisle and corridor passageways are shown in Table 9.

On the average, subjects selected a floor area of about 0.6 and 0.8 m², for the minimum and maximum starting conditions, respectively, as the minimum acceptable space they would prefer for manual lifting from floor to a table height. The selected space values can serve as a basis to develop preliminary guidelines for minimum space allowance for comfortable lifting of compact loads in industry.

Finally, the results of this study indicate that space confinement has an effect on the perceived comfort in manual lifting. The subjects in this study were willing to work harder, as estimated by the rates of perceived exertion in unrestricted space environment, compared to the selected minimum acceptable space for lifting tasks (see Table 10).

Table 8. Selected values of the minimum acceptable floor space [cm] for lifting activity.

Variable workspace	Mean*	S.D.	Range	N
Selected width [cm]				
Starting condition				
Small	a 66.45	7.33	56.55-79.13	8
Large	b 81.63	6.07	72.58-90.72	8
Selected length [cm]				
Starting condition				
Small	a 89.59	2.25	85.68-93.24	8
Large	a 97.03	15.43	84.17-131.04	8
Selected floor area [m²]				
Starting condition				
Small	a 0.60	0.07	0.48-0.70	8
Large	b 0.80	0.18	0.64-1.19	8

* Different letters indicate significant differences between means at $p \leq 0.05$.

Table 9. Recommended widths for aisle and corridor passageways (Mital, 1991).

Type of activity	Width [cm]
One person walking	65.0
One person passing, one standing against the wall	100.0
One person carrying a tray or similar load	90.0
One person with a walking frame	100.0

Table 10. Main effects for perceived overall exertion
under minimum acceptable and open spaces.

Minimum acceptable space	Mean*	S.D.	Range	N
Minimum	a 12.87	1.35	12.0-15.0	8
Maximum	b a 13.25	1.03	12.0-15.0	8
Open space	b 13.87	0.99	12.0-15.0	8

* Different letters indicate significant differences
between means at $p \leq 0.05$.

CONCLUSIONS

Assuming normal distribution of the selected values for the minimum acceptable space for lifting compact boxes with the maximum width and length less then 35 cm, it is recommended that in order to accommodate 95% of U.S. population the space allowance for manual lifting with unrestricted headroom be no less than 1.1 m².

REFERENCES

Chaffin, D. B., 1975. Ergonomics guide for the assessment of human static strength, *American Industrial Hygiene Association Journal,* 36, 505-511.

Drury, C. G., 1985. Influence of restricted space on manual materials handling, *Ergonomics,* 28, 167-175.

Mital, A., 1986. Subjective estimates of load carriage in confined and open spaces. In: *Trends in Ergonomics/Human Factors III,* edited by W. Karwowski, Amsterdam: North Holland, pp. 827-833.

Nobe, J., 1982. *Activity and Spaces, Dimensional Data for Housing Design,* The Architectural Press, London.

NIOSH, 1982. *Back Injuries Associated with Lifting,* Bulletin 2144, U. S. Government Printing Office, Washington, D.C.

Pheasant, S., 1986. *Bodyspace: Anthropometry, Ergonomics and Design,* Taylor and Francis, London.

Pytel, T. L. and Kamon, E., 1981. Dynamic strength test as a predictor for maximal and acceptable lifting, *Ergonomics,* 24, 663-672.

Ridd, J. E., 1985. Spatial restraints and intra-abdominal pressure, *Ergonomics,* 28, 149-166.

Snook, S. H., 1978, The design of manual handling tasks, *Ergonomics,* 21, 963-985.

A THREE-DIMENSIONAL MOTION MODEL AND VALIDATION
OF LOADS ON THE LUMBAR SPINE

W. S. Marras, C. M. Sommerich, and K. P. Granata
Biodynamics Laboratory
Department of Industrial and Systems Engineering
The Ohio State University
1971 Neil Avenue
Columbus, Ohio

The objective of this paper was to describe the structure and logic of
an electromyographic (EMG)-driven model that is capable of producing
estimates of spine loading under three-dimensional trunk motion
conditions. This model is intended for use during simulation of work
under laboratory conditions. The model estimates lumbar spine
compression, shear, and torsion loading of the lumbar spine as well as
the torque production of the trunk continuously throughout the
exertion. The model was tested, for validation purposes, using 11
subjects. Trunk torque estimates derived from the model were compared
with measured trunk torque. Over 85 percent of the runs produced
R-square values greater than 0.7. It was concluded that this approach
provides a reasonable means of assessing the loading of the spine
during laboratory simulations of workplace conditions.

INTRODUCTION

Models that represent the loading
of the spine under occupational
circumstances are essential for a
quantitative assessment of the work
environment. Typically, model estimates
of spine loading, during the performance
of a task, are compared with tolerance
limits of the vertebral end plates. In
this manner, one could determine how
workplace factors and methods influence
both the acute and cumulative loading of
the spine throughout the work day.

One of the significant challenges
in such efforts involves the predictions
of loads upon the lumbar spine under
realistic occupational conditions. Most
knowledge relating the reaction of the
trunk internal spine loading structures
(muscles, intra-abdominal pressure,
ligaments, and other connective tissue)
has been based upon static, two-
dimensional exertions of the trunk.
This situation has resulted in
predictive models which often use
optimization strategies of muscle
recruitment. These techniques work well
for static exertions. However,
examination of most typical industrial
workplaces indicate that these jobs
often require dynamic, three-dimensional
trunk motion and these dynamic motions
involve more co-activation of the
muscles that are not predicted well by
the optimization techniques.

The objective of this research was
to develop and validate a descriptive
simulation model of trunk loading under
dynamic trunk motion conditions. This
model is intended to be a tool that
permits one to evaluate the collective
influence of the trunk musculature upon
spine loading as the trunk moves under
controlled laboratory motion conditions.
Laboratory conditions are necessary
since information about trunk muscle
activity must be recorded via EMG which
will change as a function of muscle
length changes and muscle velocity.
Therefore, in this study, trunk muscle
EMG was monitored along with trunk
posture changes under constant trunk
velocity conditions. Constant velocity
conditions were necessary since EMG is
only related to muscle force under this
dynamic condition (Bigland and Lippold,
1954).

MODEL DEVELOPMENT

Approach

This model is an EMG-driven
simulation model that determines muscle
forces, and the subsequent spine
loading, during motion of the trunk.
The advantage of an EMG-driven model is
that it is capable of accounting for
changes in muscle activities, such as
co-activation, which are essential for
model accuracy, yet difficult to
predict. EMG recordings of the trunk
muscles are used as a base from which
muscle force is predicted. These EMG
recordings are adjusted to account for
factors that would interfere with the
EMG-force relationship such as muscle

size, motion, and length. The muscle force information is then used as continuous input to a model structure that considers the mechanical advantage of every muscle relative to the spine, and then predicts lumbar spine compression, shear and torsional forces as well as trunk torque production continuously throughout the exertion. This information may be compared with spine tolerance limits so that the risk of imposing a vertebral end-plate microfracture due to workplace requirements could be determined.

Model Overview

In order to estimate the amount of force generated by a given muscle at any point in time during an exertion, it is essential to consider the physiological aspects of the muscle's response to externally imposed torques applied over time. Figure 1 displays the basic features which compose the logic within the model for incorporating muscle physiology, as well as trunk motion components.

The key model inputs are categorized as subject characteristics, EMG signal characteristics, and trunk kinematic and kinetic elements. First, subject characteristics refer to subject anthropometry, which is utilized in estimating muscle cross-sectional areas and trunk weight. Second, two aspects of the integrated EMG signal are employed. These are the amplitude and, because of the dynamic nature of the exertion, the temporal information in the signal. In this way, the effects of peak muscle exertions and muscle coactivity can be captured, and their contributions to spinal loading can be determined. Another necessary ingredient for greater understanding of dynamic exertions is information pertaining to trunk kinematics and kinetics. This includes trunk flexion angle, angular velocity, and the output torque which the trunk is imparting to offset any external load. These elements affect one or more aspects of muscle physiology, and including their effects leads to an improved understanding of muscle response and internal spinal loading throughout the dynamic exertion.

For an individual subject and specific trial conditions, the model calculates spinal loading due to the activity of ten trunk muscles and the weight of the body above the L5 level. These values are computed throughout the extension and are based on the EMG activity of the ten trunk muscles

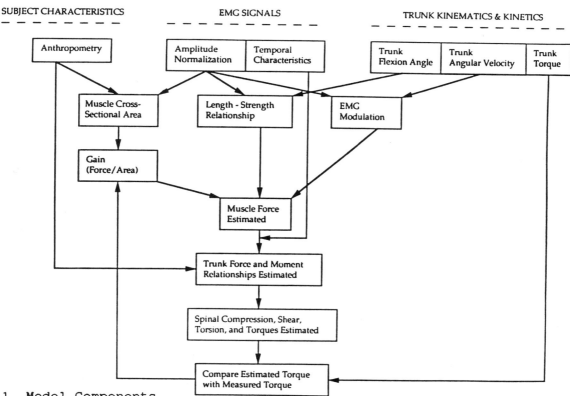

Figure 1. Model Components.

during the exertion, force and moment equations, and several assumptions regarding muscle characteristics and EMG activity.

Model Inputs

The model uses, as input, data from isokinetic trunk extension trials performed through the employment of an isokinetic dynamometer, certain anthropometric data, and maximum EMGs for use in normalizing the EMGs for the particular trial of interest. The particular task mimics the motion of the back during a manual materials handling operation which would involve only the back, that is to say that lifting is not aided through use of the legs.

Data required by the model includes trunk torque, trunk flexion angle, and trunk angular velocity data collected throughout the exertion. The model also requires EMG data from five (left/right) pairs of trunk muscles, the latissimus dorsi, erector spinae, rectus abdominus, and internal and external obliques. All of this information is collected under controlled conditions. It is important to strictly control the point of rotation about which the trunk rotates, the constant angular speed of the trunk, and any trunk twisting (trunk asymmetry) which is intentionally introduced.

Maximum EMG levels pertaining to static maximum exertion trials performed at three angles of trunk flexion are used in the EMG normalization process. The maximum value is equated with a maximum level of force generation capacity. Actual trial EMG values are divided by these maximums, in order to determine the activity level at which the muscle is operating. As an example, if the maximum capacity is considered to be 300 N and the normalized EMG was 0.35, then the force from the muscle, as indicated by the EMG would be 105 N (= 0.35 * 300 N).

Subject anthropometry, which includes height and weight measures, as well as torso depth and breadth at L5, is used in the model to determine the moment due to the weight of the upper body and in calculating muscle cross-sectional areas.

The other important input to the model is a file (for each exertion trial) containing the characteristic event times for each muscle. For a given trial, each muscle's EMG signal is reduced, in a sense, to four event

points which can be said to characterize the signal profile. Points 1 and 4 are the points in time which signal the start and end of the activity of that muscle. Points 2 and 3 are any points in between the other two which help to define the signal profile. Linear connections between these points define the shape of the EMG response.

Processing of inputs. Once the information listed above is made available to the model, the model then begins to operate on the data. The event times are ordered from first to last across all the muscles, and the time scale is converted from absolute time to relative time, where time 0.0 is the occurrence of the first event (the first muscle activating) and time 1.0 marks the final event (the last muscle returning to resting level).

Implicit in the notion of reducing the EMG signals to four event points is the idea that the values of the signal in between those points can be calculated using a straight line approximation. Based upon this assumption and given the velocity, angle, and torque at each point in time which is marked by an event (related to any muscle), the EMG values for each muscle at each point in time can be estimated. A total of 40 event times per trial are possible. This comes from four event times for each of the ten muscles. There may be fewer than 40 distinct event times if some events occur simultaneously, or if one or more of the muscles is not recruited during the exertion.

Once an EMG value is established for each muscle at every event time, those values are then modulated by velocity and length-strength effects.

The velocity modulation ratio is constructed from two values. The numerator is the average normalized EMG response from the laboratory database (see Marras and Mirka, 1990) for a particular muscle with respect to trunk angle (muscle length), external torque production, and an angular trunk extension velocity of zero. The denominator is the average EMG response, from the same database, for the same angle and torque conditions, but performed at the speed at which the trial of interest was conducted. This ratio is multiplied by the normalized EMG value for the subject in order to eliminate the velocity effect.

Each muscle has an optimal length at which its force generation capacity peaks. Deviation either way reduces that capacity. This knowledge is incorporated into the model by assuming that each modeled muscle is able to produce a certain maximum amount of tension based upon its cross-sectional area (A) and a certain inherent force per unit of area, 40 N/sq. cm. for example. If the muscle is positioned at its optimal length (Lo), the muscle is assumed to be capable of producing an optimal force (Fo=Lo*A). If, however, the muscle is not at its optimal length, it can only produce some fraction of Fo. The values used to calculate the reduced force are determined by equations derived from several graphs (Best and Taylor, 1973; Chaffin and Andersson, 1984), as well as some trial and error during model development. Muscle length is assumed to be tied to trunk angle. Perry and Bekey (1981) stated that the proportionality constant relating EMG and force level is dependent upon joint position. Asymmetry (twisting or lateral bending) will also affect muscle length, but this is not considered in the model.

Determining force from the EMG signal

While each operation is actually performed separately, the following equation sums up all of the modifications made to the EMG signal in order to derive a tension level for each muscle at each event point in time.

$$\text{Force} = \text{Gain} * \frac{\text{EMG}}{\text{EMGmax}} * \text{V Ratio} *$$

$$\text{L-S factor} * \text{Area}$$

where:

Force	= Muscle tension associated with EMG
Gain	= Factor which includes maximum muscle force per unit of area
EMG	= Recorded or interpolated value at a particular event time
EMGmax	= Maximum EMG value for particular muscle at particular angle of orientation
V Ratio	= Velocity modulation factor
L-S factor	= Length-strength modulation factor
Area	= Muscle cross-sectional area

The muscle area is calculated from the subjects' torso depth and breadth dimensions and coefficients from Schultz et al. (1982).

Once all of the other effects have been factored in, an iterative process begins during which the force level is calculated in each muscle. These forces are combined in order to calculate the resulting spinal loading at each point in time (marked by an event for any muscle). That is to say that whenever an event occurs for one muscle, the activity levels of all muscles are computed at that time and the collective influence upon spinal loading is also calculated.

Spine Force Calculations

The model structure which governs the estimation of internal force was developed from the six force and moment equations from Schultz and Andersson's (1981) static, internal trunk loading model. However, in the current model, these are not calculated once, but repeatedly as the movement progresses from the start to the finish of the isokinetic trunk extension. Reilly and Marras (1989) described an earlier version of the current model, in which they calculated relative trunk loading values for compression and shear, both lateral and anterior, based on a three-point characterization of each of the monitored muscles' EMG records.

This process begins with the assumption of an initial value of ten for the gain factor. After setting this initial value, force levels are calculated at each point in time for each muscle. Those values are combined in the equilibrium spinal force and moment equations adapted from Schultz and Andersson (1981). Once this has been done for each event point in time, a check is performed on the value of the calculated lateral torque. If the average value (area under the curve) is within five percent of the average torque value measured during testing, then the gain is judged to be sufficient and the muscle force and spinal loading values are deemed correct. If the calculated torque is too low, then the gain is increased by a preset increment, the forces are recalculated, and the torques are compared once again. This process continues until the torques are within the acceptable range.

EXPERIMENTAL VALIDATION

A highly controlled experiment was designed to test the model. The experiment required the subjects to exert a constant torque with the back throughout a 45 deg range of motion while moving at constant velocities. Subjects extended the trunk in a simulated lifting motion under both sagittally symmetric and asymmetric trunk positions. It was assumed that the point of bend about the spine was located at L5/S1 and that this motion would relate to torques experienced about the spine during lifting.

Eleven subjects experienced spinal loading during an experiment in which conditions of trunk velocity, trunk torque output, and trunk asymmetric posture were varied in a series of isokinetic velocity trunk extensions. Trunk velocities were tested at 10, 20, and 30 deg/sec, whereas trunk asymmetry was tested at 0, 15, and 30 deg of twist in the transverse plane. Trunk torque was set at 27.1 or 54.2 Nm. The exact conditions of the experiment are reported in Marras and Sommerich (1991). These conditions were controlled with a trunk dynamometer (asymmetric reference frame or ARF). The EMG activity of 10 trunk muscles, subject anthropometry, and trunk kinetics were used as input to the model. The model estimated compression, shear, and torsion loading of the lumbar spine as well as the torque production of the trunk continuously throughout each exertion.

There is no means available to directly measure compression, shear or torsion loading on the spine in-vivo. Therefore, in order to validate the model, trunk torque production was used as a standard of comparison. Continuous trunk torque estimates derived from this model were compared with continuous trunk torque measured by the ARF. R-square values of between 0.65 and 0.98 were observed in this study. Well over 85 percent of these R-square values were over 0.7.

DISCUSSION

The objectives of this study were to investigate how well this motion model simulated the action of the trunk, and to investigate loading changes of the spine as the trunk performed lifting tasks under dynamic, asymmetric conditions. Based upon the measurable performance parameters, the model appears to be robust and does reflect the action of the trunk under controlled velocity in symmetric and asymmetric conditions. The model indicated that spine compression increased directly

with trunk velocity. The model also indicated that as the trunk became more asymmetric spine compression decreased, but shear forces increased. It was also concluded that better instrumentation of the experimental task and more accurate model input information, such as MRI or CT scan information, would enhance model performance even further.

It was concluded that this approach provides a straightforward means of assessing the loading of the spine due to laboratory simulations of workplace conditions.

REFERENCES

Bigland, B. and Lippold, O.C.J. (1954). The relation between force, velocity and integrated electrical activity in human muscles. *Journal of Physiology, 123*, 214-224.

Brobeck, J. R. (ed) (1973) Best and *Taylor's Physiological Basis of Medical Practice*, 9th edition, Williams and Wilkins, Baltimore MD.

Chaffin, D.B. and Andersson, G.B.J. (1984). *Occupational Biomechanics*, John Wiley and Sons, New York.

Marras, W.S. and Mirka, G.A. (1990). A comprehensive evaluation of trunk response to asymmetric trunk motion. *Spine* (in press).

Marras, W.S. and Sommerich, C.M. (1991) A three-dimensional motion model of loads on the lumbar spine. Part II: Model Validation. *Human Factors*, in press.

Perry, J. and Bekey G.A. (1981). EMG-force relationships in skeletal muscle. *CRC Critical Reviews in Biomedical Engineering*, December, 1-22.

Reilly, C.H. and Marras, W.S. (1988). SIMULIFT: A simulation model of human-trunk motion. *Spine*, 14(1), 5-11.

Schultz, A.B. and Andersson, G.B. (1981). Analysis of loads on the lumbar spine. *Spine*, 6, 76-82.

Schultz, A.B., Andersson, G.B.J., Haderspeck, K., Ortengren, R., Nordin, M. and Bjork, R. (1982). Analysis and measurement of lumbar trunk loads in tasks involving bends and twists. *Journal of Biomechanics*, 15(9), 669-675.

LOW BACK DISORDERS IN INDUSTRY

Stover H. Snook
Liberty Mutual Insurance Company
Hopkinton, Massachusetts

ABSTRACT

Low back disorders have been around for over 5000 years, but our understanding of them remains rather primitive. Approximately 80 percent of the population experiences low back disorders at sometime during life, and the annual workers' compensation costs exceed $15 billion. Fortunately, most people recover from low back disorders in a short period of time. Low back disorders cannot be totally prevented at the present time, but they can be controlled by reducing the probability of the initial episode, reducing the length of disability, and reducing the probability of recurrence. Control is accomplished by a variety of approaches that include job redesign (ergonomics), job placement (selection), and education/training.

INTRODUCTION

Low back disorders have been around for a very long time. Bernardino Ramazzini, the founder of occupational medicine, was concerned about low back disorders in the late 17th Century (Mastromatteo, 1975). Low back disorders also afflicted the ancient Egyptians, whose physicians diagnosed sciatica about 5000 years ago (Finneson, 1975). Authorities now believe that low back disorders existed before human ancestors stood up on their hind legs, and that many four-legged animals currently suffer from this problem (Nachemson, 1975).

Although medical practitioners have been treating low back disorders for many years, our understanding of them remains rather primitive. For example, the specific cause of low back pain is still unknown. There are many opinions regarding causation but few objective data. Many authorities now believe, as Hult (1954) suggested over 35 years ago, that low back pain is basically caused by changes in the spine, usually as one gets older. It is thought that these changes lower the resistance of the spine to heavy work loads. Consequently, heavy work loads merely trigger the occurrence or onset of low back symptoms and cannot be regarded as the cause.

FREQUENCY OF LOW BACK DISORDERS

The frequency of low back disorders depends upon how disorders are defined. A distinction should be made between low back pain, low back impairment, low back disability and low back compensation.

Low back pain is generally defined as lumbosacral pain as well as buttock pain and leg pain; acute pain as well as chronic pain; and lumbago as well as lumbar insufficiency. It is widely recognized that low back pain of this nature is experienced by 80 percent of the population at some time during life (Hult, 1954; Horal, 1969; Nachemson, 1976; Bergquist-Ullman and Larsson, 1977). In the United States, it is estimated that 31 million individuals (14 percent of the population) have low back pain symptoms at any one given time (Holbrook, Grazier, Kelsey and Stauffer, 1984).

Depending upon the severity of low back pain, it may result in low back impairment. Low back impairment is defined as a reduced ability to perform various musculoskeletal activities. According to the U.S. Department of Health and Human Services (1985), there are about 10 million cases of low back impairment in the employed U.S. population between 18 and 64 years of age (representing 11 percent of that population).

Depending upon the nature of the job, low back impairment may or may not result in low back disability. Low back disability is defined as time lost from the job, or assignment to restricted duty. A person with low back impairment may not be able to perform a heavy physical job, but may be able to perform a lighter job. According to unpublished Liberty Mutual Insurance Company data (1976), low back disability affects about two workers per 100 per year. Throughout Western society, low back disability has increased dramatically in the past 30 years without an increase in the incidence of low back pain (Waddell, 1987).

Finally, low back compensation is reimbursement for medical costs and lost wages. Using data from the Bureau of Labor Statistics, Klein, Jensen and Sanderson (1984) found 0.75 claims per 100 workers per year in the U.S. working population (ranging from 0.2 in the finance industry to 1.6 in the construction industry). In 1988, 16.4 percent of all Liberty Mutual workers' compensation claims were for low back disorders (Webster and Snook, 1991).

COST OF LOW BACK DISORDERS

In 1988, 35 percent of all Liberty Mutual workers' compensation claims costs were for low back disorders. The average cost per claim was $7431; median cost, $359 (Webster and Snook, 1991). Approximately 25 percent of the cases account for 95 percent of the costs (Webster and Snook, 1990). About one-third of the costs is for medical treatment, and two-thirds for lost wages. There is great variation in costs among the different States, probably reflecting the differences in State compensation laws. It is estimated that the total compensable cost for low back disorders in the United States during 1988 was $15.3 billion.

RECOVERY FROM LOW BACK DISORDERS

Most people with low back disorders do not seek professional treatment and learn to cope with the problem by themselves. Scientific studies have shown remarkably little difference between every conceivable treatment for low back disorders. The few reported results that have reached statistical significance are usually so small as to be of no practical clinical significance (Waddell, 1987). Fortunately, most low back symptoms do not last very long, and most people return to work within a few weeks. For example, in Eastern Massachusetts, 67 percent of compensable low back cases are back to work in three weeks; 75 percent in six weeks; 80 percent in 12 weeks; 85 percent in six months; and 94 percent in one year (Webster and Snook, 1991). If an individual is out of work for over six months, the probability of returning to work is only 50 percent; over one year, only 25 percent; and over two years, practically zero (McGill, 1968). The recurrence rate is very high. After the initial episode of low back pain, an individual is four times more likely to experience another episode than someone who has never experienced low back pain (Dillane, Fry and Kalton, 1966).

CONTROL OF LOW BACK DISORDERS

It is difficult to prevent low back disorders in industry when the cause is unknown and the treatment is ineffective. However, it does appear possible to control low back disorders by reducing the probability of the initial episode (primary control), reducing the length of disability when it does occur (secondary control), and reducing the chance of recurrence (tertiary control). Industry has attempted to control low back disorders with a variety of approaches that include job redesign (ergonomics), job placement (selection), and education/training. The data supporting the use of these three approaches have been reviewed by Snook (1987). It appears quite clear that no single approach by itself will control the problem of low back disorders. An effective control program must be a combination of all three approaches (job redesign, job placement and education), applied at all three levels of control (initial episode, disability and recurrence).

Job redesign represents a more permanent, engineering approach to the problem. It reduces the workers' exposure to the risk factors of low back disorders, and consequently reduces the medical and legal problems of selecting the worker for the job, as well as finding replacements for absent workers. Job redesign also places less reliance upon the worker's willingness to follow established training procedures, such as lifting properly. Data for assessing and redesigning manual handling jobs can be found in the NIOSH Work Practices Guide for Manual Lifting (U.S. Department of Health and Human Services, 1981), and the recently revised Liberty Mutual tables of maximum acceptable weights and forces for lifting, lowering, pushing, pulling and carrying tasks (Snook and Ciriello, 1991). It is estimated that the proper design of manual handling tasks can reduce low back disorders by up to one-third (Snook, Campanelli and Hart, 1978). Job redesign is effective because it operates at all three levels of control. It reduces the probability of initial and recurring episodes of low back disorders, allows the worker with moderate symptoms to stay on the job longer, and permits the disabled worker to return to the job sooner.

Although job redesign is an effective control approach, it is sometimes difficult to apply to outdoor jobs with no defined workstation (e.g., firefighting, police and rescue work; and certain construction and delivery operations). Equipment (tools, trucks, stretchers, etc.) can be well designed, but there is little control over terrain, environment and property belonging to someone else. These jobs require greater dependence on job placement. Strength testing is the most effective of the job placement techniques. However, strength testing must simulate elements of the job as closely as possible in order to avoid possible discrimination against female and older workers. Medical examinations are more effective with older workers. Job placement is very important at the secondary and tertiary levels of control. Disabled employees can be encouraged to return to work if they are placed on jobs that are within their capabilities. Good job placement should also decrease the probability of recurrence of low back disorders.

Training the worker in strength and fitness appears to be effective in reducing the initial episode of back pain. Although fitness is difficult to enforce, management can encourage worker participation by supplying facilities and training programs. Training the worker in safe lifting does not appear to be as effective in reducing the initial low back episode as was once believed. The problem is not the training but the workers' compliance with the training. In spite of good training sessions in safe lifting procedures, workers do not lift properly -- probably because it is a harder and more difficult way to lift. Worker training should be more effective at the secondary and tertiary levels of control. In other words, it is probably more effective to train workers who have had low back disorders than to train the entire work force. These are the workers who are most likely to have the next low back episode -- and the workers who are most likely to comply with the training.

One of the most effective approaches to the control of low back disorders in industry is the education of management, unions and practitioners, because all three tend to prolong the length of disability. In general, management and unions offer little or no encouragement for workers to return to the job, and practitioners often prescribe ineffective or inappropriate treatment. The most effective control programs in industry have utilized multiple approaches that include procedures for early return to work.

REFERENCES

Bergquist-Ullman, M. and Larsson, U. (1977). Acute low back pain in industry. *Acta Orthopaedica Scandinavica* (*Supplementum No. 170*), 1-117.

Dillane, J.B., Fry, J. and Kalton, G. (1966). Acute back syndrome - a study from general practice. *British Medical Journal* 2, 82-84.

Finneson, B.E. (1975). *Dr. Finneson on low back pain*. New York: Putnam.

Holbrook, T.L., Grazier, K., Kelsey, J.L. and Stauffer, R.N. (1984). *The frequency of occurrence, impact and cost of selected musculoskeletal conditions in the United States*. Chicago: American Academy of Orthopedic Surgeons.

Horal, J. (1969). The clinical appearance of low back pain disorders in the city of Gothenburg, Sweden. *Acta Orthopaedica Scandinavica* (*Supplementum No. 118*), 1-109.

Hult, L. (1954). Cervical, dorsal and lumbar spine syndromes. *Acta Orthopaedica Scandinavica* (*Supplementum No. 17*), 1-102.

Klein, B.P., Jensen, R.G. and Sanderson, L.M. (1984). Assessment of workers' compensation claims for back strains/sprains. *Journal of Occupational Medicine*, 26, 443-448.

Liberty Mutual Insurance Company (1976). Unpublished data.

Mastromatteo, E. (1975). From Ramazzini to occupational health today from an international perspective. *Journal of Occupational Medicine*, 17, 289-294.

McGill, C.M. (1968). Industrial back problems: a control program. *Journal of Occupational Medicine*, 10, 174-178.

Nachemson, A.L. (1975). Towards a better understanding of low back pain: A review of the mechanics of the lumbart disc. *Rheumatology and Rehabilitation*, 14, 129-143.

Nachemson, A.L. (1976). The lumbar spine: An orthopedic challenge. *Spine*, 1, 59-71.

Snook, S.H. (1987). Approaches to the control of pack pain in industry: job design, job placement and educational/training. *Spine: State of the Art Reviews*, 2, 45-59.

Snook, S.H., Campanelli, R.A. and Hart, J.W. (1978). A study of three preventive approaches to low back injury. Journal of Occupational Medicine, 20, 478-481.

Snook, S.H. and Ciriello, V.M. (in press). The design of manual handling tasks: revised tables of maximum acceptable weights and forces.   Ergonomics.

U.S. Department of Health and Human Services (1981). Work practices guide for manual lifting. DHHS (NIOSH) Publication No. 81-122.

U.S. Department of Health and Human Services (1985). Current estimates from the national health interview survey, United States 1982. DHHS Pub. (PHS) 85-1578.

Waddell, G. (1987). A new clinical model for the treatment of low-back pain. Spine, 12, 632-644.

Webster, B.S. and Snook, S.H. (1990). The cost of compensable low back pain. Journal of Occupational Medicine, 32, 13-15.

Webster, B.S. and Snook, S.H. (1991). The cost of compensable low back pain: Update. (Manuscript in preparation).

PREDICTING THE MAXIMUM ACCEPTABLE WEIGHT OF LIFT FOR AN ASYMMETRICAL COMBINATION TASK

Tycho K. Fredericks and Jeffrey E. Fernandez
Department of Industrial Engineering
The Wichita State University
Wichita, KS 67260-0035

Clarence C. Rodrigues
Campbell Soup Company, Campbell Place
Camden, NJ 08103-1799

ABSTRACT

This paper presents prediction models based on a study of a combination lift and lower manual handling task that was designed to simulate the loading of grocery bags into a car trunk. Twelve male subjects performed an externally-paced task of lifting plastic grocery bags (with handles) loaded with weights from 15 cm above the floor and over a wooden sill. There were two different sill heights of 70 cm and 90 cm, and for each of these heights there were two frequencies of 3 and 6 lifts per cycle. A unique lifting sequence and a modified version of the psychophysical methodology was used to determine the maximum acceptable weight of lift (MAWOL). Prediction models were determined for MAWOL based upon task, physiological and anthropometric variables as well as a combinations of these. The results indicated that MAWOL could be predicted fairly well.

INTRODUCTION

In the United States, 25 percent of all injuries are directly related to manual material handling tasks (NIOSH, 1981). This amounts to over 25 million lost work days and between 4 and 20 billion dollars in compensation (Goldberg, 1980; Norby, 1981; and NIOSH, 1981). According to a recent study reported in the December issue of Professional Safety, (Manuel, 1991) workers' compensation costs will be approximately $60 billion, nearly double the costs of five years ago. These represent only direct costs, with indirect estimates running as high as four to six times the direct costs. Some insurance carriers have reported that close to 50% of workers' compensation (WC) claims file were ergonomic associated injuries, costing approximately 60% of the WC total dollars.

Manual Material Handling (MMH) has long been recognized as the major hazard to health and safety in business and industry, particularly in terms of lower back pain (Rowe, 1969; Ayoub, et al., 1978). Low back injuries bring substantial costs, both economically as well as in human anguish (Jones, 1971; Accident Facts, 1980). From 1938 to 1965, the number of compensable back injuries increased by 11.4 % while the

average cost increased approximately 400 percent (Snook and Ciriello, 1974). National Safety Council statistics from 1958 through 1980 depict an alarming exponential relationship between back injuries and their cost over the years.

There are several different approaches that are being utilized to prevent lower back injuries. The traditional approaches are to design the jobs to avoid excessive materials handling by the worker, train workers to follow safe lifting practices and the last option is to screen workers for musculoskeletal disorders.

While most lifting studies have been involved with single human material handling activities (lifting, lowering, holding, pushing and pulling), very little has been reported on lifting capabilities for combination activities (lift and lower, lift and carry, lift, lower and carry, etc.). The studies like Taboun and Dutta (1984, 1985 and 1986) and Jiang et al., (1986), studied combination tasks, but were carried out with boxes only. Aghazadeh (1984), suggested that loads lifted in boxes differed from loads lifted in bags and utilizing prediction models developed using boxes to predict loads lifted for bags may not be prudent. Based on the above premise,

Tycho Fredericks is now at the Department of Industrial and Manufacturing Engineering, Western Michigan University, Kalamazoo, MI.

Rodrigues, Congleton, Koppa, Huchingson, (1989), studied the simulation of loading of grocery (paper) bags (without handles) into car trunks. This task resembles lifting over an obstruction, a task commonly found in industry. This study used a unique variation of the psychophysical methodology to study eighteen male subjects lifting bags for a lifting frequency of 6 lifts per minute for 6 different obstruction height and depth combinations. The two heights (wooden sills) were 70 cm and 90 cm, and for each of these obstruction heights there were three sill depths of 28 cm, 43 cm and 57 cm. The lifting originated 15 cm above the floor. Statistical analyses on the data revealed that the obstruction heights of 70 cm and 90 cm did not effect the amount of load lifted. When the data were studied across obstruction depths however, only the 57 cm depth showed a lower lifted load than either the 28 cm or 43 cm depth. There were no significant differences in the loads lifted for either the 28 cm or 43 cm depths.

As a follow up of Rodrigues' study (1989), Fredericks (1991), used plastic bags with handles to study the simulation of loading of grocery bags into car trunks. Twelve male subjects performed an externally-paced task of lifting grocery bags loaded with weights from 15 cm above the floor over a wooden sill. There were two different sill heights of 70 cm and 90 cm, and for each of these heights there were two frequencies, 3 and 6 lifts per cycle. Figure 1 depicts the experimental set up. The dependent variable was the maximum acceptable weight of lift (MAWOL). A modified version of the psychophysical methodology was also used to determine the MAWOL. Other measures included resting heart rate, blood pressure, steady state working heart rate, and physical work capacity. Statistical analysis indicated that there was a significant difference in the weight lifted across the two sill heights as well as the weight lifted at the two different frequencies. A comparison of this study with Rodrigues (1989) revealed that individuals performing this experiment had greater lifting capacities with plastic bags with handles than that of paper bags without handles.

The elimination, or at least a reduction in the risk of injury due to lifting tasks, is the goal of researchers and practitioners from many disciplines. Therefore, the objective of this study was to determine

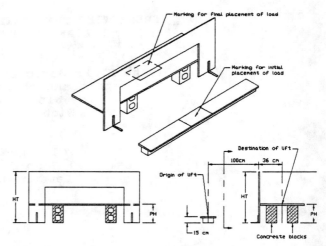

Figure 1. The experimental layout. HT= Sill height, PH= placement height.

prediction models for the maximum acceptable weight lift (MAWOL) based on task, physiological and anthropometric variables.

METHODS

All statistical procedures reported here were performed utilizing the SAS statistical package (SAS Institute, 1982) on the IBM 3081 main frame computer at The Wichita State University. A summary of the physical characteristics of the subjects are presented in Table 1.

MODEL CRITERIA

The general criteria used in the development of the data fitting models are discussed in the following section below:

1. Outlier detection. Outliers were determined by calculating the standard residuals. As a general rule of thumb, observations resulting in a standard residual with an absolute value of greater than 2.0 were checked and dealt with appropriately. These residuals were also plotted against the predicted values. When the data points appear to be randomly distributed about the residual mean, an unbiased model is indicated.

2. Multicollinearity. This refers to the situation when two or more independent variables are highly correlated with one another. When this situation exists, it can be harmful to the overall efficiency of the model. This problem can be detected by examining the correlation matrix between independent measures. If potential predictors are highly correlated, then these variables should be reduced to a

TABLE 1.   Descriptive Statistics of
          the subjects.

Body variable	Mean	STD
Age (years)	23.92	1.91
Weight (lb)	175.42	20.10
Stature (cm)	179.36	5.30
Span (cm)	182.20	5.63
Acromial height (cm)	147.63	3.72
Acromial-dactylion length (cm)	76.92	3.18
Trochanteric height (cm)	106.02	4.84
Biacromial breadth (cm)	56.43	31.74
Chest depth (cm)	24.01	2.03
Abdominal extension depth (cm)	22.61	2.74
Metacarpal III length (cm)	18.89	0.64
Knee height (cm)	56.03	2.98
Thumb-Tip Span (cm)	22.09	0.97
Grip strength at elbow angle of 180 deg (kg)	45.67	7.54
Grip strength at elbow angle of 90 deg (kg)	46.67	7.22
Resting heart rate (bpm)	66.21	8.84
Working heart rate (bpm)	95.51	10.13
Resting Blood pressure systolic (mmHg)	121.18	8.52
Resting Blood pressure diastolic (mmHg)	73.38	10.37
Estimated absolute PWC (L/minute)	4.12	0.73

number of independent variables that best represent the original set. Tests for multicollinearity are also available on SAS.

3.   Evaluation of F values. An F statistic is generated for every model and each independent variable within a model. These indicate the significance of the relationship between the independent variable(s) and the dependent measure.

4.   Evaluation of R-squared. R-squared represents the coefficient of multiple determination. The value of R-square for a model indicates the percentage (R-square * 100) of variance within the dependent measure that is accounted for by the independent variables in the model.

5.   Mallows Cp. If Cp is graphed with p, Mallows recommends the model where Cp first approaches p. When the most adequate model is reached, the parameter estimates tend to be unbiased, and Cp will approximate p. Models with substantial basis will tend to fall considerable above the line Cp=p. Values below the line are interpreted as showing no basis; that is, they are below the line due to sampling error.

MODEL DEVELOPMENT

A stepwise multiple regression procedure (PROC STEPWISE) and (PROC RSQUARE) were used to develop equations to predict MAWOL. Stepwise selection is a variation of forward selection where tests are performed at each step, significance of F, to determine the contribution of each predictor already included in the model. It was then possible to identify predictors that were significant at an earlier stage but then lost their usefulness when additional predictors entered the equation. Such a predictor would then be removed from the model.

The RSQUARE procedure finds subsets of independent variables that best predict a dependent variable by linear regression in the given sample. Regression coefficient and a variety of statistics useful for model selection can also be generated (SAS, 1982). The independent measures used were sill height, frequency of lift, age, weight, stature, span, acromial-dactylion length, trochanteric height, biacromial breath, chest depth, abdominal extension, metacarpal III length, knee height, thumb-tip reach, grip strength at elbow (90 degrees), grip strength at elbow (180 degrees), resting heart rate, steady state heart rate, resting blood pressure systolic, resting blood pressure diastolic, average blood pressure, and physical work capacity for using the treadmill. Both procedures generated similar and dissimilar models. Models were selected based on the previous mentioned criteria. Each model was checked for outliers and multicollinearity. Five prediction equations were developed to predict MAWOL and are presented in Table 2.

DISCUSSION

The models presented in Table 2 offer the industrial ergonomist a means of predicting MAWOL so as to determine a safe weight and possibly decrease the risk of back injuries. Model 1 was an attempt at documenting that task variables alone are not sufficient in predicting MAWOL. Model 2 was developed using physiological measures exclusively. Again, it is evident that the physiological parameters used in this experiment were not an adequate means of predicting MAWOL. Model 3, on the other hand, appears to predict MAWOL rather well (R-squared = 0.798). This model was developed using anthropometric variables exclusively. The anthropometric variables presented in

TABLE 2. Regression equations.

Model				R-squared	Mallows Cp	
MAWOL =	63.641	− 0.245(SH)		0.059	2.87	(1)
MAWOL =	77.574	− 0.507(RHR)		0.196	4.49	(2)
MAWOL =	− 137.226	+ 4.672(AGE) +	0.551(WT)			
	− 2.851(SP)	+ 1.367(TH) −	0.183(BB)			
	+ 30.998(MT)	+ 3.216(KH) −	18.472(TT)	0.798	10.16	(3)
MAWOL =	− 113.491	− 0.245(SH) −	0.913(FQ)			
	+ 4.672(AGE)	+ 0.551(WT) −	2.851(SP)			
	+ 1.367(TH)	− 0.183(BB) +	30.998(MT)			
	+ 3.216(KH)	−18.472(TT)		0.876	14.84	(4)
MAWOL =	22.287	− 0.245(SH) −	0.913(FQ)			
	+ 3.107(AG)	− 0.253(TH) −	0.981(KH)			
	− 4.126(TT)	+ 0.008(PWC)		0.868	18.83	(5)

where,

SH = sill height
AGE = age in years
SP = arm span (cm)
BB = biacromial breadth (cm)
MT = metacarpal III length (cm)
TT = thumb-tip reach (cm)
PWC = physical work capacity (ml/minute)

FQ = frequency of lift per cycle
WT = weight in pounds
TH = trochanteric height (cm)
ABD = abdominal extension depth (cm)
KH = knee height (cm)
RHR = resting heart rate (bpm)
MAWOL = maximum acceptable weight of lift

this model coincide with variables already prevalent in the literature (Rodrigues, 1989; Rodrigues et al. 1992). Prediction models based on anthropometric variables offer industry a distinct advantage for they are quick and easy to measure. Model 4 is a combination model. It combines task variables and anthropometric variables to predict MAWOL. It is evident by the high R-squared value that this model could also be very useful in an industrial setting. The difference between models 3 and 4 is that model 4 takes into consideration the sill height and the frequency of the load lifted. These two additional variables are very easy to measure and offer a significant improvement in R-squared over the previous model. Model 5 is a combination of task, anthropometric and physiological variables and appears to predict MAWOL rather well. Due to the physical work capacity variable in this model, it would be easier to use this model in a laboratory setting rather than an industrial setting.

It must be mentioned that many models with higher R squared values (as high as 0.91) were developed but excluded due to problems with multicollinearity. Multicollinearity appears to be an unavoidable obstacle when trying to develop prediction models that include anthropometric measures.

CONCLUDING REMARKS

The models developed offer the industrial ergonomist a means to determine a safe weight for an asymmetrical lifting task and possibly decrease the risk of back injuries. These equations could be beneficial to those who are involved in task design, task modification, or updating existing lifting capacity standards.

Further studies are needed in this area. Exploration of other anthropometric variables and ratios, physiological measures and other task variables could lead to stronger prediction models. Static and dynamic strength measurements in combination with anthropometric measurements could also aid in the prediction of psychophysically acceptable weight of lift.

REFERENCES

Accident Facts. 1959-1981 Editions. Chicago IL: National Safety Council.

Aghazadeh, F., (1984). Prediction models for manual handling of bags. In: A.Mital (Ed.), Trends in Ergonomics/Human Factors I. Elsevier, Amsterdam, pp.155-160.

Ayoub, M. M., Dryden, R. D., McDaniel, J.W., Knifer, R.E., and Aghazadeh, F., (1978). "Modeling of Lifting Capacity as a Function of Operator and Task Variables". In Safety in Manual Material Handling, Colin G. Drury, Editor, DHEW(NIOSH) Publications, No. 78-185, pp. 120-130.

Fredericks, T.K. (1991). The Maximum Acceptable Weight of Lift for an Asymmetrical Combination Task. Unpublished Master's Thesis, The Wichita State University, Wichita, KS.

Goldberg, H.M., (1980). Diagnosis and Management of Low Back Pain. Occupational Health and Safety, June, pp. 14-15.

Jiang, B.C., Smith, J.L., and Ayoub, M.M., (1986). Psychophysical Modeling for Combined Manual Materials Handling Activities. Ergonomics, 29(10), pp. 1173-1190.

Jones, D.F., (1971). Back Strains: The State of the Art. Safety Research, 3(1), pp. 28-34.

Krager, D. and Hancock, W., (1982). Advanced Work Measurement. New York: Industrial Press.

Manuel, F.A., (1991). Workers' Compensation cost control through ergonomics. Professional Safety, pp. 37-32.

NIOSH, 1981. Work Practices Guide for Manual Lifting (Publication No. 81-122). U.S. Department of Health and Human Services, Centers for Disease Control, Cincinnati, OH.

Norby, E.J., (1981). Epidemiology and Diagnosis in Low Back Injury. Occupational Health and Safety, January, pp. 38-42.

Rodrigues, C., Congleton, J.J., Koppa, R.J., Huchingson, R.D., (1989). Maximum Acceptable Weight of Lift for an Asymmetrical Combination Manual Handling Task. International Journal of Industrial Ergonomics, 4, pp. 245-253.

Rodrigues, C., Fredericks, T.K., Fernandez, J.E. (1992). Prediction Models For Asymmetrical Lifting Combination Tasks. In S. Kumar (Ed.), Advances in Industrial Ergonomics and Safety IV, pp.875-882. London: Taylor & Francis.

Rowe, M.L., (1969). Low Back Pain in Industry: A Position. Journal of Occupational Medicine, 11, pp. 161-169.

SAS Institute, Inc., (1982). SAS User's Guide, 1982 Ed.. Cary, NC.

Snook, S.H., and Ciriello, J.M., (1974). The Effects of Heat Stress on Manual Handling Tasks. American Industrial Hygiene Association Journal, 35, pp. 681-695.

Snook, S.H., (1978). The Design of Manual Material Handling. Ergonomics, 21(2), pp. 963-985.

Taboun, S.M., and Dutta, S.P., (1984). Prediction Models for Combined Tasks in Manual Material Handling (CTMMH). In: Proc. 1984 International Conference on Occupational Ergonomics, Toronto, Canada. Human Factors Association of Canada, pp. 551-555.

Taboun, S.M., and Dutta, S.P., (1985). Effect of Task Variables in Simultaneous Manual Lifting/Lowering and Carrying Loads. In: Proc. 18th Annual Meeting of the Human Factors Association of Canada. Human Factors Association of Canada, pp. 55-58.

Taboun, S.M., and Dutta, S.P., (1986). Modeling Psychophysical Capacities for Combined Manual Materials Handling Activities. In: W. Karwowski (Ed.), Trends in Ergonomics/Human Factors II. Elsevier, Amsterdam, pp. 785-791.

THE EFFECTS OF LIFTING POSTURE ON TRUNK MUSCLE ACTIVITY

Christopher A. Hamrick and Sean Gallagher
U.S. Department of the Interior, Bureau of Mines
Pittsburgh Research Center
Pittsburgh, Pennsylvania

Trunk muscle activity of twelve healthy males with coal mining experience was examined while each subject lifted a box under various conditions. The independent variables were four levels of posture (kneeling, stooped under a 1.2 m roof, stooped under a 1.6 m roof, and standing), height to which the box was lifted (35 cm or 70 cm), and weight of the lifting box (15 kg, 20 kg, or 25 kg). The dependent variables were the peak EMG values recorded during a lift for each of eight trunk muscles (left and right erectores spinae, left and right latissimus dorsi, left and right external oblique, and left and right rectus abdominis). Posture and weight of lift significantly affected peak activity of the left and right erectores spinae, the left and right latissimus dorsi muscles, and the right external oblique muscle. The latissimus dorsi muscle activity was highest in the low stooping posture, and was lowest in the kneeling posture, while erectores spinae activity was highest in the kneeling posture and decreased as the trunk became more flexed. Thus, the muscle activity during lifting tasks is affected by restricting a worker's posture. Consequently, many lifting guidelines and recommendations currently in use may not be directly applicable to work being performed in restricted postures.

INTRODUCTION

The height of an underground coal mine is generally determined by the thickness of the seam. About half of the coal seams in the United States are below 1.2 meters (48 inches) in thickness. Furthermore, as the mineral reserves are depleted, mining in these low seams will become more common. Because of the restricted roof heights, miners are often forced to adopt restricted postures during their work activities, and these restricted postures undoubtedly are a major factor in the high incidence rate of musculoskeletal injuries in the mining industry.

According to Bobick (1987), the two most common postures used by miners who work in low coal are stooping and kneeling on two knees. Since workers must often perform heavy lifts in these restricted postures, traditional recommendations for lifting in unrestricted postures may not apply to the underground mining environment. Thus, it is important to compare stresses on the body in restricted and unrestricted postures. One measure of such stresses on the low back is trunk muscle activity, which is captured using electromyography (EMG).

METHOD

Subjects

The subject population consisted of twelve healthy male subjects with coal mining experience. [Age(years) = 42.8 ± 5.2 s.d., Weight (kg) = 82.9 ± 11.8 s.d., Stature (cm) = 173.9 ± 6.7 s.d.] All subjects were paid volunteers who operated under terms of informed consent. Each subject received a thorough medical examination and a graded exercise tolerance stress test before taking part in the testing.

Experimental Design

The independent variables that were manipulated in this experiment were: four levels of posture [kneeling, stooped under a 1.2 m roof (low stooping), stooped under a 1.5 m roof (high stooping), and unrestricted standing], height to which the box was lifted (35 cm or 70 cm), and weight of the lifting box (15 kg, 20 kg, or 25 kg). Thus, there were twenty-four conditions, and each subject completed all lifting conditions. The dependent variables were the peak EMG values recorded during the lifts for each of eight trunk muscles [left erector spinae (les), right erector spinae (res), left latissimus dorsi (lld), right latissimus dorsi (rld), left external oblique (leo), right external oblique (reo), left rectus abdominis (lra), and right rectus abdominis (rra)]. A diagram of these muscles and their geometry with respect to the spine is presented by Schultz and Andersson (1981). A within subjects split-plot design was used, and the order of treatments was randomized for each subject.

Apparatus

Bipolar surface electrodes were used to obtain integrated EMG's of the eight trunk muscles of interest. Belt-wearable preamplifiers were used to amplify the EMG signals, which were sent through shielded cables to an amplifier/integrator where the signals were conditioned using an 80 Hz high pass filter and a 1000 Hz low pass filter. The integrator constant was 500 ms. During the experiment, the integrated EMG signal of all eight trunk muscles for each lifting task were plotted on a computer screen to ensure integrity. The fully conditioned signals were then sent via an A/D board to a MicroVax II microcomputer for storage and analysis. (Reference to specific products does not imply endorsement by the Bureau of Mines.)

Vertical space constraints for the stooped lifting conditions were controlled through an aluminum frame that supported a 1.2 m x 2.4 m plexiglass roof. The roof height could be set to either 1.2 m or 1.5 m. An aluminum lifting box (50.8 cm x 33.0 cm x 17.8 cm) with two covered compartments was used in the lifting task. The box had a rounded, recessed handhold on either side so that it would be easy to grasp. An automatic lowering device was designed and built for the experiments so the load never had to be lowered by the subject or the experimenters.

Experimental Procedure

First, each subject was prepared for the experimental session. The eight trunk muscles of interest were identified, and the skin above each muscle was prepared by shaving (if necessary) and thoroughly cleaning with alcohol. Two electrodes filled with electrolyte gel were placed 3 cm apart (center to center) over each selected muscle. A single electrode, which served as a ground, was placed at a remote site on the body. After all electrodes had been attached, the integrity of the signal was checked by monitoring the raw EMG signal on an oscilloscope while the subject submaximally activated each of the muscles in isolation.

The subject was then placed into a frame attached to a Cybex II isokinetic dynamometer and performed maximum voluntary isometric flexion and extension exertions in four postures: 1) kneeling

with 22½° of trunk flexion, 2) standing with 22½° of trunk flexion, 3) standing with 45° of trunk flexion, and 4) standing with 67½° of trunk flexion. EMG data were collected during these exertions so the lifting data could be normalized relative to the maximum EMG signals for the particular lifting posture.

Each subject then completed the series of twenty-four lifting tasks in random order. The subject was required to lift or lower the box to the proper height every fifteen seconds for a period of one minute. All lifting was performed symmetrically with respect to the sagittal plane. The EMG data were collected during a specified lift of the sequence. The subject was given a two minute rest period between each condition.

Analysis

The EMG values for each subject were normalized for each condition relative to the maximum value obtained for that posture. The peak EMG value during the duration of the lift was then determined for each condition performed by each subject. These values were then subjected to a 4 x 2 x 3 (posture x height x weight) Analysis of Variance (ANOVA) procedure to determine which of the independent variables significantly influenced the peak EMG signals. Separate ANOVA's were performed for the EMG signals obtained from each of the eight trunk muscles. A significance level of 0.05 was used for all tests.

RESULTS

Table 1 contains a summary of the significant effects for each dependent variable from the ANOVA's. No significant interactive effects were found to be present. Figures 1 and 2 display the muscle activity versus lifting weight and lifting height, respectively. Averages and standard error for the mean of the peak EMG values for each posture are presented for each trunk muscle in Table 2, and the average values are shown graphically in Figure 3.

The results of the ANOVA indicate that both posture and weight significantly affected peak EMG values for the left and right erectores spinae muscles, the left and right latissimus dorsi muscles, and the right external oblique muscle. Lifting height was found to influence left and right latissimus dorsi activity. The activity of both rectus abdominis muscles was significantly affected by posture.

Table 1. Summary of significant effects.

	Posture (P)	Height (H)	Weight (W)
Left Erector Spinae	***		***
Right Erector Spinae	***		***
Left Latissimus Dorsi	***	*	***
Right Latissimus Dorsi	***	***	***
Left External Oblique			
Right External Oblique	**		**
Left Rectus Abdominis	***		
Right Rectus Abdominis	*		

```
  *   p≤0.050
 **   p≤0.010
***   p≤0.001
```

Table 2. Averages of peak EMG values for four postures (% of maximum). Standard errors for the mean are given in parentheses.

	Standing	High Stooping	Low Stooping	Kneeling
Left Erector Spinae	60.83 (1.78)	56.42 (2.20)	52.23 (2.30)	63.47 (1.47)
Right Erector Spinae	63.32 (1.93)	57.88 (2.21)	54.51 (2.22)	65.62 (1.12)
Left Latissimus Dorsi	51.18 (2.69)	54.44 (2.61)	57.53 (2.74)	38.30 (2.37)
Right Latissimus Dorsi	51.86 (2.40)	52.66 (2.55)	56.39 (2.74)	34.65 (1.88)
Left External Oblique	29.03 (2.20)	28.53 (1.84)	31.78 (2.34)	26.34 (1.83)
Right External Oblique	22.94 (1.15)	25.53 (1.48)	28.46 (1.56)	21.12 (1.15)
Left Rectus Abdominis	34.53 (2.70)	38.84 (2.60)	38.66 (2.17)	25.60 (1.98)
Right Rectus Abdominis	43.35 (2.77)	42.45 (2.56)	41.18 (2.55)	33.68 (2.36)

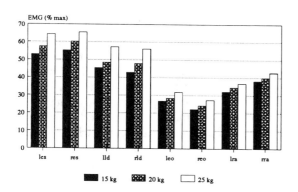

Figure 1. Muscle activity versus weight of box lifted.

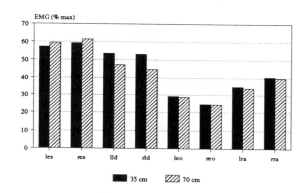

Figure 2. Muscle activity versus height of lift.

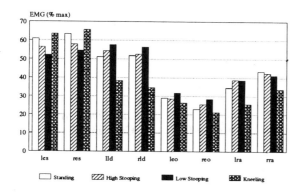

Figure 3: Muscle activity versus lifting posture.

DISCUSSION

Effects of Weight

Muscle activity increased with an increased weight of lift, although this increase was not significant for the left external oblique or for the left and right rectus abdominis muscles. This relationship is not surprising, since the trunk muscles must increase their activity to generate forces that counteract the additional external forces of the load. Furthermore, this EMG - force relationship has been extensively examined in the literature (e.g. Lippold, 1952; Komi and Viitasalo, 1976).

Effects of Lifting Height

Erectores spinae activity generally increased as the lifting height increased from 35 cm to 70 cm, although this change was not statistically significant. Conversely, latissimus dorsi activity generally decreased with the elevated lifting height. An informal examination of the latissimus dorsi activity for each height-posture combination reveals that the activity of these muscles while kneeling increases slightly with increasing lifting height, the opposite effect of that seen in the other postures. According to Miely, *et al.* (1990), the latissimus dorsi acts as a "prime mover of the shoulder and in the process exerts some force on the spine." Thus, during lifting in the stooped and standing postures, the arm may be used more when lifting the box to the lower shelf than to the higher shelf, thus resulting in increased latissimus dorsi activity.

Effects of Posture

Left and right latissimus dorsi activity appeared highest in the low stooping posture and lowest while kneeling. The same relationship holds for the external oblique muscle activity, although statistical significance was found only for the right muscle. In contrast to the latissimus dorsi and external oblique muscles, the left and right erectores spinae muscles were most active in the kneeling and standing postures, and the activity decreased as the trunk became more flexed. Hence, the erectores spinae activity was least active in the stooping positions. Rectus abdominis activities were generally lowest in the kneeling posture.

Forces generated by the erectores spinae muscles are known to contribute substantially to compressive forces on the spine, forces generated by the external oblique muscles add to anterior-posterior shear and compression, and forces generated by the latissimus dorsi muscles contribute to both compressive and lateral shear forces (Schultz and Andersson, 1981). Therefore, the EMG data suggest that compressive forces due to the erectores spinae muscle forces should be highest in the standing and kneeling postures and become lower as the trunk becomes more stooped. Furthermore, anterior-posterior shear forces in the spine should be highest in the stooping postures and decrease as the trunk becomes more erect. These force relationships are consistent with findings obtained in previous Bureau of Mines studies (Gallagher, *et al.*, 1992).

There are two possible reasons for the decreased erectores spinae muscle activity in the stooped postures. The first reason is that the erector spinae muscles are relatively inactive during extreme forward flexion, as in the low stooped posture (Floyd and Silver, 1955). During forward flexion, the ligaments are used to counteract external moments and stabilize the trunk. In the more erect postures, however, the erectores spinae muscles provide the forces that counteract the moment created by the weight of the lifting box.

The second reason for the decreased erectores spinae muscle activity during stooping is that in this posture the forces from the box due to gravity and the upward acceleration, which act vertically, are no longer acting in line with the spine, as is the case with the standing and kneeling postures. Thus, other muscles that can counteract anterior shear forces are relied upon to provide these forces. According to a model presented by Schultz and Andersson (1981), the erector spinae muscles provide forces parallel to the spine, while the external oblique muscles can counteract anterior-posterior shear forces.

The findings are in agreement with both of the previous explanations, since as the trunk flexion angle increased, the erectores spinae muscle activity decreased and the external oblique activity generally increased. Thus, while lifting in flexion, spinal ligaments may have provided the restorative moment provided by the erectores spinae muscles in the erect postures. Furthermore, the external oblique muscles were relied upon in the flexed postures to combat the increased anterior-posterior shear forces in the stooped postures.

CONCLUSIONS

Based upon these findings, the trunk muscle activity during lifting tasks does depend upon lifting posture. Consequently, the forces on the low-back associated with the trunk musculature certainly vary with posture, as well. Hence, current lifting guidelines and recommendations based upon biomechanical data gathered in unrestricted postures may not be directly applicable to work being performed in restricted postures. Such work includes manual materials handling tasks being performed in low-seam coal mines.

REFERENCES

Bobick, T.G. (1987). Analyses of materials-handling systems in underground low-coal mines. In: *Human Engineering and Human Resources Management in Mining*. Bureau of Mines Information Circular 9245, pp. 13-20.

Floyd, W.F. and Silver, P.H.S. (1955). The function of the erectores spinae muscles in certain movements and postures in man. *Journal of Physiology*, **129**:184-203.

Gallagher, S., Hamrick, C.A., Love, A.C., and Marras, W.S. (1992) Dynamic biomechanical modeling of symmetric and asymmetric lifting tasks in restricted postures. *Ergonomics* (accepted).

Komi, P.V. and Viitasalo, J.H.T. (1976) Signal characteristics of EMG at different levels of muscle tension. *Acta Physiol. Scand.*, **96**:267-276.

Lippold, O.C.J. (1952) The relation between integrated action potentials in a human muscle and its isometric tension. *Journal of Physiology*, **117**:492-499.

Miely, W.R., McLain, R., Weinstein, J.N., Goel, V.K., and Found, E.M. (1990) Anatomy of the lumbar spine. In *Biomechanics of the Spine: Clinical and Surgical Perspective* (Vijay K. Goel and James N. Weinstein, *eds.*), CRC Press, Inc., Boca Raton, Florida, p. 33.

Schultz, A.B. and Andersson, G.B.J. (1981). Analysis of loads on the lumbar spine. *Spine*, **6**(1):76-82).

INDUSTRIAL QUANTIFICATION OF OCCUPATIONALLY-RELATED LOW BACK DISORDER RISK FACTORS

William S. Marras[1], PhD, Steven A. Lavender[2], PhD, Sue E. Leurgans[3], PhD, Sudhakar L. Rajulu[4], PhD, W. Gary Allread[1], Fadi A. Fathallah[1], MS, and Sue A. Ferguson[1], MS,

[1] The Ohio State University, Department of Industrial and Systems Engineering,Biodynamics Laboratory, 1971 Neil Avenue, Columbus, Ohio 43210 USA

[3] The Ohio State University, Department of Statistics, 1958 Neil Avenue, Columbus, Ohio 43210 USA

[2] Rush-Presbyterian-St.Luke's Medical Center, Department of Orthopaedic Surgery, 1653 West Congress Parkway, Chicago, Illinois 60612 USA

[4] Lockheed Engineering and Sciences Company, Ergonomics Section, 2400 NASA Road, C95, Houston, Texas 77058 USA
(713) 333-7821

Few assessment techniques have attempted to define the role of occupational trunk motion in the risk of occupationally-related low back disorder (LBD) even though laboratory studies have indicated that motion significantly increases spine loading. An in-vivo study was performed to assess the contribution of three- dimensional dynamic trunk motions to the risk of LBD during occupational lifting in industry. Over 400 industrial lifting jobs were studied in 48 industries. Specific manual materials handling jobs historically identify as either high risk or low risk for LBD were identified. A tri-axial electrogoniometer was worn by workers and documented the three-dimensional trunk motion characteristics associated with these high risk or low risk jobs. Workplace characteristics such as load moment arm, load weight, etc. were also documented for each of the repetitive lifting tasks. A multiple logistic regression model indicated that a combination of five trunk motion and workplace factors (lifting frequency, load moment, trunk lateral velocity, trunk twisting velocity, and trunk sagittal angle) predicted occupational-related LBD risk well. The analyses have enabled us to determine the LBD risk associated with combined changes in the magnitudes of the five factors. This model could be used as a quantitative, objective measure to redesign the workplace so that the risk of occupationally-related LBD is minimized.

INTRODUCTION

Epidemiologic studies (Andersson, 1981; 1991; Kelsey et. al. 1984; Magora, 1973) acknowledge that there are several risk factors associated with lifting and manual materials handling that increase the risk of suffering an occupationally related low back disorder (LBD). These studies often cite heavy work, static bending postures, frequent bending and twisting, lifting, pushing-pulling, and repetition as significant risk factors.

The literature has also attempted to explain the biomechanical significance of these actions under occupational circumstances. Studies are often limited to static, two-dimensional assessments of compressive load on the lumbar spine and are often unable to associate increases in predicted spine compression with increases in occupationally related low back disorder incidence rates.

It has been suggested (Andersson, 1991) that this lack of association between risk factors and LBD may be due to "confounding factors" as well as the inability to determine the level at which physical work becomes dangerous. It is hypothesized that an overlooked mechanism for explaining biomechanical increases in spine loading during lifting may be the action of the three-dimensional dynamic motion characteristics of the spine and its effect on the individual motion segments. Tracking these dynamic three-dimensional biodynamic motions of the spine during lifting may help explain how the spine is loaded and may help define dangerous limits of spine loading. The objective of this study, therefore, was to quantitatively evaluate, in-vivo, the trunk motions and physical workplace factors associated with high risk (of occupationally-related LBD) and low risk (of occupationally-related LBD) industrial jobs.

METHODS

This study was a cross-sectional study of 403 industrial jobs from 48 manufacturing companies throughout the midwestern United States. Only repetitive jobs without job rotation were used in this study. This was necessary to prevent the confounding effects created by alternate jobs. Jobs examined in this study were divided into two groups, high and low risk of LBD, based upon examination of the injury and medical records. Whenever possible, company medical reports were used to categorize risk. In some cases only injury logs (OSHA 200 logs) were available.

Independent Variables

The independent variable in this study consisted of two levels of job-related LBD risk categories. Low risk jobs were defined as those jobs with at least three years of records showing no injuries and no turnover. High risk jobs were those jobs with at least 12 injuries per 200,000 hours worked.

Dependent Variables

The dependent variables in this study consisted of workplace, individual and trunk motion characteristics which were indicative of each job. The workplace and individual characteristics consisted of variables typically considered in current workplace guidelines for materials handling. Specifically, these variables were: 1) the maximum horizontal distance of the load from the spine; 2) the weight of the object lifted; 3) the height of the load at origin of the lift; 4) the height of the load at the destination of the lift; 5) the frequency of lifting (liftrate); 6) the asymmetric angle of the lift; 7) worker anthropometry (12 measures); 8) worker injury history; and 9) worker satisfaction.

Trunk motion characteristics were those variables obtained from a tri-axial electrogoniometer (lumbar motion monitor) which was worn by workers. The lumbar motion monitor was essentially an exo-skeleton of the spine that emulated the motions of the worker's back during the work cycle. The lumbar motion monitor is shown in figure 1. This device documented the three-dimensional angular position, velocity, and acceleration characteristics of the lumbar spine while workers lifted in these high risk or low risk jobs.

Analyses

The analyses consisted of univariate and multivariate logistic regression assessments which enabled us to identify those individual variables as well as combinations of variables that were most sensitive to differences between the low and high risk groups. Selected trunk motion factors along with selected workplace factors were used to develop a quantitative model of occupational risk factors.

Figure 1. Lumbar motion monitor worn by the workers.

RESULTS

Univariate logistic regressions were used to identify those variables that significantly differentiated between the low risk and high risk jobs. This analysis indicated that of the workplace factors, the object weights, moments, and vertical load destination locations all yielded significant odds ratios. Of the trunk motion factors, the average and maximum velocity observed during the work cycle were the only trunk motion factors that resulted in significant odds rations in all planes of the body.

A multiple logistic regression model indicated that a combination of five trunk motion and workplace factors predicted occupational-related LBD risk well and is shown in table 1. The risk factors included: 1) lifting frequency, 2) load moment, 3) trunk lateral velocity, 4) trunk twisting velocity, and 5) the trunk sagittal angle.

Increases in the magnitude of these factors significantly increased the risk of LBD. The analyses have enabled us to determine the LBD risk associated with combined changes in the magnitudes of the five factors. The results indicate that increases in the magnitudes of these variables can increase the probability of occupationally-related LBD by almost eleven fold. The predictive power of this model was also found to be over three times greater than that of current lifting guidelines (NIOSH, 1981).

Based upon the logistic regression analysis trunk motion and workplace factor benchmarks were obtained for the various probabilities of LBD risk. Benchmarks permit graphical presentation of the way in which multiple risk factors combine to contribute to the probabilities of LBD. These benchmarks and the associated probability of high risk group membership are shown in figure 2.

Table 1. Multiple logistic regression model used to assess occupationally- related risk factors.

Variables	Coefficients	Standard Error
Constant	-3.80	0.67
Liftrate	0.0014	0.0006
Avg. Twist Velocity	0.061	0.041
Max. Moment	0.024	0.004
Max. Sagittal Flexion	0.20	0.012
Max. Lateral Velocity	0.036	0.014
Estimated Odds Ratio	10.7	
Confidence Interval	(4.9 - 23.6)	

DISCUSSION

This work represents the first study to relate epidemiological findings with quantitative biomechanical findings in a large and varied industrial data base. A deliberate attempt was made to include repetitive jobs representative of a broad range of industries so that the data and LBD risk model would be as generalizable as possible. For the first time, we have been able to quantify biomechanical factors in-vivo during industrial work. This has provided a data base that will be useful for research purposes as well as for ergonomic application purposes. There

are several immediate implications of this study.

First, we have been able to identify and describe the trunk motions that are present in industry. This information has shown that there is considerable three-dimensional trunk motion occurring in most industrial tasks. These findings indicate that assumptions, such as those in current lifting guidelines, of sagittally symmetric, slow, smooth lifting are not consistent with the types of motions that are routinely experienced by the workers in the workplace.

Second, we have been able to identify key factors that are indicative of occupationally-related LBD risk. LBD risk is identified as a function of a multivariate vector of workplace factors and trunk motion factors. Many of these factors are highly correlated. Thus, by tracking just five factors (moment, lift rate, lateral velocity, sagittal trunk angle, and twisting velocity) used in the multiple logistic regression risk model, we are able to determine how a particular job relates to occupational LBD risk. Individually, each of these factors is unable to discriminate reliably between a high risk situation and a low risk situation. However, when these factors are considered in combination as a representation of a multivariate vector, the risk model is capable of predicting situations that would result in a greater probability of LBD. This model also permits one to determine the effects of changing the specific values of the risk factors. This model has excellent predictive power and could identify nearly an eleven-fold increase in odds of occupationally-related LBD. The model has immediate application potential for the ergonomic design and redesign of workplaces involving MMH. This study has also demonstrated that the best discriminators of risk in the workplace are generally achieved by studying the maximum or peak measures of trunk motion and workplace factors. Biomechanically extreme job elements are often not considered in current ergonomic evaluations. The notion that the maximum values of the workplace and trunk motion factors impact risk also has implications for the assessment of cumulative trauma disorders. Perhaps, the variance in load demands on a biomechanical structure is a confounding factor in current attempts to understanding how the accumulation of stressors affects the tolerance of the structure.

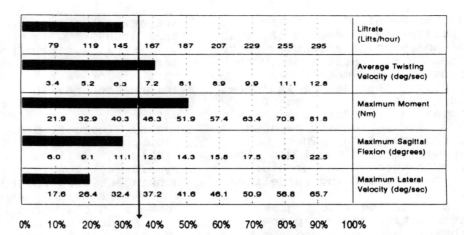

79	119	145	167	187	207	229	255	295	Liftrate (Lifts/hour)
3.4	5.2	6.3	7.2	8.1	8.9	9.9	11.1	12.8	Average Twisting Velocity (deg/sec)
21.9	32.9	40.3	46.3	51.9	57.4	63.4	70.8	81.8	Maximum Moment (Nm)
6.0	9.1	11.1	12.8	14.3	15.8	17.5	19.5	22.5	Maximum Sagittal Flexion (degrees)
17.6	26.4	32.4	37.2	41.6	46.1	50.9	56.8	65.7	Maximum Lateral Velocity (deg/sec)

0%   10%   20%   30%   40%   50%   60%   70%   80%   90%   100%

Probability of High Risk Group Membership

Figure 2. Risk factor benchmarks which predict probability of occupationally-related low back disorder high risk group membership.

Third, these data can be used to help understand the conditions under which various biomechanical mechanisms operate during MMH. Since this data base is the only quantitative in-vivo industrial data base of trunk motions and workplace factors that we are aware of, it can provide information about the expected in-vivo conditions associated with trunk motion loading during work. Such information is necessary so that biomechanical properties can be tied to realistic workplace conditions. This is becoming increasingly important in light of new findings regarding in-vivo versus in-vitro experimentation.

Finally, this information could be used to further trunk biomechanical modeling efforts. Biomechanical models which are sensitive to three-dimensional positions of the trunk as well as those that are sensitive to dynamic trunk motions could use this data to define the limits of trunk motion characteristics that would be expected during work. Furthermore, in-vivo studies of muscle and intra-abdominal pressure responses during lifting could use this information to define test conditions, thereby providing valuable insight as to how the musculoskeletal system responds to realistic industrial lifting conditions.

ACKNOWLEDGMENT

Funding for this project is provided by the Ohio Bureau of Workers' Compensation, Division of Safety and Hygiene.

REFERENCES

Andersson GB: Epidemiologic aspects of low back pain in industry. *Spine*, 6, 53-60, 1981

Andersson GBJ: The epidemiology of spinal disorders, *The Adult Spine: Principles and Practice*, Chapter 8, (JW Frymoyer, ed) Raven Press Ltd. New York, 1991

Kelsey KL, Githens PB, White AA III, Holford TR, Walter SD, O'Conner T, Ostfeld AM, Weil U, Southwick WO, Calogero JA: An epidemiologic study of lifting and twisting on the job and risk for acute prolapsed lumbar intervertebral disc. *J Ortho Res*, 2(1), 61-66, 1984

Magora A: Investigation of the relation between low back pain and occupation. 4. Physical requirements: Bending, rotation, reaching and sudden maximal effort. *Scand J Rehab Med*, 5(4), 186-190, 1973

National Institute for Occupational Safety and Health (NIOSH): Work practices guide for manual lifting. Department of Health and Human Services (DHHS), National Institute for Occupational Safety and Health (NIOSH) Publication No. 81-122, 1981

MODELLING THE STOCHASTIC NATURE OF TRUNK MUSCLE FORCES

Gary A. Mirka[1]
William S. Marras

The Biodynamics Laboratory
Ohio State University
Columbus, Ohio

Now at North Carolina State University
Department of Industrial Engineering
Raleigh, North Carolina

In an effort to understand the mechanism of low back disorders, researchers have developed EMG driven biomechanical models which estimate the magnitude of the internal reaction forces of the spine (compression and shear), by using information about the activity of the muscles of the trunk. But, because the trunk is multi-dimensional in nature, there is variability in the relative contribution of the various muscles, which implies variability in the reaction forces of the spine. Therefore, it may be more appropriate to discuss the range of spine reaction forces during a lift as opposed to the mean spine reaction force. The present research was an attempt to model the muscle forces stochastically and to develop a simulation model which predicts trunk muscle EMG that could occur during a lift. The simulated EMGs which resulted were then input into an EMG driven biomechanical model so the variability in spine reaction forces could be quantified. Under simple sagittally symmetric isometric conditions, compression which occurred at three standard deviations above the mean was 12% higher than that of the mean. The results for anterior shear (24% higher) and lateral shear (50% higher) showed even larger increases.

INTRODUCTION

The incidence of occupational low back disorders (LBDs) has grown to epidemic proportions. In a paper which reviewed epidemologic studies, Andersson et al (1984) cited studies which showed that between 12 and 35 percent of the working population are experiencing low back pain (LBP) at any point in time. It has also been estimated that between 48 and 70 percent of all people will experience LBP at some time in their lifetime.

In an effort to understand and control this problem, researchers have developed techniques to estimate the internal muscle forces that occur during a lift to better understand the types of forces which the spine must withstand. Among the methods used to estimate these forces are mechanical analysis, optimization and electromyography (EMG). The estimations from EMG analyses are often more accurate because the technique involves indirect measurement instead of prediction of muscle forces (Marras and Mirka, 1991, 1992).

Some of the more advanced EMG analysis techniques have made quantitative assessments of the internal reaction forces of the spine by developing "EMG driven" biomechanical models. These models utilize EMG inputs from the various muscles of the trunk and calculate the spine reaction forces which the spine must resist.

The fundamental principle of these EMG driven biomechanical models is that muscle tissue has a maximum capacity to exert force per unit area and by quantifying the relative EMG activity of the muscle an estimate of the muscle force results. These models contain factors which modify the EMG signal based on the velocity of contraction and the location along the muscle's length-tension curve. Two of the most notable EMG driven models are that of McGill and Norman (1986) and that of Reilly and Marras (1989). Subsequent iterations of these models include McGill (1991, 1992) and Marras and Sommerich (1991a,b).

Focus of this Research

One of the common problems among mechanical analysis, optimization and EMG driven biomechanical models is that they approach the problem deterministically. The multi-dimensional nature of the trunk indicates that the problem could be stochastic in nature. That is, in order to perform a given lift, the relative contribution of each muscle to the total external torque may vary. This variability will have profound effects on the spine reaction forces because each muscle has its own moment arm and line of action. By using stochastic values as inputs to these EMG driven biomechanical models the spine reaction forces which result will more accurately reflect the types of forces which occur during lifting.

The specific goals of this research were to model the variability present in muscle forces and to develop a simulation model which could generate EMG signals which reflect this innate variability. In order to achieve these goals several steps had to be accomplished. First, an experiment was conducted which had a group of subjects perform a lift of precise specifications repeatedly. Then, using this data, the probability density functions (PDFs) that describe the stochastic activity of the ten muscles of the trunk under the experimental conditions were developed. Using these PDFs as a data base, an EMG generator was developed which produces simulated EMG values that could occur during a lift of given specifications. To gain an understanding of the spine reaction forces that result, the simulated data was input into an EMG driven biomechanical model.

METHODS

Subjects

Four male college students served as subjects in this study. In order to minimize learning effects, subjects were chosen from a pool of subjects from previous similar studies.

Experimental Design

Independent variables. There were two independent variables in this study: torque level and angular trunk position. Torque was controlled by the subject and the levels of this variable were 30 and 60 ft-lbs. Trunk position was set at two levels in the sagittal plane, 5 and 40 degrees of forward flexion. Each subject had to perform each combination of independent variables 5 times.

Dependent variables. The dependent variables in this study were the normalized integrated EMG values for ten trunk muscles. These muscles included: the right and left erector spinae (RES, LES), the right and left latissimus dorsi (RLAT, LLAT), the right and left rectus abdominis (RAB, LAB), the right

and left external oblique (REX, LEX) and the right and left internal oblique (RIN, LIN).

In addition to these measures, two new variables SES and DES were studied and these variables correspond to the sum of the erector spinae (RES+LES) and the difference between the erector spinae (RES-LES). The need for these two variables will be shown later in the simulation model section.

Equipment

The equipment used for this experiment consisted of a Kin/Com dynamometer, a trunk motion reference frame, an EMG data processing system and a data collection system. The dynamometer and the reference frame were combined to make a back machine that allowed for the precise control of the angular trunk position and the external torque (about the L5/S1 joint) being exerted by the subject. See Figure 1.

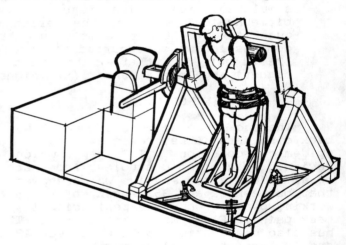

Figure 1. Trunk Reference Frame

The EMG signal collected by the electrodes was amplified 52000x and was conditioned by a high pass filter set at 80 Hz and a low pass filter set at 1000 Hz. This filtered signal was rectified and averaged using a 20 msec window. The EMG data along with the torque and motion data from the Kin/Com were collected at 100 Hz by the data collection system.

Experiment

Upon arrival, subjects had surface electrodes applied to their skin through standard preparation procedures. When all of the EMG signals were verified, the subject was placed in the reference frame such their L5/S1 joint was aligned with the rotating axis of the Kin/Com dynamometer. Once the subject was in the apparatus, he performed maximum voluntary contractions (MVCs) at each of the two trunk positions (5 and 40 degrees of sagittal bend). For a more complete description of the method employed in gathering of MVC data see Marras and Mirka (1992).

After the MVCs were obtained the subjects performed the submaximal contractions defined by the combination of independent variables. The order of presentation of these combinations was completely randomized and the subjects were given a rest period of 1 minute between exertions. The position was controlled by the dynamometer while the subject controlled his torque output using a video feedback mechanism. A VDT was located directly in front of the subject which displayed the subject's torque output as well as the designated amount of torque for that trial.

ANALYSIS

Data Processing

Smoothing. The nature of the integrated signal using a 20 ms averaging window is quite noisy. Therefore, in order to reduce signal noise, while simultaneously maintaining the time dependent nature of the muscles' activity, further processing of the EMG signals was done in software, using a Hanning weighted filter. This smoothing function had a bandwidth of 17 and was repeated 3 times.

Normalization. The smoothed signal was normalized by the angle dependent maximum and resting EMG values. In addition, in order to control for the differences between subjects, the EMG values were normalized across subjects within an experimental condition.

Distribution Development

The goal of this phase was to arrive at probability density functions for each of the ten muscles studied for each set of experimental conditions. In this way the distributions would show how the environmental factors affected the muscles. Empirical distributions were formulated using SAS software which plotted the empirical PDFs.

This empirical data was then input into a software package called FITTR1 (Venkatraman and Wilson, 1987) which fit the data into a four parameter distribution using Johnson's translation system. The four parameters are xi (a

location parameter), lambda (a scale parameter), gamma (a shape parameter) and delta (a shape parameter). This translation system allows for fitting to a variety of distributions from normal to lognormal to a general bounded distribution to a general unbounded distribution. The output from this software were the values for the four parameters which describe the best fit distribution. It was these values which were utilized later in the simulation model to generate the simulated EMG signals. An example of a fitted empirical distributions is shown in figure 2.

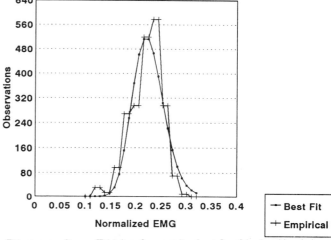

Figure 2. Fitted empirical distribution for the right erector spinae

Simulation Model

The basic principle of this simulation model is that the erector spinae muscles are the primary lifting muscles and the other muscles are primarily trunk stabilizers. Given that as the fundamental premise, the sampling from distributions was done in two steps. The first step involved the sampling for the erector spinae muscles using the distributions which were conditional on the environmental parameters. The second step was to sample for values for the other muscles conditional on both the values derived from step one as well as the environmental parameters.

It would be inappropriate to sample for the two erector spinae muscles from their individual conditional distributions because the synergism between the right and left erector spinae would be lost. Therefore, two new variables which describe these muscles interactions were developed and were called SES and DES. SES describes the sum of the erector spinae normalized

EMG values while DES describes the difference between these two values (RES-LES). The exact same procedure described above for fitting the empirical data to distributions was used to describe the distributions of these new two variables.

Once the distributions for SES and DES were formulated, subsequent conditional distributions were developed for each of the ten muscles. The method chosen for development of these conditional distributions was to partition the SES/DES space into nine sections and conditional distributions were developed for each muscle in each section. Therefore, the data points which made up the conditional distributions for each muscle were not only from the same environmental condition but were also from the same SES/DES section. The final step in the simulation model was to perform a linear transformation of the EMG values as a function of subject isometric strength.

RESULTS

The results of this study will be shown at several levels. The first level will be to show the type of distributions developed and how those distributions changed as a function of the environmental factors. The second level will show the simulated EMG signals and how the variability from the distributions translates into variability of the simulated EMG signals. The last level of results will show how the variability of the simulated EMGs translates into variability in the spine reaction forces as computed using an EMG driven biomechanical model.

An example of the distributions which were developed as a function of environmental conditions is shown in figure 3. This figure shows how the shape and location of the distribution changed as a function of the various environmental factors. This figure shows the environmental conditions tested in the experiment as well as those developed by interpolating the values of the four parameters which describe the distributions.

The second level of results describes how these sampling distributions translates into variable EMG signals through the simulation model. An example of these simulated EMG results is shown in figure 4. These represent the output of fifty runs of the simulation model.

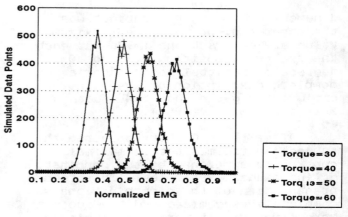

Figure 3. PDFs used in sampling for the SES value during simulation (ANGLE=5)

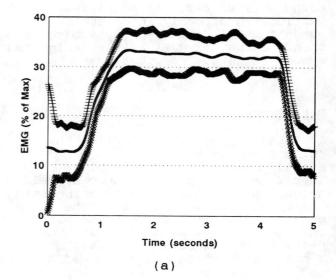

(a)

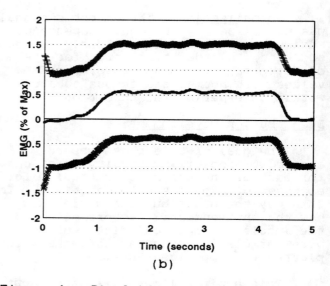

(b)

Figure 4. Simulation model output describing mean and +/-3 standard deviations for the (a) right erector spinae and (b) left rectus abdominis, (TORQUE 60, ANGLE 5).

The last level of results is the model output describing the spine reaction forces which are the sum total of all of the muscles working together. Figure 5 shows several examples of the types of reaction forces which occur during a variety of lifting conditions.

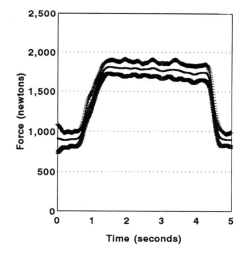

(a)

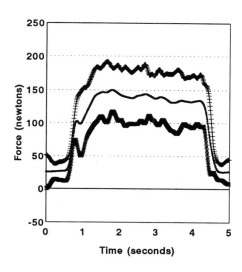

(b)

Figure 5. Compression (a) and anterior/posterior shear (b) forces (mean, +/- 3 standard deviations) computed by biomechanical model (TORQUE 60, ANGLE 5)

CONCLUSIONS

The results of this study have shown that there is significant variability in the amount of muscle force exerted by the muscles of the trunk to perform a lift of precise specifications. The result is variable spine reaction forces. It is hoped that by gaining a better understanding of this variability a more complete understanding of the etiology of low back pain/disorders will be obtained.

REFERENCES

Andersson, G. B. J., Pope, M. H. and Frymoyer, J. W. (1984) "Epidemiology" Chapter 4 from **Occupational Low Back Pain**, Praeger, NY.

Marras, W. S. and Mirka, G. A. (1990) "Muscle Activities During Asymmetric Trunk Angular Accelerations", <u>J Orth Res</u>, 8, 824-832.

Marras, W. S. and Mirka, G. A. (1992) "A Comprehensive Evaluation of the Trunk Response to Asymmetric Trunk Motion", <u>Spine</u>, 17, 318-326.

Marras, W. S. and Sommerich, C. M. (1991a) "A Three Dimensional Motion Model of Loads on the Lumbar Spine: I. Model Structure", <u>Human Factors</u>, 33, 123-138.

Marras, W. S. and Sommerich, C. M. (1991b) "A Three Dimensional Motion Model of Loads on the Lumbar Spine: II. Model Validation", <u>Human Factors</u>, 33, 139-150.

McGill, S. M. and Norman, R. W. (1986) "Partitioning of the L4-L5 Dynamic Moment into Disc, Ligamentous, and Muscular Components During Lifting", <u>Spine</u>, 11, 666-678.

McGill, S. M. (1991) "Electromyographic Activity of the Abdominal and Low Back Musculature During the Generation of Isometric and Dynamic Axial Trunk Torque: Implications for Lumbar Mechanics", <u>J Orth Res</u>, 9, 91-103.

McGill, S. M. (1992) "A Myoelectrically Based Dynamic Three-Dimensional Model to Predict Loads on the Lumbar Spine Tissues During Lateral Bending", <u>J Biomech</u>, 25, 395-414.

Reilly, C. H. and Marras, W. S. (1989) "Simulift: A Simulation Model of Human Trunk Motion", <u>Spine</u>, 14, 5-11.

SOME ERGONOMIC CONSIDERATIONS IN THE DESIGN
OF MATERAL HANDLING DEVICES

Marc L. Resnick, Don B. Chaffin
Center for Ergonomics
University of Michigan

Material handling devices (MHD's) are being proliferated in factory workplaces to prevent workers from being injured due to the lifting of heavy loads. These devices require exertions which have not been adequately studied from an ergonomic perspective. Jobs with MHDs often require complex 3-dimensional movements and loaded axial rotation. One type of MHD, an articulated arm, was used to investigate the effects of inertial load, arm joint friction, and positioning accuracy requirements. The kinematic variables of peak push and pull hand forces, velocities, and accelerations were measured or computed in both a task that allowed sagittally symmetric postures as well as one in which loaded axial torso rotation was required. Greater inertial loads increased the peak push and pull hand forces in all cases by an average of 20%. The activation of a 40 psi brake at both joints of the articulated arm increased the peak hand forces by about 40% on average and decreased the peak velocities and accelerations in both tasks by about 20% and 15% respectively. The effects of positioning accuracy required were not as universal. There was a 10% decrease in peak velocity and acceleration for smaller target size in the sagittally symmetric task but no significant effect in the torso twisting task. The study generated some guidelines for the implementation of MHD's, and suggests some areas where further research is required.

INTRODUCTION

The incidence and costs associated with low back pain (LBP) have been increasing to staggering proportions for many years. Snook (1978) compiled data indicating that 79% of all manual materials handling injuries were due to low back injuries. Manual materials handling activities are a major source of worker absence and high costs due to compensation claims (NIOSH 1981), costing 170 to 240 million work days per year at a cost of $4.6 billion (Khalil 1991) The total direct costs of low back pain reached $30 billion in 1985 (Stephens 1991).

In order to eliminate these lifting requirements, manufacturing engineers are implementing material handling devices (MHDs). These devices require operators to exert forces horizontally instead of vertically. The redesigned jobs now often require asymmetric loaded torso axial rotations. These types of movements have not been sufficiently studied to allow informed decisions regarding the implementation of MHDs. Axial rotations cause workers to assume more awkward postures which have already been identified as a risk factor in low back injury. Drury (Drury, Debb, Hartman, Woolley, Drury and Gallagher 1989) cites several disadvantages to asymmetric lifting. Workers are unable to use muscles effectively because the load is concentrated on only a few muscles. Asymmetric lifting adds stress to the vertebral column and discs. Energy costs are also increased with asymmetric lifting.

Woldstad, Langolf and Chaffin (1988) studied the kinematic performance of subjects, including both dynamic postures and inertial effects, by using an industrial hoist to simulate a real factory task. Peak pushing forces exerted ranged from 200 N to 500 N, and peak pulling forces ranged from 150 N to 300 N. Peak accelerations ranged from 0.30 g to 0.10 g and peak decelerations ranged from 0.25 g to 0.07 g.

Resnick, Chaffin and Erig (1991) had subjects push a laden cart to simulate the low friction and high inertia found in MHD jobs. The kinematic results corroborated Woldstad et al (1988). With loads of up to 1000 lbs (220 kg), they found peak horizontal pushing forces of 250N for women and 500N for men. Accelerations ranged from 0.35 g for light loads to 0.05 g for heavier loads. With the heavy loads, velocities were as low as 0.2 m/sec, indicating that these conditions are too difficult for a broad range of prospective workers if speed of motion is required.

Resnick et al (1991) also calculated biomechanical stresses at the subjects' L5/S1 spinal disc. The results showed that loads of 500 lbs (227 kg) and above produced unacceptable compression forces at the low back of 3500 to 5000 N. These results indicate that higher performance and less biomechanical stress occured when the subjects pushed at elbow height.

By determining how MHD design and layout parameters contribute to safe and effective performance of MHD jobs, a set of guidelines to aid in the implementation of the systems in the workplace can be compiled. In the course of determining the performance attributes of each of these parameters, it will be possible to gain a greater understanding of the motor and kinematic abilities of the population.

METHOD

Subjects

Ten subjects, seven male and three female, were selected to perform the experiment. No subjects reported any musculoskeletal disorders and were healthy at the time of testing. Subjects were paid for their participation.

Apparatus

An MHD Simulator designed by Ergomatic systems of Pontiac, MI was used to simulate MHD use. The Simulator is an articulated arm with adjustable load balancing. Encoders on each joint enabled a kinematic model of the arm to be programmed, and the movements of the arm to be recorded on an IBM PC. A load cell was built into the handle to measure hand forces in three directions. The target was an aluminum plate located at the end of a five foot movement zone. The plate had two holes, of diameters 1.5" (3.8 cm) and 2.0" (5.1 cm), establishing two levels of positioning difficulty for the task. The articulated arm was equipped with a 40 psi brake on both joints. This increased the frictional component at each joints such that the forces required to accelerate the arm increased by 25 N.

Experimental Design

For each task, subjects manipulated the arm under 16

conditions corresponding to levels of inertial load, level of motion difficulty, and joint friction (see Table 1). Inertial load and level of positioning difficulty were randomized within blocks of joint friction. Each combination was repeated five times consecutively.

Dependent Variables measured were the motions of each arm joint and hand forces exerted at the force handle in three directions as a function of time. Peak and average MHD handle velocities and accelerations were calculated from the position data.

Procedure

Subjects were informed of the purpose and possible risks involved in the study. They each signed a voluntary consent form. The tasks required were explained to subjects in detail at the beginning of the session. Subjects were instructed to work at a comfortable rate which could be maintained for eight hours.

In the sagittally symmetric trials (see Figure 1a), the subjects were instructed to begin with the arm facing the target. In the torso twisting trials (see Figure 1b), they began with the arm in a target facing 180° from the opposite target. At the beginning of each trial, a tone was sounded, at which time the subject was instructed to proceed. They pushed the mandrel to the target five feet away, and pulled it back to the starting position. This was performed at a rate of once every fifteen seconds, five times for each load/brake/target size combination. After half of the trials, subjects were given an additional fifteen minute rest. A minimum of two days separated the sagittally symmetric trials from the torso rotation trials.

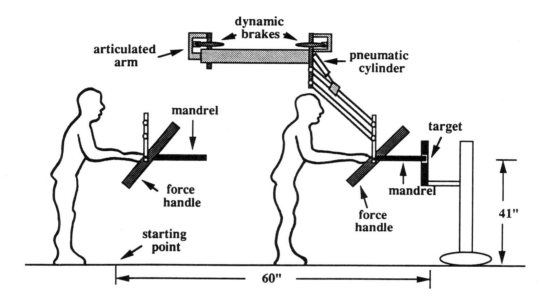

Figure 1a. Experimental Layout for Sagittally Symmetric Study

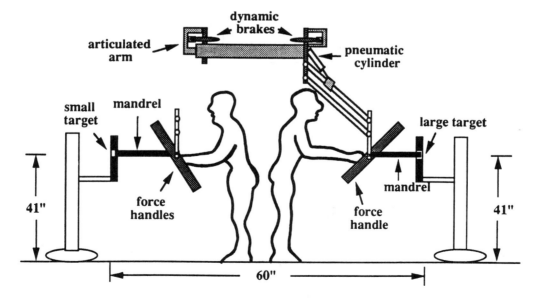

Figure 1b. Experimental Layout for Torso Twisting Study

Table 1. Levels of each Experimental Condition

condition	levels
inertial load	0, 50, 100, 150 (lbs)
diameter of target	1.5, 2.0 (inches)
arm friction of brake	0, 40 (psi)

RESULTS

Tables 2 and 3 present the results of the Analysis of Variance for all independent and dependent measures for the sagittally symmetric and torso twisting tasks. As predicted, greater inertial loads produced higher peak push and pull forces in all cases (see Figure 2). When the brake was activated, subjects exerted larger peak hand forces. In the sagittally symmetric task, the magnitude of the peak push hand forces ranged from 20 N with no brake and no load to 100 N with the brake activated and a 150 lb (68 kg) load. In the torso rotation task, the magnitudes ranged from 20 N to 300 N, 20 to 30 N greater than the sagittally symmetric trials on average.

The magnitudes of the peak pull forces are greater than the peak push forces. In the sagittally symmetric task, they ranged from 40 N with no load and no brake to 160 N with the brake activated and 150 lb (68 kg) load. In the torso rotation task, peak pull forces ranged from 40 N to 220 N in the same conditions, about 10 N greater on average.

Two strategies for determining how much force to exert in each condition were observed. Some subjects pushed with greater force to overcome higher inertial loads and retain a consistent acceleration, while others exerted a consistent force which yielded less acceleration and a longer time to complete the task. This could be a function of the strength of the subject. When subjects reach their maximum force exerting capability, they are unable to increase the forces they exert and therefore must use the latter strategy.

Peak accelerations and decelerations are lower when the brake is activated. With the exception of the heaviest load in the torso twisting trials, acceleration and deceleration appear to decrease with increased inertial load (see Figure 3). The magnitudes range from 0.2 g to 1.3 g for the sagittally symmetric task and 0.2 g to 1.6 g for the torso twisting task for both peak acceleration and peak deceleration. Averaged across brake and target size levels, the variations were much lower, ranging from 0.7 g to 0.8 g in the sagittally symmetric trials and 0.85 g to 1.00 g in the torso twisting trials. The accelerations and decelerations calculated for each load were remarkably consistent, with the decelerations 0.2 g to 0.3 g less in all cases.

The inclusion of the brake decreased peak velocity by about 20% on average. Though not statistically significant, the greater inertial loads appear to yield lower peak velocities (see Figure 4). This factor is not significant because of the large variances found with several of the subjects. Subjects were 5-10% faster when using the large target compared to the small target. The magnitudes of the peak velocities range from 0.6 m/sec to 1.8 m/sec in the sagittally symmetric task and 0.8 m/sec to 2.0 m/sec in the torso rotation task. Averaged across brake and target size levels, the variations were much lower, ranging from 1.15 to 1.30 m/sec in the sagittally symmetric trials and 1.30 to 1.45 m/sec in the torso twisting trials.

Table 2. P-values from the Analysis of Variance for all independent
measures and all significant interactions in the Sagittally Symmetric Task

factor \ measure	Borg CR-10 rating	peak push force	peak pull force	peak velocity	peak acceleration	peak deceleration
inertial load	0.12	0.94	0.75	0.30	0.78	0.70
brake	0.03+	0.01+	0.06	0.02+	0.03+	0.03+
target size	0.03+	0.86	0.83	0.01+	0.03+	0.06
subject	0.02+	0.03+	0.43	0.29	0.04+	0.03+
S*B	0.01+	0.00+	0.01+			

+ significant at the p<.05 level

Table 3. Results of the Analysis of Variance for all independent
measures and all significant interactions in the Torso Twisting Task

factor \ measure	Borg CR-10 rating	peak push force	peak pull force	peak velocity	peak acceleration	peak deceleration
inertial load	0.12	0.01+	0.01+	0.53	0.96	0.84
brake	0.02+	0.01+	0.02+	0.01+	0.01+	0.06
target size	0.11	0.60	0.51	0.23	0.31	0.47
subject	0.06	0.02+	0.02+	0.01+	0.01+	0.01+
S*B	0.02+	0.02+	0.04+			
B*L		0.01+	0.03+			

+ significant at the p<.05 level

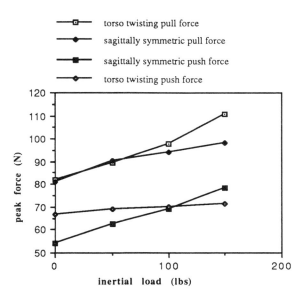

Figure 2. Peak push and pull forces for the sagittally symmetric and torso twisting trials

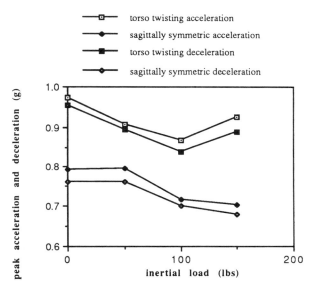

Figure 3. Peak acceleration and deceleration for the sagittally symmetric and torso twisting trials

DISCUSSION

The manipulation of several task parameters had significant effects on the kinematic performance of the subjects on the articulated arm used in this pilot study. The peak push and pull forces of around 100 N measured for most subjects are lower than those found in Woldstad et al (1988) with an industrial hoist and Resnick et al (1991) with a cart. In both of these studies, peak forces of 300 to 500 N were measured. However accounting for the greater loads used in both Woldstad et al (1988) and Resnick et al (1991) eliminates these differences. Furthermore, in both previous studies, peak forces were measured while subjects worked at peak performance, whereas in the present study, the subjects were instructed to work at a comfortable rate which could be maintained for eight hours.

The peak velocities observed in this study range from 0.6 to 1.9 m/sec. These are slightly higher than the 0.6 to 1.2 m/sec reported by Resnick et al (1991) for similar loads and postures. As the arm moves freely in all directions, it is not surprising to find higher velocities. Forces in the lateral and transverse directions will contribute to the velocity whereas with the cart of Resnick et al (1991), they were restricted. The velocities reported in Woldstad et al (1988) range from 1.0 to 1.5 m/sec. There is less variability in these numbers due to less motion variation in the conditions of the study (ie no brake and no arm configuration). The similarity between the findings of all three studies, however, adds to the conclusion that these paradigms are approached similarly by the subjects. The lower peak velocities measured with

the brake activated indicate that performance times will be increased when there is more friction in the joints of an articulated arm.

The trends of the peak push forces measured in the torso rotation task are similar to those obtained in the sagittally symmetric trials, though on average about 20 to 30 N greater. In the sagittally symmetric task, there was more of a trend towards increasing peak forces for higher inertial loads. This is due to smaller peaks for lower inertial loads than in the torso twisting trials. The magnitudes of the peak pull forces also exhibit an increasing trend for higher inertial loads in only the sagittally symmetric trials, but have the same magnitudes in both sets of trials. The trends of the velocities, accelerations and decelerations are similar to the sagittally symmetric trials, though the accelerations are on average 0.3 g larger.

The performance characteristics of the subjects suggest several issues to be addressed in the design and selection of MHD's, specifically articulated arms, for industrial jobs. The magnitudes of the forces exerted in the sagittal and lateral directions were largely dependent on the geometry of the arm joints. When the arm links are normal to the hand force vector, the moment arm around the joint is greatest, and smaller amounts of force are required to move the arm. Likewise, when the arm links are more parallel with the force vectors, it is difficult to move the arm, especially when the brake is set. When designing the layout of an MHD job in the factory, it is important to determine where the peak forces will be exerted, and orient the MHD in such a manner as to cause the arms to be perpendicular at these points.

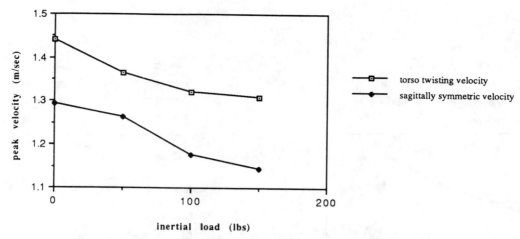

Figure 4.   Peak velocity for sagittally symmetric and torso twisting trials

Activation of the 40 psi brake on the joints of the arm caused a 50-100% increase in the amount of peak force subjects exerted in pushing and pulling the arm, both for the sagittally symmetric trials, and the torso twisting trials. A braking system should only be included if it can be automated to work when the operator is decelerating, or the brake could be put under operator control.

There are several implications of this study for the design of MHD jobs. When axial rotation is required, inertial loads should be minimized. Positioning accuracy requirements should be as low as possible to improve both the speed and accuracy at which the job will be performed. Any braking system must not increase the forces required for positioning or lateral movement. The geometry of the links of an articulated arm should be configured to give the least resistance when the most force is required.

ACKNOWLEDGEMENT

This research was sponsored in part by Ford Motor Company, Contract NP 8800-55169. The authors would like to express their gratitude to Brad Joseph at Ford and Jim Foulke and Ulrich Raschke at the Center for Ergonomics for their instrumental assistance.

REFERENCES

Drury C. G., Deeb J. M., Hartman B., Woolley S., Drury C. E.(1989), and Gallagher S. Symmetric and Asymmetric Manual Materials Handling Part 1: Physiology and Psychophysics. *Ergonomics*, 32(5), 467-489.

Khalil T. M. (1991), Ergonomic Issues in Low Back Pain: Origin and Magnitude of the Problem.*Proceedings of the Human Factors Society National Conference*, San Francisco, 820-823.

National Institute for Occupational Safety and Health, (1981),"Work Practices Guide for Manual Lifting" Technical Report 81-122, NIOSH, Cincinatti.

Resnick M. L., Chaffin D. B., and Erig M. (1991), Biomechanical Analysis of Dynamic Horizontal Movement Strategies in the Sagittal Plane. *Proceedings of the Human Factors Society National Conference*,San Francisco, 785-789.

Snook S. H. (1978), The Design of Manual Handling Tasks. *Ergonomics*, 21(12), 963-985

Stephens R. (1991), Plenary Address. *Human Factors National Conference*, San Francisco, 820-823.

Woldstad J. C., Langolf G. D., and Chaffin D. B. A Psychomotor and Biomechanical Analysis of the Use of Small Material Handling Assist Devices, Submitted to IIE Transaction 11/91.

Effects of gender, lift height, direction, and load on the ability to estimate weight

Valerie J. Rice, Marilyn A. Sharp, Tania L. Williamson, and Bradley C. Nindl

U.S. Army Research Institute of Environmental Medicine
Natick, MA 01760-5007

The study evaluated the effects of gender, lift height, direction (lift/lower), and load on the ability to correctly estimate weight handled. Seven women and six men lifted and lowered boxes to and from knuckle, waist, and shoulder heights. Subjects were asked to estimate weights corresponding to 50, 40, 30, and 20% of gender specific lifting strength to 152 cm. The difference between the actual and estimated weight (DIFF) was 100% greater for men than for women ($F = 6.27$, $p = 0.03$). When the percent difference was analyzed, there was no significant gender effect. The least accurate estimates occurred when lowering a weight from knuckle height ($p < 0.05$). The majority of subjects underestimated the weight and men underestimated more frequently than women ($Chi^2 = 12.57$, $p = 0.0004$). Subjects overestimated the weight more often at higher weights. The results suggest that both men and women tend to underestimate weight, especially when lowering from knuckle height, possibly putting them at risk for injury.

INTRODUCTION

The total compensable cost for industrial low back pain in the United States has been estimated to be $11.1 billion, with a mean cost per case of $6,807 (Webster and Snook, 1990). Indirect costs also exist, such as lost work time, training of new employees, administrative costs, personal suffering, and low morale. Methods of injury prevention include ergonomic job or worksite evaluation, job redesign, training, and use of predetermined lifting limits. Safe lifting limits have been evaluated by biomechanical estimation of stress on the lumbar discs (Anderson, 1982), psychophysical measurement techniques, such as maximum acceptable weight limits (MAWL) (Snook, 1978; Snook and Irvine, 1967; Snook and Ciriello, 1974; Snook, Irvine, and Bass, 1970), or having subjects estimate the stress level they feel on their backs (Waikar, Lee, Aghazadeh, and Parks, 1991).

When using psychophysical measurement techniques, subjects overestimate the amount of weight which can be lifted during an eight hour shift (Mital, 1983). When asked to lift their selected weight for eight hours, men reduced their load to about 65% of their MAWL over the course of the day, while women reduced their load to 85%.

Karwowski (1988, 1991) conducted a series of studies to evaluate human perception of load heaviness. Subjects were required to classify weights into categories of very light, light, less-than-medium, medium, more-than-medium, heavy, and very heavy. Subsequently, subjects selected acceptable loads that they would be willing to lift over an eight hour shift. He found that men selected loads which they had independently judged to be very heavy or heavy. Women selected weights they had judged to be more-than-medium to heavy. Karwowski concluded that women are better able to select a safe lifting load than are men, when using a psychophysical methodology (Karwowski, 1991).

Mital and Kromodihardjo (1986, as reported by Waikar, et al., 1991) found a high correlation between the psychophysical determination of MAWL and biomechanically estimated compressive forces. In contrast to this, Waikar et al. (1991) found that tasks determined to be biomechanically stressful were not considered, by subjective evaluation, to be stressful to the lower back.

Thus, it appears that workers inadequately judge the lifting work they can perform over the course of a day and the work causing the most stress on their backs. It has also been suggested that men tend to underestimate weights, while women tend to overestimate weights (Sloane, 1991).

In an epidemiological study conducted by the National Institute for Occupational Safety and Health, injuries occured most frequently when the object was initially located on the floor (52%), at waist height (17%), or at knee height (13%)(NIOSH, 1982). Additionally, thirty six percent of the injured workers felt they were lifting objects which were 'too heavy' and 14% felt they had underestimated the weight of the object before lifting (NIOSH, 1982). Little controlled research has been conducted to determine the effect of height and direction (lift/lower) on weight estimation.

The main objective of this study was to examine the effects of gender, lift height, direction of lift, and load on weight estimation. An additional objective was to examine the relationship between the ability to estimate weight and the subjects' body weight, fat free mass, and strength.

METHOD

Seven female and six male volunteers participated in this study. Subjects did not lift weights on a regular basis. Physical characteristics are included in Table 1. Percent body fat was measured by underwater weighing. Lifting strength was assessed using an incremental dynamic lift machine (IDL), a deadlift, and a 38 cm upright pull (Teves, Wright, and Vogel, 1985).

A covered metal box (47 x 23 x 31 cm in width, height, and depth, respectively) with side handles was used for all lifts. The box was lifted/lowered (direction), to/from three heights (knuckle, waist, and shoulder). Four weights were used,

Table 1. Physical Characteristics of Subjects

	Mean	SD	Range
Height (cm)			
Males	178.0	7.1	17.4 - 190.5
Females	161.9	4.4	154.0 - 168.1
Weight (kg)			
Males	74.8	13.8	60.6 - 100.4
Females	58.0	9.3	49.9 - 76.6
Age (yrs)			
Males	20.3	1.8	19.0 - 23.0
Females	27.1	6.7	18.0 - 34.0
Percent Body Fat			
Males	16.2	6.2	10.2 - 19.6
Females	24.6	6.5	16.3 - 34.7
Fat free Mass (kg)			
Males	61.8	7.9	53.5 - 60.8
Females	46.3	7.1	40.1 - 52.5
IDL			
Males	73.5	14.5	54.5 - 95.4
Females	37.7	5.1	31.8 - 45.4
Deadlift			
Males	132.1	24.0	105.0 - 175.0
Females	80.0	10.4	60.0 - 95.4
38 cm upright pull			
Males	140.4	30.4	101.4 - 178.4
Females	76.0	11.9	62.5 - 91.4

approximately 50, 40, 30, and 20% of the mean incremental dynamic lift strength mean for male and female subjects (Teves, Wright, and Vogel, 1985). The box loads were varied (± 2.5 lbs) to provide a range of weights, thus preventing subjects from identifying and responding to four specific loads. Lift heights were adjusted specifically for each subject, using a pneumatically driven lifting/lowering device (Teves, McGrath, Knapik, and Legg, 1986). Subjects were trained in lifting procedures and proper posture was encouraged for all lifts. The order of lifts was randomized. The dependent measure used was the difference between actual and estimated weights (lbs) (DIFF).

Analysis of variance (ANOVA) procedures were performed on dependent variables. The Duncan Multiple-Range test was used to compare means. A Chi-square was used to determine whether under- or over-estimation occurred at greater than random frequency, and whether gender affected the frequency of under- or over-estimation. Pearson Product-moment correlations were used to examine the association between dependent and descriptive variables.

RESULTS

Women's estimates were closer to the actual weight (10.2 lbs) than were mens (20.0 lbs) (F = 6.27, p = 0.03). When the percentage of actual to estimated weight was examined, there was no significant difference between men (51.6%) and women (59.4%) (Figure 1).

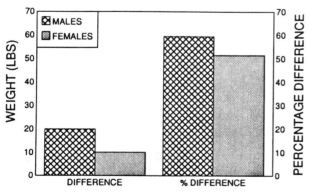

Figure 1. DIFF and % DIFF for men and women.

In both male and female subject's, estimates were less accurate at lower weights than at the highest weight (p < 0.05, Figure 2). Estimates were also less accurate at knuckle height than at waist or shoulder height (p < 0.05, Figure 3). Subject's estimates were least accurate when lowering a weight from knuckle height than for any other condition (p < 0.05, Figure 4).

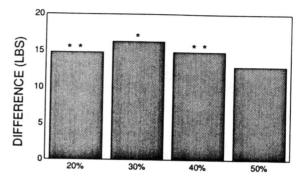

* SIGNIFICANTLY GREATER THAN 50% (p < 0.01)
** SIGNIFICANTLY GREATER THAN 50% (p < 0.05)

Figure 2. DIFF by weight lifted.

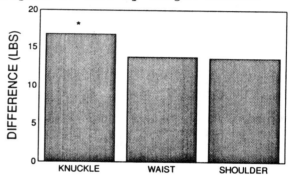

* SIGNIFICANTLY GREATER THAN OTHER CONDITIONS (p < 0.05)

Figure 3. DIFF by lift height.

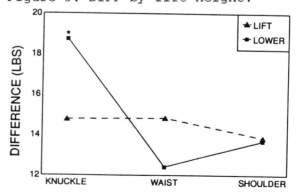

* SIGNIFICANTLY DIFFERENT FROM LIFT (p < 0.05)

Figure 4. DIFF by direction and height.

A significant gender x weight interaction was found (F = 5.61, p = 0.03). The most accurate estimates were achieved by women lifting at the highest weight (50% max, p < 0.01) and the least accurate estimates were achieved by men at the lowest weight (20% max, p < 0.01).

Both men and women tended to underestimate the load and men underestimated more frequently (96% of the time) than did women (86% of the time)(Chi^2 = 12.57, p = 0.0004). The limited occurrences of overestimation were more frequent at the highest weight (15% of the time) than at the two lower weights combined (8% of the time and 4% of the time, respectively) (Chi^2 = 5.89, p = 0.02).

The difference between actual and estimated weight was more highly correlated with body weight (r = .81, p < 0.001), than with fat free mass (r= 0.66, p = 0.02). Positive correlations were found between the difference between actual and estimated weight and strength measurements measured by a 38 cm upright pull (r = 0.68, p = 0.01) incremental dynamic lift (r = 0.72, p = 0.006), and dead lift (r = 0.71, p = 0.007). These results indicate that lifting strength is inversely associated with the ability to accurately estimate weight. The scatter plots showed the heavier, stronger males were the least accurate in their estimations.

DISCUSSION

Discrimination between actual and estimated weight was least accurate when lifting to or lowering from knuckle height and at the lowest weights. The least accurate combination was lowering a box from knuckle height to the floor, which required the least amount of effort. It may be that subjects perceive the weight as less when they are required to expend the least effort.

Both men and women appear to be inaccurate judges of weight, and men underestimated weights more frequently than women. The most accurate estimates were obtained by women at the highest weight. The fact that subjects who weighed more and were stronger were less accurate in their estimations also appears to be the result of gender differences. Although a different procedure was used, these findings are consistent with Karwowski's results in which women more accurately selected their own lifting capacity for an eight hour day (Karwowski, 1991).

As noted earlier, workers injured during lifting reported they had underestimated the weight of the object lifted (NIOSH, 1982). Although the greatest discrepancies were found at the lowest weights, mis-perceptions of weights might lead individual to attempt to lift more than they should, thereby putting themselves at greater risk for injury.

REFERENCES

Andersson, G.B.J. (1982). Measurements of load on the lumbar spine. Paper presented at the Symposium on Idiopathic Low Back Pain, Mosby, St. Louis, pp 220-251.

Cross, D.V., and Rotkin, L. (1975). The relation between size and apparent heaviness. Perception and Psychophysics, 18(2), 79-87.

Karwowski, W. (1991). Psychophysical acceptability and perception of load heaviness by females. Ergonomics, 34(4), 487-496.

Karwowski, W. (1988). Human perception of load heaviness. In K.H.E. Kroemer, J.D. McGlothlin, and T.G. Bobick (Eds.), Manual Material Handling: Understanding and Preventing Back Trauma (pp. 9-14). Akron, OH: American Industrial Hygiene Association.

Jones, L. (1986). Perception of force and weight: Theory and Research. Psychological Bulletin, 100(1), 29-42.

Mital, A. (1983). The psychophysical approach to manual lifting: A verification study. Human Factors, 25, 485-491.

Mital, A., and Kromodihardjo, S. (1986). Kinetic analysis of manual lifting activities: Part II-Biomechanical analysis of task variables. International Journal of Industrial Ergonomics, 1, 91-101.

NIOSH (1982). Back injuries associated with lifting (Bulletin 2144). Washington, D.C.: U.S. Government Printing Office.

Sloane, P. (1991, June). Principles of Ergonomics Course. Presented at the U.S. Army Environmental Hygiene Agency, Aberdeen, Maryland.

Snook, S.H. (1978). The design of manual handling tasks. Ergonomics, 21, 963-985.

Snook, S.H. and Irvine, C.H. (1967). Maximum acceptable weight of lift. American Industrial Hygiene Association Journal, 28, 322-329.

Snook, S.H. and Ciriello, V.M. (1974). Maximum weights and workloads acceptable to female workers. Journal of Occupational Medicine, 16, 527-534.

Snook, S.H., Irvine, C.H., and Bass, S.F. (1970). Maximum weights and workloads acceptable to male industrial workers. American Industrial Hygiene Association Journal, 31, 579-586.

Teves, M.A., McGrath, J.M., Knapik, J.J., and Legg, S.J. (1986). An ergometer for maximal effort repetitive lifting. In Proceedings of the Eighth Annual Conference of the IEEE/Engineering in Medicine and Biology Society (pp. 592-593).

Teves, M. A., Wright, J. E., and Vogel, J. A. (1985). Performance on selected candidate screening test procedures before and after army basic and advanced individual training. (Tech. Report T13/85). U.S. Army Research Institute of Environmental Medicine, MA: Exercise Physiology Division.

Waikar, A., Lee, K. Aghazadeh F., and Parks, C. (1991). Evaluating lifing tasks using subjective and biomechanical estimates of stress at the lower back. Ergonomics, 34 (1), 33-47.

DETERMINATION OF THE MAXIMUM ACCEPTABLE WEIGHTS OF LIFTING AND LOWERING USING THE DIRECT ESTIMATION METHOD

F. Aghazadeh, H. Lu
Department of Industrial and Manufacturing Systems Engineering
Louisiana State University, Baton Rouge, LA 70893

S.J. Morrisey
Oregon, OSHA, Portland, OR

The maximum acceptable weights of lifting and lowering were determined using the direct estimation method. Five task frequencies (1, 2, 4, 6, and 8 handlings per minute) were studied. Both lifting and lowering were fixed at knuckle-shoulder height. A base load for each individual subject was first established using the psychophysical approach. The direct estimation method was then applied. The results revealed that task frequency exerted a significant role in the variation of the data. The type of tasks (lifting and lowering) also significantly affected the maximum acceptable capacity performing the tasks. The study shows that the direct estimation method is a faster and feasible way to determine the maximum acceptable weight that an individual can handle.

1. INTRODUCTION

Improving human performance requires that workloads be matched with worker capacities. That is, given a reasonable work schedule which includes rest periods the worker should be able to perform his tasks without undue strain. This matching process depends on load capacity measurements on individuals over a prescribed set of conditions. The objective is to determine the load or task which does not exceed the maximum capacity of the individual. When the task is lifting objects, the measurement will center on finding the maximum acceptable weight of lift (MAWOL). This can be carried out by two methods of measurement: (1) the psychophysical method, and (2) the direct estimation method. In the former, subjects are required to lift loads for twenty to forty minutes during which they adjust the loads until they chose the maximum load without undue strain. Subjects are informed in advance that the test load is assumed to be the working load for a number of hours per day of work. In the latter, subjects perform a standard task and a base load is determined for each subject by the psychophysical method. A numeric ratio, usually 100%, is assigned to the strain of performing the standard task. Subject then perform other new task using the same load and rate the strain as being more or less stressful compared with the standard task. The designer can then predict or directly estimate new loads based on the subjects' rating for different tasks. The direct estimation method is faster compared to the psychophysical method. It has been proven to be as reliable as the psychophysical method for symmetrical and asymmetrical lifting tasks (Chen, et. al., 1992), but it has not been applied to other manual material handling tasks.

The objective of this study was to apply the direct estimation method to both lifting and lowering tasks at different task frequencies to predict the maximum acceptable handling capacity and to examine the relationship between lifting and lowering.

2. METHODOLOGY

Dependent and Independent variables

The dependent variable is the maximum weight of load subjects can handle for different manual handling tasks. The independent variables are: (1) type of tasks (lifting and lowering), and (2) task frequency (1, 2, 4, 6, and 8 handlings per minute). The height for lifting and lowering tasks was fixed at knuckle-shoulder height.

Subjects

Eleven male graduate students participated this study. Their age ranged from 22-43 and the average is 26.4. All subjects were free from back pain and musculoskeletal abnormalities.

Apparatus

A wooden box with a size of (45.8cm) * (30.5cm) * (20.3cm) (Length*Width*Height) was used for the experiment. Various steel weights were used to increase or decrease the weight of the box. A wooden platform with adjustable pinhole settings was set for the height of lifting/lowering. It was set at 75 cm for the knuckle height and 136 for the acromial height for knuckle-shoulder lifting and lowering.

Experimental design

Two types of task (lifting and lowering) and five task frequencies (1, 2, 4, 6, and 8 handling per minute) were examined in this study. A randomized complete block design with 2*5 factorial treatment combinations was utilized. Each subject was considered as a block. The total treatments assigned to each subject were ten. The trials were randomized to avoid the effects of interaction.

Experimental Procedure

The subjects were briefed on the purpose of the experiment and the procedure in general. Then individual task routines were explained in detail. A one-page instruction of psychophysical approach for subjects (Jiang, 1986) served as a guideline for subjects participating in the experiment. The requirement of the ratio estimation was also explained in detail.

The experiment was conducted in two phases: (1) determination of base load, and (2) ratio estimation for other designed handling tasks. In Phase I, subjects performed a standard task and the base loads were determined by using the psychophysical method. Each subject lifted a load from knuckle to shoulder height at a frequency of 4 lifts per minutes for 20 minutes. During the 20-minute session, the subject adjusted the load until he felt that he could handle it with maximum effort without undue strain for eight hours on a normal working day. In Phase II, subjects performed other prescribed handling tasks and rated the stain compared to the standard load. The strain for handling base load determined in the first phase was instructed to rate as 100 percent. For example, if a subject felt 25 percent more stressful when performing 8 lifts per minute compared to the base load (4 lifts per minute), he would rate the handling effort (or strain) as 125 percent. On the other hand, if the subject felt 25 percent less stressful when performing a lowering task, he would rate the handling effort as 75 percent.

After collecting the data, the ratio estimations were converted to weight based on the base load according to the following formula:

Maximum acceptable weight(Kg)

$$= \text{BASE(kg)} * [\, 1 - \frac{\text{RATING} - 100}{100} \,]$$

where BASE is the base load, and RATING is the task rating.
The data was then analyzed using SAS package (Statistical Analysis Software) to examine possible effect of task and frequency.

3. RESULTS

Table 1 shows the means and standard deviation of task rating as percentage of the standard task for various handling tasks. Note that a 100 percentage rating is at 4 lifts per minute since this is the standard task. Figure 1 shows the result of converting the ratio estimations to load weights.

Analysis of variance shows significant effect of task frequency (p=0.0001) and type of tasks (lifting and lowering) (p=0.001) on the ratio estimation of the handling effort and therefore the maximum acceptable weight of load for lifting and lowering tasks. No interaction of type of task and task frequency was found.

Table 1. Means (standard deviations) of the task rating using direct estimation method (%)

| | Task Frequency (lifts/min or lowering/min) | | | | |
	1	2	4	6	8
Lifting	70.46 (12.14)	79.55 (7.57)	100.00 (0.00)	120.46 (9.34)	136.36 (17.19)
Lowering	61.36 (14.68)	73.64 (13.25)	90.00 (10.25)	115.91 (11.14)	128.18 (11.89)

When examining the effect of task frequency, it is shown by stepwise comparison that the task ratings are significantly different from each other. Further polynomial contrast shows a linear relationship among the task ratings and the converted maximum acceptable weight of handling for different task frequencies. For lifting tasks, as the frequency deviates from 4 to 2, from 4 to 1, from 4 to 6, and from 4 to 8 lifts per minute the maximum acceptable weight of lift changes +21%, +29%, -20%, and 34%, respectively. For lowering tasks, as the frequency deviates from 4 to 2, from 4 to 1, from 4 to 6, and from 4 to 8 lowering per minute, the maximum acceptable weight of lowering changes +15%, +26%, -22%, and 34%, respectively.

When further examining the effect of type of tasks (lifting and lowering), it was found that overall the frequencies, the maximum acceptable weight of lowering is 7% more than that of lifting.

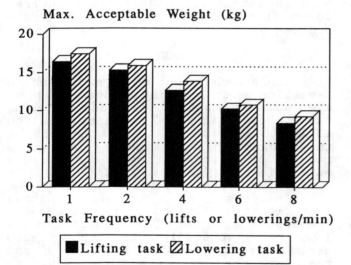

Figure 1. Maximum acceptable weight of load using the direct estimation method

4. DISCUSSION

As task changes from lifting to lowering, the average maximum acceptable weight increases by 7%. This result is similar to the data of the revised tables of maximum acceptable weight of lifting and lowering for 90% percent population, with vertical distance 51 cm and box width 34 cm, which shows 7% increase from lifting to lowering for lifting frequency 4.29 lifts and lowering per minute (Snook and Ciriello, 1991). This indicates that the direct estimation method is feasible to apply to other type of manual handling tasks using one standard task (lifting task, 4 lifts per minute).

Comparing to the revised tables of maximum acceptable weight for lifting and lowering developed by Snook and Ciriello (1991), it is found that the percentage changes of the maximum acceptable weight in this study is larger. For example, when the frequency changes from 4.29 lifts/min to 1 lift/min, the percentage change of the maximum acceptable weight of lift is 21% in the revised tables by Snook and Ciriello (1991); this study shows 29% change when the frequency changes from 4 lifts/min to 1 lift/min. This result may indicates that subjects over-estimated the maximum acceptable weight when the frequency increases and under-estimated when the frequency decreases using the direct estimation method.

The acceptable weight values in this study were lower than those reported in literature (e.g. Snook and Ciriello, 1991). Two factors may account for this discrepancy: (1) student subjects has significantly lower capability than industrial workers; and (2) vertical distance of lifting/lowering (61 cm) is longer than that in the literature (51 cm) (e.g. Snook and Ciriello, 1991).

It was observed that when the task frequency was low (1 and 2 lifts or lowering per minute), subjects encountered difficulty in making judgement for task ratings. Subjects had to spend more time on re-familiarizing the standard task before performing the task rating for the new task. The same observation is also noted by Chen et al. (1992).

The two-phase study of determining maximum acceptable weight of load of lifting and lowering for eleven subjects using direct estimation method took about 60 minutes for each subject. It is faster than psychophysical method which would take 200 minutes for the ten treatments for each subject in this study.

In summary, the direct estimation method is a faster and feasible way to predict the maximum acceptable weight of handling for lifting and lowering tasks. The experiment needs to be well controlled and subjects should fully understand the method in order to obtain accurate result. The relationship between the accuracy of the method and frequency of manual handling tasks needs further study.

5. REFERENCES

Chen, F., Aghazadeh F. and Lee, K.S., 1992. Utilization of direct estimation method to predict the maximum acceptable weight of symmetrical and asymmetrical lift. Proceedings of the human factors society 36th annual meeting. pp 679-683.

Jiang, Bernard C., 1986, Psychophysical modeling for combined manual materials-handling activities. Ergonomics, 29(10): 1173-1190.

Snook, Stover H., and Ciriello, Vincent M., 1991, The design of manual handling tasks: revised tables of maximum acceptable weights and forces. Ergonomics, 34(9): 1197-1213.

A CATASTROPHE THEORY-BASED MODEL FOR QUANTIFICATION OF RISK OF LOW BACK DISORDERS AT WORK

Waldemar Karwowski
Center for Industrial Ergonomics
Department of Industrial Engineering
University of Louisville
Louisville, KY 40292, USA

and William S. Marras
Biodynamics Laboratory
Department of Industrial and Systems Engineering
Ohio State University
Columbus, OH 43210, USA

This paper discusses applications of the catastrophe theory in the dynamic modeling of occupational low back disorders, and offers a framework for conceptualization of such disorders in view of the elementary cusp catastrophe models. It was proposed that low back disorders due to manual lifting should be considered as a discontinuous phenomenon, reflecting dynamic changes in the state of human musculoskeletal system, which are dependent upon the combination of human strength abilities, muscular fatigue and endurance, spinal loading tolerance, as well as dynamic equilibrium between these variables. The behavior of the proposed cusp-catastrophe based model for the risk of LBDs due to manual lifting jobs was examined based on empirical data collected in industry.

INTRODUCTION

Low back disorders (LBDs) at work are recognized as one of the main occupational health problems that adversely affects the quality of life in the United States. As pointed out by the National Safety Council, in 1988 the overexertion injuries across all industries accounted for 28.2% of all work injuries involving disability. Despite many years of research efforts, the occupational exposure limits of different risk factors for development of LBDs have not yet been established. One of the problems in setting such guidelines is the limited understanding of how different risk factors of LBDs interact in causing the injury. Furthermore, even the nature and mechanism of LBDs are still relatively unknown phenomena.

Current ergonomic techniques for controlling the risk of occupationally-related low back disorders (LBDs) consist mainly of static assessments of spine loading during lifting activities. However, as discussed by Marras (1992), biomechanical models and epidemiologic studies suggest that the dynamic characteristics of lifting increase spine loading and the risk of occupational LBD.

OBJECTIVES

As discussed by Karwowski, et al. (1991), the underlying assumption of a system's behavior continuity in many processes related to human factors research cannot be supported when considering events of non-continuous nature. This is the case of low back syndrome and traumatic low back injury, which may occur quite suddenly and in a non-linear fashion (Karwowski et al., 1992). In general case, the nature of such changes may depend upon the combination of human strength abilities, muscular fatigue and endurance, spinal loading tolerance, as well as dynamic changes in the state of equilibrium between these variables.

The objective of this study was to propose a framework for a mathematical model, where a set of occupational risk factors for development of LBDs can be considered interactively, through a dynamic process which illustrates the nature of a transition from one state of the equilibrium in the human musculoskeletal system to another. In general case, these changes in the system depend upon combination of occupational LBDs risk factors, such as motion repetition, the extent of force exertion, deviations from neutral trunk or some other body part posture, rest requirements, fatigue patterns, and discal endurance characteristics, as well as the dynamic changes in a state of equilibrium between these variables.

According to Karwowski et al. (1992), at a certain stage of the musculoskeletal system's loading, the already strained elements of the considered subsystem can no longer tolerate any further stresses, and the system rapidly changes its state. This phenomena occurs even though the changes in system properties leading to the point of *system instability* are gradual, and they cumulatively affect the system tolerability over time. This sudden change will lead to the new state of *system equilibrium*, most likely accompanied by the LBDs-related system damage.

In view of the above, the mechanism of LBDs can be modeled as a dynamic but discontinuous process, reflecting the dynamic changes in the state of human muscular, skeletal and neural subsystems. Furthermore, the risk of LBDs can be modeled as the effect of sudden changes (discontinuity or elementary catastrophe) in the state of equilibrium between the LBDs risk factors and the specific musculoskeletal subsystem's tolerability.

THE CATASTROPHE THEORY

The process of discontinuity

The catastrophe theory (Thom, 1975), allows to model

sudden changes in the state of the system which are of nonlinear and discontinuous nature. The elementary models of changes in a system behavior describe specific dynamic processes and relationships among several system variables. According to Zeeman (1977), the catastrophe theory provides a level of insight and understanding in very complex systems which were not previously thought possible, especially when applied in biology and medicine, behavioral and social sciences, psychology, or engineering.

The catastrophe theory is primarily useful in modeling phenomena where gradually changing forces produce sudden effect(s), which can be mathematically described using few archetypal geometrical forms, i.e. elementary catastrophes. The elementary catastrophes (Thom, 1975; Zeeman, 1977) are mathematically defined as real-valued polynomial functions, where k denotes dimension of the state space or behavior space, while q denotes dimension of the control space, i.e. the space of external parameters. The critical equilibrium set or the catastrophe manifold can be found by equating the appropriate differential of function (F) to zero.

METHODS AND PROCEDURES

Experimental basis for model definition

The experimental data used for development of the cusp catastrophe-based model was collected earlier by Marras (1992). That study involved an industrial surveillance of the trunk motions and workplace factors involved in high and low risk for LBDs repetitive tasks. A system called the lumbar motion monitor (LMM) was used for the purpose of documenting the three-dimensional components of trunk motion in the work environment.

The workplace-related and individual characteristics observed by Marras (1992) consisted of the following variables: 1) the maximum horizontal distance of the load from the spine; 2) the weight of the object lifted; 3) the height of the load at the origin of the lift; 4) the height of the load at the destination of the lift; 5) the frequency of lifting (lift rate); 6) the asymmetric angle of the lift; 7) worker anthropometry and 8) worker injury history.

Trunk motion variables consisted of the trunk angular position, velocity, and acceleration characteristics in each of the cardinal planes. Each job was weighted proportionally to the number of person-hours from which the injury and turnover rates were derived. The odds ratio for LBDs was defined as the ratio of the probability that a LBD occurs (probability of being in the high risk LBD group) to the probability that LBD does not occur (probability of being in the low risk LBD group).

The data from 403 industrial jobs from 48 manufacturing companies was used. These jobs were divided into two groups, high and low risk of LBD, based upon examination of the injury and medical records. Low risk group jobs were defined as those jobs with at least three years of records showing no injuries and no turnover. High risk group jobs were those jobs associated with at least 12 injuries per 200,000 hours of exposure. The high risk group category incidence rate corresponded to the 75th percentile value of the 403 jobs examined.

The cusp catastrophe model

The canonical model of the cusp catastrophe (Zeeman, 1977) can be described after as follows. Let $\mathbf{M}$ be the cubic surface in three-dimensional space (R^3), given by:

$$-x^3 + bx + a = 0 \qquad (1)$$

Parameters $\mathbf{a}$ and $\mathbf{b}$ are two control factors, which form the control (horizontal) space $\mathbf{C}$; while $\mathbf{X}$ denotes the behavior space with vertical coordinate x. The *fold curve* $\mathbf{F}$ is where vertical lines are tangent to $\mathbf{M}$, and is given by the differential of $\mathbf{M}$ with respect to x: $3x^2 = b$. The projection of the fold curve $\mathbf{F}$ into the control space $\mathbf{C}$ is called the *bifurcation set* $\mathbf{B}$, which has a cusp at its origin. The bifurcation set $\mathbf{B}$ is the two-dimensional projection of the cusp manifold, and can be found by solving for $\mathbf{a}$ or $\mathbf{b}$, as follows: $27a^2 = 4b^3$. The fold curve $\mathbf{F}$ divides the surface $\mathbf{M}$ into two parts, i.e.: 1) the upper space outside the cusp given by $3x^2 > b$ and 2) the lower surface inside the cusp, defined by $3x^2 < b$. In the control space $\mathbf{C}$, parameter $\mathbf{b}$ is defined as the *splitting factor*, while $\mathbf{a}$ is the *normal factor*.

The butterfly catastrophe model

In order to allow for more than two control factors (risk parameters) in modeling risk of LBDs, the higher order elementary catastrophe models can be used. For example, the butterfly catastrophe model, which allows to include up to four control factors, can be mathematically represented by the following polynomial:

$$F = \pm\, 1/6x^6 + 1/4dx^4 + 1/3cx^3 + 1/2bx^2 + ax \qquad (2)$$

The four control parameters of the butterfly catastrophe are as follows: $\mathbf{a}$ is the normal factor, $\mathbf{b}$ is the splitting factor, $\mathbf{c}$ is the bias factor, and $\mathbf{d}$ is the butterfly factor. The derivative of the potential function for the butterfly elementary catastrophe model is given as follows:

$$\mathbf{F}^B(c, x) = ax + 1/2bx^2 + 1/3cx^3 + 1/4dx^4 + 1/6x^6 \qquad (3)$$

according to the state variable $\mathbf{x}$ (see Figure 2) is given as:

$$\partial \mathbf{F}^B(c, x)/\partial x = f^B(c, x) = a + bx + cx^2 + dx^3 + x^5. \qquad (4)$$

The cusp catastrophe-based model of low back injury

Karwowski et al. (1992) proposed that the risk of low back overexertion injury due to manual lifting can be conceptualized within the framework of two risk equilibrium states (*low, high*), and the potential for sudden system behavior changes. The risk behavior ($\mathbf{x}$) was influenced by changes in the state of equilibrium between the outside forces (due to specific lifting task factors) and the spinal system stress tolerance. These, in turn, were expressed by an interplay between the two internal forces (control factors) acting on the L5/S1 joint, i.e. the compression force (parameter $\mathbf{a}$), and the shear force (parameter $\mathbf{b}$), which define the fold cure $\mathbf{F}$.

Empirical data for model development

The empirical findings by Marras (1992) were used as

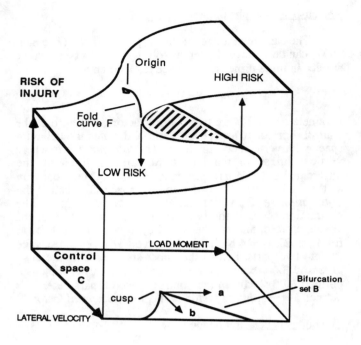

Figure 1.   The cusp catastrophe model of the risk of LBD
with two dynamic control space factors.

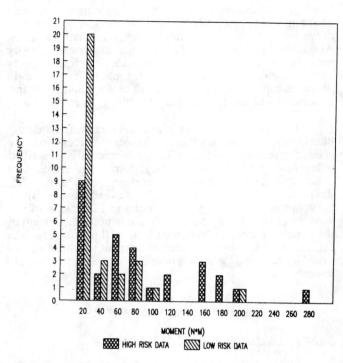

Figure 3.   Distribution of load moment data for
low and high risk lifting jobs.

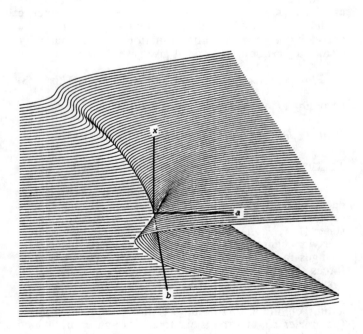

Figure 2.   The butterfly catastrophe manifold with
c > 0, d > 0 (after Murata et al., 1984).

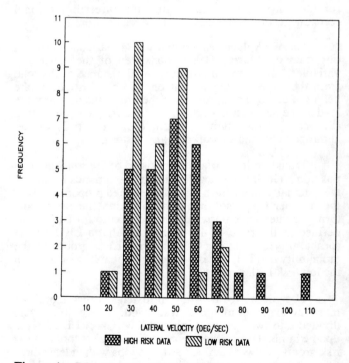

Figure 4.   Distribution of lateral velocity data for
low and high risk lifting jobs.

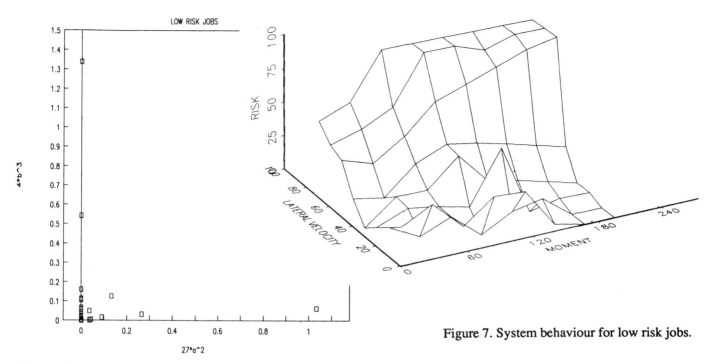

Figure 5. Cusp catastrophe fold distribution for the low risk jobs.

Figure 7. System behaviour for low risk jobs.

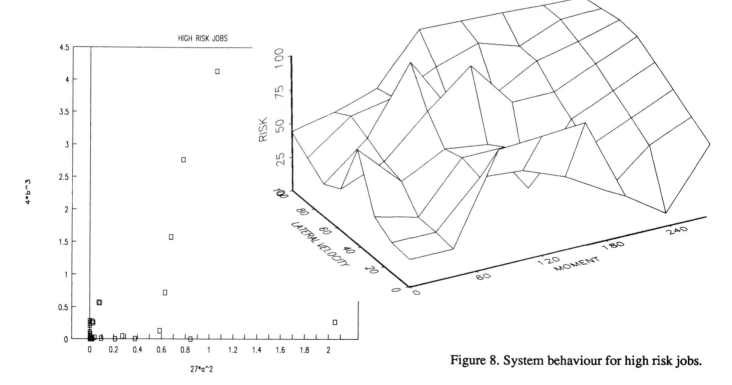

Figure 6. Cusp catastrophe fold distribution for the high risk jobs.

Figure 8. System behaviour for high risk jobs.

the basis for the proposed cusp catastrophe-based model. These findings pointed out that there large trunk range of motion, trunk velocity, and trunk accelerations were associated with manual handling jobs. Furthermore, the trunk motion patterns were primarily a function of the job environment. The occupationally-related low back disorder risk can be associated with a combination of five measures representing both workplace and trunk motion factors. These include two workplace factors: 1) load moment and 2) lifting frequency, and three trunk motion factors, i.e. 3) the lateral trunk velocity, 4) twisting trunk velocity, and 5) sagittal flexion angle.

Marras (1992) has also shown that as the magnitude of each of these variables increased the risk increased. It was also concluded that when these factors were considered in combination, the risk model indicated greater probability of high risk of LBDs. By tracking the five above listed occupationally-related factors, Marras (1992) was able to predict the probability of high risk group membership for any repetitive job. However, individually, each of these task factors did not allow for reliable discrimination between a high risk situation and a low risk situation for LBDs.

In this study, the two control parameters chosen are the moment load (parameter **a**), and the lateral trunk velocity (parameter **b**). According to the model, at certain state of the risk potential for low back injury, the already strained elements of the lumbosacral joint and/or muscular system will no longer tolerate any further stress changes, and the system will rapidly change its state. This will lead to the new risk equilibrium, most likely accompanied by some neural, end-plate, vertebral, or other aspects of lumbosacral disk damage, as well as the low back muscle strain or sprain, resulting in low back syndrome or low back injury.

For the purpose of examining the proposed cusp catastrophe-based model of the LBD risk behavior, a total of 60 industrial lifting jobs were selected based on previously collected data (Marras, 1992). Thirty of the examined jobs were categorized as the low risk category, while another thirty were categorized as high risk category, using the above stated classification criteria. The distributions of values for the load moment and lateral velocity of trunk motion for the low and high risk groups of the examined lifting jobs are shown in Figures 3 and 4.

PRELIMINARY RESULTS AND DISCUSSION

The manifold distribution values, as defined by the elementary cusp catastrophe model, for jobs with the low and high risk of LBDs are shown in Figures 5-6. It can be seen that these two distributions are distinctively different from each other, indicating qualitatively different states in the behavior of the risk potential of LBDs system. As discussed above, such risk potential for LBDs (**x**) is influenced by changes in the state of equilibrium between the specific occupational LBDs risk factors, and a given musculoskeletal subsystem's stress tolerance level.

Additional insights into the risk potential behavior are offered by the geometries of the LBD risk values, treated as the function of a load moment and lateral trunk velocity (see Figures 7-8).

Extension of control space for quantification of the risk of low back disorders

In the butterfly catastrophe-based model, the control space can be extended to four control variables, i.e.: **a, b, c** and **d**, defined in C = R⁴, and shown below as follows:

a - load moment (normal factor),
b - lifting frequency (splitting factor),
c - lateral trunk velocity (bias factor),
d - sagittal flexion angle (butterfly factor), and
x - the risk potential for LBDs.

It should be noted that the interplay between the control parameters modifies the system's risk potential (behavior) for LBDs (**x**), rather than the effect of either one variable taken separately.

CONCLUSIONS

This study outlined a preliminary framework for modeling of the LBDs using the concepts of catastrophe theory. According to the proposed framework, the risk of LBDs can be modeled as the effect of sudden changes (discontinuity or catastrophic changes) in the state of equilibrium between the LBDs risk factors and human musculoskeletal system tolerability. At a certain stage, the already strained elements of the system can no longer tolerate any further stresses, and the system will rapidly change its state, leading to a new system equilibrium, accompanied by the LBD-related damage. Current research efforts involve estimation of the proposed model parameters and their empirical validation.

REFERENCES

Karwowski, W., 1978. On controllability of systems in view of the catastrophe theory. *Unpublished Technical Report (in Polish), I-23,* Institute of Organization and Management, Technical University of Wroclaw, Poland.
Karwowski, W., Hancock, P., Zurada, J. M. and Ostaszewski, K., 1991. Risk of low back overexertion injury due to manual load lifting in view of the catastrophe theory. In: *Proceedings of the 11th Congress of the International Ergonomics Association,* London, Taylor and Francis, pp. 66-68.
Karwowski, W., Ostaszewski, K and Zurada, J., 1992. Applications of the catastrophe theory in modeling the risk of low back injury in manual lifting tasks. *La Travail Humain, (The Journal of French Ergonomics Society; in English),* 55 (3), 259-275.
Marras, W. S., 1992. Toward an understanding of dynamic variables in ergonomics. *Occupational Medicine: State of the Art Reviews,* 7(4), October-December, 655-677.
Marras, W. S., Wongsam, P. E. and Rangarajulu, S. L., 1986. Trunk motion during lifting: the relative cost, *International Journal of Industrial Ergonomics,* 1, 103-112.
Murata, A., Kume, Y., and Hashimoto, F., 1984. Geometry of catastrophe model, *Bulletin of University of Osaka Prefecture,* Series A, 33(2), 145-157.
Thom, R., 1975, *Structural Stability and Morphogenesis,* New York, Addison-Wesley.
Zeeman, E.C., 1977, *Catastrophe Theory, Selected Papers: 1972-1977,* Reading, MA, Addison-Wesley.

EVALUATION OF LIMITING STRENGTH CONSTRAINTS
IN A COMPREHENSIVE BIOMECHANICAL MODEL

Carter J. Kerk
Industrial Engineering Department
Texas A&M University
College Station, Texas 77843-3131

Don B. Chaffin
Center for Ergonomics
University of Michigan
Ann Arbor, Michigan 48109

The strength constraints in a two-dimensional static human force exertion capability model (HFEC) have been evaluated using eight male and female subjects of varying anthropometry and strength capability. The model comprehensively estimates feasible exertion capability under symmetric conditions using a set of fifteen linear constraint equations from three constraint classes: strength, stability, and coefficient of friction (COF). This evaluation examines the nature of the limiting strength constraints. The computer model aided in designing tasks (combining posture with force exertion direction) that isolated upper extremity strength constraints and hip/torso strength constraints from stability and COF constraints. Subject performances of maximum exertions were recorded using force platforms and a multi-axis load cell to record external reaction forces at the hands and feet. Body posture was recorded with a 2D motion analysis system. The observed hand force exertions were compared to the exertions predicted by the model. The identity of the limiting constraints was well predicted by the model. The location of the constraints was logical and predictable. The results are discussed in the context of other modeling approaches as well as implications for future research. The HFEC approach shows excellent potential as an ergonomic engineering tool for teaching, evaluation, and design.

INTRODUCTION

The importance of static, whole-body, two-dimensional biomechanical models is well established in ergonomics. These models provide, at the minimum, a threshold of understanding in studying occupational exertions and resultant stresses by both practitioners and researchers. The strength related aspects of these models are fundamental (Chaffin and Baker 1970, Martin and Chaffin 1972). Strength equations predict percentile strength capability for a given hand force vector using methods described in Chaffin and Andersson (1984). These strength equations draw on strength data from Clarke (1966), Schanne (1972), and Burggraaf (1972), with population corrections by Stobbe (1982).

The development of a two-dimensional static human force exertion capability (HFEC) model (Kerk, Chaffin, Page, and Hughes 1991) created a formulation that used constraints from three classes (strength, stability and coefficient of friction (COF)) in a comprehensive manner. This new model reformulates the Two-Dimensional Static Strength Prediction Program (2DSSPP^{TM}). The original strength prediction equations predict a specific joint strength moment for a given joint function (e.g., flexion or extension), joint angle, appropriate adjacent joint angles, gender, and population strength percentile. This joint strength moment is compared to the moment produced at the joint by a specific hand force vector and appropriate body segment masses and reports the percentage of the gender population that has the necessary strength capability.

The new constraint formulation predicts the complete set of feasible horizontal and vertical components of force at the hands that satisfy the joint strength moment given by a specific joint function, joint angle and appropriate adjacent joint angles, gender, and population strength percentile. Likewise for stability and coefficient of friction, the constraint formulation predicts the complete set of feasible horizontal and vertical components of force at the hands that satisfy stability and COF requirements as described in Kerk et al. (1991). Predicting the hand force capability over the entire region is an improvement over the *single–point* prediction of the traditional model because it allows the user the opportunity to learn about the complex interactions of the factors and provides a better tool for engineering design.

The graphical solution of the HFEC model produces a result similar to the concept of the Postural Stability Diagram (PSD), developed by Grieve (1979ab), especially for stability and COF.

Carter Kerk is now at the South Dakota School of Mines and Technology, Industrial Engineering Program, Rapid City, SD.

However, in contract to PSD, HFEC incorporates the strength aspect and provides a predictive model.

Three independent studies were designed in which the objectives were to isolate and evaluate specific constraints in the model. This paper focuses on the strength constraints. Although it may cause confusion, some mention of the stability and COF constraints are made because it is difficult to isolate one portion from such a comprehensive model. The comprehensive model is thoroughly discussed in Kerk et al. (1991). The stability constraints are examined in Kerk (1992b) and the COF constraints in Kerk (1992c).

A computer version of the HFEC model simulated combinations of the input parameters to design tasks in which selected strength constraints theoretically determine the force exertion capability. It was extremely difficult, if not impossible, to individually isolate each of the strength constraints. However, it was possible to isolate a logical grouping of the strength constraints: upper extremity (elbow and shoulder) strength constraints and torso/hip strength constraints. The objective of this study was to evaluate the accuracy of the model predictions constrained by strength.

METHOD

Two types of trials were conducted, each with a unique posture and force exertion directions: upper extremity strength trials (to test the upper extremity joint strength constraints) are shown in Figure 1a, and hip/torso strength trials (to test the hip and L5/S1 joint strength constraints) are shown in Figure 1b. A complete description of the methods is found in Kerk (1992a).

Eight healthy male and female subjects were selected to represent a broad perspective of anthropometric range and strength capability. Those that approximately fit one of four general classes (small/weak, small/strong, large/weak, large/strong) by gender were selected for extensive strength testing as described in Stobbe (1982). Condensed subject attributes are given in Table 1.

The data acquisition system was constructed to control the timing aspects of the trials and to collect the necessary posture and force information. A pair of force platforms were used to monitor three-dimensional forces and moments beneath the feet. The hands were symmetrically coupled to vertically oriented, non-slip cylinders (14 cm long, 3.2 cm diameter, 45.7 cm separation) rigidly fixed to a three-dimensional hand force transducer. A video camera provided a right-side sagittal plane view. Subjects wore retro-reflective spheres on their approximate joint centers (right elbow, shoulder, hip, knee, and ankle). Fixed markers were also placed on the right-foot force platform and the handle. Computer programs guided calibration and synchronized data collection of the kinetic (87 Hz) and kinematic data. The program calculated the average forces, moments, and foot centers of pressure during seconds four through six after the threshold trigger. Subjects wore industrial-type shoes with skis (50.8 cm length) and bindings rigidly fastened. A rubber pad was placed between the skis and the force platform surface to create a non-skid interface. These precautions were

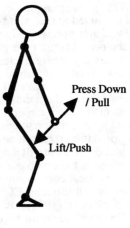

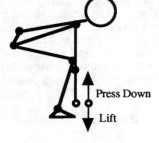

(a)   (b)

Figure 1. The requested postures and force exertion vectors for the upper extremity (a) and hip/torso strength trials (b).

Table 1. Subject attributes.

Subject #	Age (years)	Height (%tile)	Weight (%tile)	Strength (%tile)	Anthropometry/Strength Category
Female 1	36	8.1	16.9	19.5	small / weak
Female 2	22	79.7	79.7	17.3	large / weak
Female 3	35	30.5	34.5	55.3	small / strong
Female 4	37	72.6	50.0	45.8	large / strong
Male 1	26	14.7	4.3	4.5	small / weak
Male 2	20	71.2	20.6	15.1	large / weak
Male 3	20	72.2	19.8	36.7	small / strong
Male 4	23	90.5	88.3	73.1	large / strong

Note 1: %tile = percentile.
Note 2: Strength is average percentile.

taken to minimize the possibilities of exerted hand forces being limited by stability or COF and to maximize the opportunity for the exertions to be strength limited. Prior to each trial, the subject assumed the requested posture. The subject was requested to perform a maximal horizontal pushing or pulling exertion while maintaining the static posture.

The kinematic data for each trial was digitized and input into the HFEC model with the subject's average percentile joint moment strength and anthropometry. This provided the necessary input information for a model prediction for hand force exertion capability for each specific trial.

RESULTS

Each trial was analyzed to determine the accuracy of the model prediction, or modeling error. This was done by overlaying the observed exertion vector on the feasible region predicted by the model. The limiting constraint was defined as the nearest constraint (in distance on the plot) to the observed exertion vector in units of force. This determined the identity of the limiting constraint. The occurrence of these limiting constraints was compiled by force direction for both the upper extremity strength trials and the torso/hip strength trials and is shown in Table 2. The light gray outlines indicate the hypothesized limiting constraints.

For the upper extremity strength trials with the press down/pull exertion, all of the trials were limited by elbow extension strength as hypothesized. In these cases, shoulder extension strength was

always the second or third nearest limiting constraint.

For the upper extremity strength trials with the lift/push exertion, nearly all of the trials were limited by elbow and shoulder flexion strength as hypothesized.

For the torso/hip strength trials with the pressing down exertion, few of the trials were limited by L5/S1 and hip flexion strength, the expected limiting constraints. Examination of the individual solution space graphs indicated that the L5/S1 and hip flexion strength constraints were usually on or quite near the feasible region of the graph. Because observed exertion magnitudes fell generally short of the predicted values, the strict rules used in selecting the minimum distance to the nearest constraint often chose the elbow extension strength constraint when there was an element of pulling component and the shoulder flexion strength when there was a pushing component.

A similar result occurred in the torso/hip lift exertions, although the average modeling error was very close to zero and the model actually under-predicted the observed exertion in many trials. It appears that in these trials, that although L5/S1 and hip extension strength constraints were very near the feasible region, that upper arm strength constraints were more restrictive. The observed knee posture deviated from the requested posture. This postural deviation effected the slopes of the two constraints. The effect made it less likely for the two constraints to affect the solution space. The subjects tended to exert a slight pulling component along with the orthogonal lifting component. This drew the observed exertion vector closer to the elbow and shoulder extension strength constraints.

Table 2. Occurrence of limiting constraints.

Limiting Constraint	Upper Extremity Press Down/Pull	Upper Extremity Lift/Push	Torso/Hip Press Down	Torso/Hip Lift
(1) Elbow Flexion	0	2	0	0
(2) Elbow Extension	17	0	10	3
(3) Shoulder Flexion	0	15	5	0
(4) Shoulder Extension	0	0	0	14
(5) L5/S1 Flexion	0	0	2	0
(6) L5/S1 Extension	0	0	0	0
(7) Hip Flexion	0	0	0	0
(8) Hip Extension	0	1	0	0
(9) Knee Flexion	0	0	0	0
(11) Ankle PFlexion	0	1	0	0
(10) Knee Extension	0	0	0	0
(12) Fall Back	0	0	0	0
(13) Fall Forward	0	0	0	0
(14) Slip Back	0	0	0	0
(15) Slip Forward	0	0	0	0

Notes: Each cell with a gray outline represents the theoretically expected limiting constraints. Column sums are not equal.

DISCUSSION

The identity of the limiting constraints was well predicted by the model. The magnitudes of the modeling error in some cases were large. Possibly the best explanation was the effects of the novel postures on performance as well as the small sample population of extremely diverse individuals. This type of model is not designed to predict individual performance, but better suited to make predictions for general populations. It is also important to keep in mind that the HFEC model is a reformulation of an existing model (2DSSPP^{TM}) that has been successfully evaluated using an extensively larger subject population utilizing commonly used postures and force exertions (Chaffin, Freivalds, and Evans 1987, Chaffin and Erig 1991).

In this study in which the strength constraints were isolated from the stability and COF constraints, the identity of the limiting constraints was well predicted by the model. The location of the constraints for a given posture was logical. Even when the modeling errors were relatively large, the presence of the expected constraints with respect to the observed exertion vector was predictable. This provides positive support for the HFEC modeling concept.

The ability to predict major aspects of hand force exertion capability can explain and complement the usefulness of the Posture Stability Diagrams (Grieve 1979ab). The Posture Stability Diagrams provide physics-based limits to hand force

exertion capability due to stability and coefficient of friction. The HFEC modeling concept provides predictive limits from muscle generated joint strengths that are so vital in force exertion capability. Therefore a union of the HFEC and PSD concepts is a powerful tool.

It is difficult to compare the results of this study to that of Stobbe (1982) which developed a set of seven standardized, whole-body strength tests from which joint moment strengths could be predicted. For one set of subjects, joint moment strength tests were conducted, as well as an augmented set of standardized, whole-body strength tests. A regression analysis identified the set of seven whole-body tests that best predicted the joint strengths. This regression model was then validated with a second set of subjects and the predicted joint strengths were compared with the regression model predictions. This HFEC study used the same protocol as Stobbe in quantifying joint moment strength, but would need to apply the experimental procedure to each of Stobbe's subjects, given inputs of the joint strengths and anthropometry. If the Stobbe study postures were available, the HFEC model could make an exertion prediction. The whole-body strength tests required precise location of the hands and feet, but subjects were free to choose a posture. This suggests a future study using Stobbe's whole-body strength tests while recording postures and comparing the HFEC prediction to the actual exertion, as well as comparing Stobbe's regression predictions to measured joint strength data. This could provide an evaluation of the HFEC model for more industrial-type, practical exertions, as opposed to the novel postures used in the current study to test individual constraints.

It is a fundamental contention of the HFEC modeling approach that hand force exertion capability is limited, not only by strength, but also by stability and COF in a comprehensive manner. Use of the terms *whole-body strength* and *whole body strength tests*, must be used and interpreted carefully. If these are in fact limited by stability or COF, then better terminology is *whole-body exertion capability*.

Extensive analysis of the data from this study, including the absolute and relative magnitudes of the observed exertions, the variation in the observed force vector directions, and the trends in the quantitative modeling prediction errors is the subject of an expanded paper. Implications for future studies include improved design and control of subject posture, and control of force exertion direction. The HFEC model is a reformulated version of a model (2DSSPP^{TM}) that has been shown to be unbiased for large, normally distributed population strengths. The most important contribution from the HFEC model lies in the comprehensive manner in which it accounts for important input factors and the manner in which the output is graphically displayed. This output has been shown to be an excellent teaching tool, providing insights previously not available. There exists excellent potential for HFEC as a valuable ergonomic engineering tool for teaching, evaluation, and design.

ACKNOWLEDGMENTS

This research was supported by Paul McMahan, Research and Test Department, Association of American Railroads.

REFERENCES

Burggraaf, J.D. (1972) *An Isometric Biomechanical Model for Sagittal Plane Leg Extension.* M.S. thesis (Industrial Engineering), University of Michigan, Ann Arbor.

Chaffin, D.B. and Andersson, G. (1991) *Occupational Biomechanics.* New York: John Wiley & Sons, Inc. 2nd Ed. p. 170-263.

Chaffin, D.B. and Baker, W.H. (1970) A Biomechanical Model for Analysis of Symmetric Sagittal Plane Lifting. *IIE Transactions* 2(1):16-27.

Chaffin, D.B. and Erig, M. (1991) Three-Dimensional Biomechanical Static Strength Prediction Model Sensitivity to Postural and Anthropometric Inaccuracies. *IIE Transactions* 23(3):215-227.

Chaffin, D.B., Freivalds, A., and Evans, S.M. (1987) On the Validity of an Isometric Biomechanical Model of Worker Strengths. *IIE Transactions* 19(3):280-288.

Clarke, H.H. (1966) *Muscle Strength and Endurance in Man.* Prentice-Hall, Englewood Cliffs, pp. 39-51.

Grieve, D.W. (1979a) The Postural Stability Diagram (PSD): Personal Constraints on the Static Exertion of Force. *Ergonomics* 22(10):1155-1164.

Grieve, D.W. (1979b) Environmental Constraints on the Static Exertion of Force: PSD Analysis in Task-Design. *Ergonomics* 22(10):1165-1175.

Kerk, C.J., Chaffin, D.B., Page, G.B., and Hughes, R.E. (1991) A Comprehensive Biomechanical Model Using Strength, Stability, and COF Constraints to Predict Hand Force Exertion Capability Under Sagittally Symmetric Static Conditions. *IIE Transactions* (accepted for publication, July 1991).

Kerk, C.J. (1992a) *Evaluation of a Comprehensive Biomechanical Model: The Strength Constraints.* Chapter III, Ph.D. Thesis, Department of Industrial and Operations Engineering, University of Michigan, Ann Arbor.

Kerk, C.J. (1992b) *Evaluation of a Comprehensive Biomechanical Model: The Stability Constraints.* Chapter IV, Ph.D. Thesis, Department of Industrial and Operations Engineering, University of Michigan, Ann Arbor.

Kerk, C.J. (1992c) *Evaluation of a Comprehensive Biomechanical Model: The Coefficient of Friction Constraints.* Chapter V, Ph.D. Thesis, Department of Industrial and Operations Engineering, University of Michigan, Ann Arbor.

Martin, J.B. and Chaffin, D.B. (1972) Biomechanical Computerized Simulation of Human Strength in Sagittal-Plane Activities. *IIE Transactions* 4(1):19-28.

Schanne, F.J. (1972) *A Three-Dimensional Hand Force Capability Model for the Seated Operator.* Ph.D. thesis (Industrial Engineering), University of Michigan, Ann Arbor.

Stobbe, T.J. (1982) *The Development of a Practical Strength Testing Program for Industry.* Ph.D. thesis (Industrial Engineering), University of Michigan, Ann Arbor.

MAXIMUM LIFTING CAPACITY IN SINGLE AND MIXED GENDER THREE-PERSON TEAMS

Marilyn Sharp, Valerie Rice, Brad Nindl, Tania Williamson
U.S. Army Research Institute of Environmental Medicine
Natick, MA 01760-5007

Little information is available regarding isoinertial lifting ability in teams of two or more people. The relationship between the sum of individual lifts and the lifting capacity of a three person team was examined. Eleven men and ten women were randomly combined into 18 teams for each of the following four combinations: three men (3M), three women (3W), two men with one woman (2M&1W), and one man with two women (1M&2W). While the absolute load lifted from floor to knuckle height decreased with a decrease in the number of males on the team, team lifting strength as a percentage of the sum of individual lifting strength was generally higher for single gender teams (91.0% for 3W and 85.0% for 3M) than for mixed gender teams (82.7% for 1M&2W and 74.4% for 2M&1W).

INTRODUCTION

In industry (Johnson and Lewis, 1989) and the military (Department of the Army, 1990) there are many lifting tasks that require individuals to work in groups of two or more. Examples of these are loading large rolls of material onto machines, lifting injured persons on stretchers and moving furniture. While there are some reports of team lifting in the literature (Karwowski and Mital, 1986; Karwowski and Pongpatanasuegsa, 1988; Karwowski, 1988), there are no reports of isoinertial lifting in teams of three persons. In addition, there are no reports of combined gender teams. Combined gender teams are a reality in the military and civilian workforce, therefore it is important to know the relationship between individual strength and team lifting strength for combined gender teams.

Karwowski and Mital (1986) found that the isometric and isokinetic team lifting strength of two men was less than the sum of their individual lifting strengths and this difference increased when the number of men lifting increased from two to three. Karwowski and Pongpatanasuegsa (1988) found that the isometric and isokinetic team lifting strength of two women was less than the sum of individual strengths, but did not consistently decrease with the addition of a third woman. No gender comparisons were made for the percentage difference between the sum of individual lifting strengths and team lifting strength.

The percentage difference between the sum of individual lifting strengths and team lifting strength was greater in isokinetic trials than in isometric trials and lack of team coordination was cited as a possible explanation (Karwowski and Mital, 1986; Karwowski and Pongpatanasuegsa, 1988). If team coordination decreases during an isokinetic movement, where velocity is controlled, further decreases might be expected during an isoinertial lift, where velocity is not controlled. This hypothesis, however, was not supported in an investigation by Karwowski (1988). The maximum load lifted by pairs of men and pairs of women was 87.5% and 91.0%, respectively, of the sum of their individually determined maximum lifts. This is similar to the data from isometric composite strength in pairs of men and pairs of women (Karwowski and Mital, 1986; Karwowski and Pongpatanasuegsa, 1988).

The objective of this study was to determine the relationship between individual lifting strength and lifting strength in teams of three persons, and to determine the effects of same and mixed gender groupings on team lifting strength.

METHODS

Eleven men and ten women participated in the study. Subjects were assigned to teams on a random basis to obtain equal numbers of male, female and mixed gender teams. The four gender groupings were: all men (3M), all women (3W), 2 men with 1 woman (2M&1W) and 1 man with 2 women

(1M&2W).

Three person team lifting strength was measured using the weight lifting device shown in Figure 1. The triangular shaped device, which was similar to a standard weight lifting bar, had reinforcing bars in the center and extensions to hold standard weight plates.

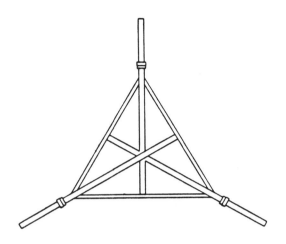

Figure 1. Three person team lifting device

On command, subjects lifted the device using a smooth continuous motion from floor level to the full standing position of the shortest team member. Weight was added to each arm of the device following each successful lift, until the team was unable to complete the lift. Subjects used a bent-knee, straight-back technique, and the test was terminated if proper form was not adhered to. The last successful lift was recorded as the maximal team lifting strength for that trial. Adequate rest was provided between lift attempts (3 minutes minimum). One determination of maximum lifting strength was conducted per day, with 48 hours rest between testing sessions. Lifting trials with same gender teams were conducted before mixed gender trials. The individual maximum lifting strength procedure was identical to that of the team lifting procedure, but was conducted using a standard weight-lifting bar. Individual lifting strength was measured at the beginning and middle of the study.

The percentage of the sum of individual lifting strengths represented by the team lifting strength (% sum) was calculated as follows:

$$\% \text{ sum} = \frac{\text{team lift}}{\text{sum individual lifts}} * 100$$

The maximum machine lift was also assessed for each subject by lifting the handles of a weight stack machine from floor to 152 cm. The initial load of 18 kg was increased in 4.5 kg increments until the subject was unable to safely complete the lift (Sharp and Vogel, 1992). Body composition was estimated using hydrostatic weighing (Fitzgerald, Vogel, Milleti, and Foster, 1987).

Paired t-tests were used to compare team lift with the sum of individual lifts for each gender grouping. Analysis of variance was used to compare the % sum for the four team gender combinations. Tukey HSD post-hoc tests were performed to examine differences between means.

RESULTS

The mean and standard deviation of descriptive measures are listed by gender in Table 1. Males were significantly younger, taller, weighed more, had more fat free mass and less body fat ($p < .01$), and were significantly stronger on all strength measures ($p < .01$).

The results of the team lifting trials are listed in Table 2 and include the team lift, the sum of individual lifts (i.e. subject 1+ subject 2 + subject 3), the team lift as a percentage of the sum of individual lifts (% sum), and the absolute difference between the sum of individual lifts and the team lift. The absolute load lifted decreased as the number of men in the team decreased; three men lifted the heaviest load, while 2M&1W, 1M&2W and three women lifted 18.8% ($p < .05$), 29.2% ($p < .01$) and 37.8% ($p < .01$) less, respectively.

The % sum for 2M&1W teams was lower than for any other gender grouping ($p < .01$). Teams of three women had a significantly greater % sum than teams of 1M&2W ($p < .01$). The % sum for teams of three men was not

Table 1. Descriptive data for men and women (mean ± SD)[1].

	Men	Women
n	11	10
Weight (kg)	75.2 ± 13.7	59.1 ± 9.5
Height (cm)	177.8 ± 6.1	163.1 ± 3.7
Age (yrs)	20.0 ± 1.3	29.2 ± 6.7
% body fat	15.0 ± 6.8	24.5 ± 5.4
Fat free mass (kg)	62.7 ± 8.0	44.3 ± 5.4
Individual lift (kg)	138.2 ± 23.0	78.8 ± 8.9
Machine lift (kg)	76.0 ± 13.5	38.6 ± 6.3

[1] There was a significant gender difference for all measures (p<.01).

Table 2. Maximum team lifting strength in three person teams.

		3M	3W	1M&2W	2M&1W
n		18	18	18	18
Team lift (kg)[1]	X̄	345.1 ± 39.5	214.6 ± 17.6	244.3 ± 19.1	280.3 ± 35.4
Sum of individual lifts (kg)[2]	X̄	405.9 ± 30.2	235.8 ± 15.1	297.7 ± 34.7	377.2 ± 31.7
% sum[3]	X̄	85.0 ± 7.6	91.0 ± 5.7	82.7 ± 7.5	74.4 ± 7.5
Difference (kg)[4]	X̄	-60.9 ± 30.9	-21.3 ± 13.7	-53.4 ± 27.8	-96.9 ± 28.6

[1] All gender groups significantly different (p<.01), except 3M vs 2M&1W (p<.05).
[2] All gender groups significantly different (p<.01), except 3W vs 1M&2W (p<.05).
[3] 2M&1W significantly different from all other teams (p<.01) and 1M&2W significantly different from 3W (p<.01).
[4] All gender groups significantly different (p<.01) except 3M vs 1M&2W.

significantly different from teams of three women or from teams of 1M&2W. For each gender grouping, the sum of individual lifting strength was significantly greater (p<.01) than the team lift. The female to male ratio for lifting in three person single gender teams was 62.2%.

DISCUSSION

The absolute load lifted decreased with a decrease in the number of men lifting. Teams of three women scored closest to the sum of their individual lifts (% sum), but this was not significantly different from teams of three men. The

percentage sum for the team of 2M&1W was significantly less than all other teams. If the objective of grouping the teams were to have each person lifting a load closest to their maximum capacity, the gender mix of 2M&1W is least effective.

Data from studies by Karwowski and colleagues (1986, 1988) are listed in Table 3 for comparison with the present data. The percentage of the sum lifted during isometric composite strength testing in trios of male (Karwowski and Mital, 1986) and female (Karwowski and Pongpatanasuegsa, 1988) subjects was comparable to the single gender three person lifts of the

Table 3. Team lift as a percentage of the sum of individual lifts (% sum): Comparison of present data with 3 person isometric and isokinetic lifting strength.

	Current study	Karwowski & Mital		Karwowski & Pongpatanasuegsa	
	isoinertial	isometric	isokinetic	isometric	isokinetic
3M	85.0	88.6	60.3	NA	NA
3W	91.0	NA	NA	90.1	73.4
1M&2W	82.7	NA	NA	NA	NA
2M&1W	74.4	NA	NA	NA	NA

present study. The % sum for three person isokinetic lifting strength appears to be lower than for either isometric or isoinertial lifting in three person teams. If lack of team coordination during the lifting effort were responsible, one would expect isoinertial lifting to show a greater decrease in % sum than the more controlled effort of isokinetic lifting. It may be that the velocity of the isokinetic lifting test was slower than that of free lifting, and therefore a more difficult or unfamiliar task to perform.

The % sum for box lifting in two, three or four person teams from Karwowski (1988), Sharp et al (1993) and the present study are listed in Table 4. The % sum levels off after moving from an individual lifter to a pair with no significant losses due to further increases (two to three or four persons) in team size. This trend supports the work of Ringelmann (as reported by Karwowski and Pongpatanasuegsa, 1988) for multiple person rope pulling in which "the addition of a fourth, fifth or sixth member to the team produced insignificant decrements in strength exertion".

The three-person compact load limit for occasional lifting in Military Standard 1472-D is 109 kg for all male teams, and 55 kg for mixed gender teams (Department of Defense, 1989). The mixed gender recommendation represents 16% (3M) to 26% (3W) of the maximum three-person lifting capacity of the healthy young subjects in this study. The all male lifting standard is only 50% of the maximum lifting capacity of 3W. Based on actual lifting capacity, Military Standard 1472-D appears to be a safe, conservative recommendation (Department of Defense, 1989).

This study demonstrated that single gender teams are able to lift a greater percentage of their potential

Table 4. Effects of increasing team size on team lift as a percentage of the sum of individual lifts (% sum).

Team Size	2	2	3	4
Source	(Sharp et al, 1993)	(Karwowski, 1988)	(Present data)	(Sharp et al, 1993)
men	90	88	85	86
women	92	91	91	90
mixed	80	NA	79	84

lifting strength than mixed gender teams, and that a greater % sum is achieved when one man lifts with two women than when two men lift with one woman. When these data are compared to similar data for different sized teams, the % sum is not significantly decreased from that of two-person lifting in teams of three or four. Military Standard 1472-D (Department of Defense, 1989) recommends doubling the individual load for two-person lifting, and adding 75% of the individual load for each additional lifter (Department of Defense, 1989). As no additional decrease in lifting strength was found with more than two lifters, perhaps this additional 25% reduction in load is not necessary.

REFERENCES

Department of Defense (1989). *Military Standard Human Engineering Design Criteria for Military Systems, Equipment and Facilities*. Philadelphia, PA: Naval Publications and Forms Center.

Department of the Army. (1990). *Military Occupational Classification and Structure*. Washington, DC: U.S. Government Printing Office.

Fitzgerald, P. I., Vogel, J. A., Milleti, J.and Foster, J. M. (1987). *An Improved Portable Hydrostatic Weighing System for Body Composition*. (Tech. Report T4/84). Natick, MA: U.S.Army Research Institute of Environmental Medicine.

Johnson, S.L. and Lewis, D.M. (1989). A psychophysical study of two-person manual materials handling tasks. *Proceedings of the Human Factors Society 33rd Annual Meeting* (pp. 651-653). Santa Monica, CA: Human Factors Society.

Karwowski, W. (1988). Maximum load lifting capacity of males and females in teamwork. *Proceedings of the Human Factors Society 32nd Annual Meeting* (pp. 680-682). Santa Monica, CA: Human Factors Society.

Karwowski, W. and Mital, A. (1986). Isometric and isokinetic testing of lifting strength of males in teamwork. *Ergonomics, 29*, 869-878.

Karwowski, W. and Pongpatanasuegsa, N. (1988). Testing of isometric and isokinetic lifting strengths of untrained females in teamwork. *Ergonomics, 31*, 291-301.

Sharp, M., Rice, V., Nindl, B., and Williamson, T. (1993). Effects of gender and team size on floor to knuckle height one repetition maximum lift. *Medicine and Science in Sports and Exercise, supplement 25*, S137.

Sharp, M. A.and Vogel, J. A. (1992). Maximal lifting strength in military personnel. In S. Kumar (Ed.), *Advances in industrial ergonomics and safety IV* (pp. 1261-1268). Washington,D.C.: Taylor & Francis.

CAN BIOMECHANICALLY DETERMINED STRESS BE PERCEIVED?

Deborah D. Thompson* and Don B. Chaffin
Center for Ergonomics
The University of Michigan
Ann Arbor, MI

Back and overexertion injuries are a costly and debilitating problem in industry. It has been suggested that the best protective action in the prevention of back injuries is to rely on a person's perception of the risks, and allow them to operate within them. However, this assumes that a person is aware of the sensory information from the body concerning unsafe levels of stress, particularly in the back. Unfortunately, there is some question as to whether this assumption is valid. The purpose of this study was to determine how well physical stress resulting from performing occasional lifting exertions could be perceived. This required an evaluation to determine how perception (psychophysical approach) relates to physical tolerances (biomechanical approach). The results showed that back stress resulting from occasional lifting exertions is not well perceived in general. The fact that the stress was not well perceived by some may indicate why low back injuries are so pervasive in the population, and why engineering and ergonomic changes are needed to reduce the exposure to conditions that would overstress the back.

INTRODUCTION

Back and overexertion injuries are a costly and debilitating problem in industry today. A National Institute for Occupational Safety and Health report (NIOSH, 1981) states that approximately one third of the US workforce is required to exert significant strength as part of their job demands. It also states that overexertion injuries account for about one fourth of all reported occupational injuries in the US. Further, Klein et al. (1984) reported that back injuries and low back pain are responsible for 26% of all worker compensation claims.

There are those who have suggested that the best protective action in manual materials handling is that a person know his/her limits, and operate within them (Brown, 1971; Jones, 1972; Snook and Ciriello, 1972). However, this assumes that the individual is aware of the somatic signals about unsafe levels of physical stress. Is it valid to assume that people have the ability to perceive unsafe levels of stress? Is it possible to rely on a person's subjective tolerance in defining load lifting capabilities to reduce back and overexertion injuries? In order to address these questions, the purpose of this study was to evaluate how the dynamic biomechanical tolerances of the low back (back compression force) relate to the subjective perception of physical stress to determine if a lifting stress on the back is adequately perceived. Back compression force was used since it has been accepted as one of the primary means of stress on the spine during lifting activities, and force limits

which identify potential injury risk levels have been developed (NIOSH, 1981). Since one's perception can only be measured through the use of self-reporting techniques, the Borg CR-10 rating scale (Borg, 1982 and 1990) was used to determine the perception of the lifting stress. The ability to predict potential biomechanical tolerances of the low back through the perception of stress could provide a simple mechanism to minimize the risk of injury associated with manual materials handling, particularly to the back.

METHOD

Subjects

Nine male subjects volunteered to participate in the study. They were of good health, with no prior history of back pain nor injury. The subject's health status was determined through screening interviews. Their ages ranged from 23-33 years (mean age = 27.7 yr). Their body weights ranged from 62.2-115.7 kg (mean = 80.3), and the heights ranged from 175.3-188 cm (mean = 181.4).

Apparatus

The subjects were required to lift a moderately sized box ($21\frac{1}{2}$ X $10\frac{1}{4}$ X $9\frac{3}{4}$ inches) with $5\frac{1}{2}$ inch handles located in the middle on each end of the box. The box was attached to a pneumatic overhead suspension system which was designed such that the subject was only required to lift the box.

The LiftTrak[TM] Motion Analysis System (Version 2.10) was used to record the postures and the trajectory of the critical joints during the lifting activity by way of reflective markers. LiftTrak[TM] used the postures, body type (height and weight) of the subject, the amount of weight involved (box weight), along with the kinematic data (velocity and acceleration) to automatically calculate the L5/S1 compression force, using the 2D Dynamic Strength Prediction Program[TM] (Version 4.2). The program allowed for the evaluation of the dynamic exertions by using a pseudodynamic biomechanical analysis wherein a frame by frame analysis was conducted (isometric effect) with the dynamic inertial effects added in order to compute the compression forces.

Procedure

In order to place a load moment on the back, the subjects were asked to perform floor to standing knuckle height lifting tasks. The subjects were trained in using a back lift posture (back bent, knees straight lifting posture) which relies on the back muscles to execute the lift. The floor to knuckle height lifts were performed at three different horizontal distances from the body, which were based on the anthropometric reach capabilities of the subject. The lifts were performed on a occasional basis (less than one lift per minute) since Chaffin and Park (1973) found that occasional, or infrequent, maximum lifting may be more hazardous, in terms of the incidence rates of low back injuries, than more frequent lifting (except at the higher frequencies - greater than 150 lifts per day).

In determining dynamic lifting capability, the subjects were allowed to adjust the load weight until they reached their maximum dynamic limit (MDL). The MDL represented the maximum amount of weight that the subject could lift. They were given up to 20 minutes to adjust the load by adding to or subtracting weight from the box. Once the MDL had been established, submaximal loads representing a percentage of the MDL (60, 40, and 25%) were presented to the subjects to perform the lifting tasks at horizontal distances from the body.

After lifting each maximum dynamic limit (MDL) and submaximal load, the subjects were instructed to rate their perception of the exertion using the Borg CR-10 rating scale (Figure 1). They were instructed to select the number on the scale which most accurately corresponded to their perception of the exertion on the back. The scale provides ratings ranging from 0, which indicates no effort, to 10, which is associated with heavy or extremely stressful exertions, or above if the subject desires to indicate supermaximal exertions. The specific instructions were based on the recommendations of Borg (1990) and Kroemer et al. (1990).

Borg's CR-10 scale

0	Nothing at all	
0.5	Extremely weak	(just noticeable)
1	Very weak	
2	Weak	(light)
3	Moderate	
4		
5	Strong	(heavy)
6		
7	Very strong	
8		
9		
10	Extremely strong	(almost max)

Maximal

Figure 1 - Borg CR-10 rating scale.

RESULTS

The initial analysis involved determining if the horizontal distance of the load had an affect on the rating of perception of effort/exertion on the back. The nonparametric statistical evaluation found that the horizontal distance of the load did not have an affect on the ratings (Kruskal-Wallis test: $H(2) = 0.63$, $p > 0.05$). Since it was found that the horizontal distance did not affect the ratings, a logistic response function was used, on the combined data, to determine whether the subjects were able to perceive the magnitude of the stress relative to the computed L5/S1 compression forces. A logistic response function was fitted to the data since it was expected that the ratings would not continue to increase to infinity as the magnitude of the stress increased. At some point, one's perception would become saturated resulting in the ratings asymptoting at some level. Although, with the Borg CR-10 scale, latitude is given to rate the perceived effort beyond 10, it could be hypothesized that since the scale is anchored at 0 and 10, there may be a tendency to not rate an effort greater than 10. Thus, the ratings would begin at 0 and asymptote at 10. The logistic response function would adequately model the asymptoting behavior.

The graphical results of the dynamic exertions (Figure 2) revealed a random pattern between the rating of perceived exertion on the back and the peak L5/S1 compression forces, suggesting that no consistent relationship exists. The logistic response function could not explain any of the observed variation ($r^2 = 0$), resulting in no correlation ($r = 0$) between the ratings and compression force. With none of the variation explained, it could be concluded that either the subjects were inconsistent in their perception of the magnitude of the stress relative to the computed L5/S1 compression forces, or that the Borg CR-10 rating scale was not appropriate, while performing occasional dynamic lifting exertions.

The inability of the subjects to consistently perceive the magnitude of the stresses tends to indicate that the inter-subject variability was large. The ANOVA results confirmed this fact by indicating that there were statistically significant differences between the subject's ratings ($F(8,160) = 8.06$, $p < 0.05$). With the subjects designated as the blocking variable, the variability between subjects was responsible for 30% of the variation in the data. The variability in perception indicates that the subjects did not perceive levels of physical stress equally. A question arises as to how much better the modeling of the perception would have been in explaining the variation in the data if the inter-subject variability was excluded. With the inter-subject variability responsible for 30% of the variation in the data, 70% remained to be explained. So at best, the logistic response function could only explain 70% of the variation in the data, which could be interpreted as the function having the ability to be more successful in explaining the variation than initially indicated. However, the results showed that the function was not able to explain any of the remaining variation (Figure 2). Even if the function was able to explain a significant portion of the remaining variation, from a practical standpoint, the exclusion of inter-subject variability is not appropriate since the source of the variability is not explained. The factors affecting inter-subject variability (psychological, movement strategies, anthropometry, etc.) are unknown. Until these factors can be identified, the exclusion of the inter-subject variability would be considered inappropriate, in terms of modeling the behavior in order to reduce the risks of back and overexertion injuries.

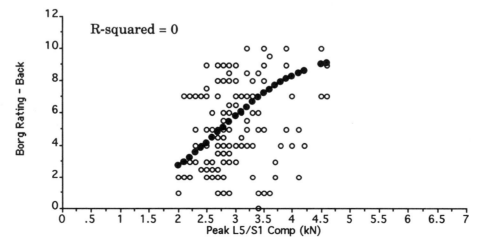

Figure 2 - Logistic response model of the Borg ratings at the low back vs. peak dynamic L5/S1 compression force during back lifts for all subjects.

DISCUSSION

The results indicated that when the back is used to perform a lifting exertion, the magnitude of the lifting stress is not consistently perceived. The fact that the stress is not well perceived may indicate why low back injuries are so pervasive in the population. These results indicate why engineering and ergonomic changes are needed to combat the problem. The goal of ergonomics is to design the job to fit the individual performing the job, including their mental and physical capabilities, limitations and tolerances. Snook et al. (1978) have suggested that the ergonomic redesign of tasks to reduce the manual handling exposure is the only currently effective control for low back injuries. This is substantiated by Ridyard (1990) who indicated that ergonomic improvements have accounted for a 62% reduction in the incidence of overexertion and back injuries.

The fact that the lifting stress was not perceived also brings into question the utility of the psychophysical approach. The assumption is that in choosing the load, a person will integrate the resulting stress and ultimately choose loads that are below their physical tolerance (Hafez et al., 1982; Karwowski, 1982, 1983). The results of this research suggests that with occasional (infrequent) lifting tasks, people are not necessarily aware of, and do not perceive, levels of physical stress equally. There is the potential to choose loads which exceed their physical tolerance; therefore, loads which can lead to manual handling injuries may be chosen.

ACKNOWLEDGMENTS

* The first author is now at Chrysler Corporation, Auburn Hills, MI.

REFERENCES

Borg, G. A. V. (1982 a). A category scale with ratio properties for intermodal and interindividual comparison. In *Psychophysical Judgement and Process of Perception*. Berlin.

Borg, G. A. V. (1982 b). Psychophysical bases of perceived exertion. *Med. Sci. Sports, 14*, 377-381.

Borg, G. A. V. (1990). Psychophysical scaling with applications in physical work and the perception of exertion. *Scand. J. Work Envirom Health, 16 (suppl 1)*, 55-58.

Brown, J. R. (1971). *Lifting as an industrial hazard*. Labour Safety Council of Ontario, Ontario Department of Labour, Toronto, Ontario, Canada.

Chaffin, D. B., and Park, K. S. (1973). A longitudinal study of low back pain as associated with occupational lifting factors." *AIHA Journal, 34*, 513-525.

Hafez, H. A., Gidcumb, C. F., Reeder, M. J., Beshir, M. Y., and Ayoub, M. M. (1982). Development of a human atlas of strength. In *Proceedings of the 26th Annual Meeting of the Human Factors Society*, 575-579.

Jones, D. F. (1972). Back injury research: a common thread. *AIHA Journal*, 596-602.

Karwowski, W. (1982). *A fuzzy sets based model on the interaction between stresses involved in manual lifting tasks*. PhD dissertation, Texas Tech University.

Karwowski, W. (1983). A pilot study of the interaction between physiological, biomechanical and psychophysical stresses involved in manual lifting activities." In K. Coombes (Ed.), *Proceedings of the Ergonomics Society's 1983 Conference*. Taylor & Francis: London, 95-100.

Klein, B. P., Roger, M. A., Jensen, R. C., and Sanderson, L. M. (1984). Assessment of workers' compensation claims for back sprain/strains." *J. Occ. Med., 26*, 443-448.

Kroemer, K. H. E., Kroemer, H. J., and Kroemer-Elbert, K. E. (1990). *Engineering Physiology Bases of Human Factors/Ergonomics*. Van Hostrand Reinhold: New York.

NIOSH. (1981). *Work practices guide for manual lifting*. NIOSH: Cincinnati, OH.

Ridyard, D. T. (1990). A successful applied ergonomics program for preventing occupational back injuries." In B. Das (Ed.),*Advances in Industrial Ergonomics and Safety II*. Taylor & Francis: London, 125-132.

Snook, S. H., and Ciriello, V. M. (1972). Low back pain in industry. *ASSE Journal, 17*, 17-23.

Snook, S. H., Campanelli, R. A., and Hart, J. W. (1978). A study of three preventive approaches to low back injury. *J. Occ. Med., 20*, 478-481.

AGE EFFECTS IN BIOMECHANICAL MODELING OF STATIC LIFTING STRENGTHS

Don B. Chaffin, Charles B. Woolley, Trina Buhr, Lois Verbrugge
The University of Michigan

ABSTRACT

There is growing awareness that age results in reduced strengths in the population, and that significant decreases start in the 5th decade. The magnitude of the decrease in strength depends on the specific muscle function being tested. Because of differential effects it is not clear how various decreases could alter whole-body strength performance. This paper describes how specific strength decreases measured in an older population of men and women could affect their whole-body exertion capabilities in selected scenarios. A computerized strength prediction program is used to both predict the whole-body strength changes with age, and to study how older populations can alter their postures to achieve maximum exertion capability. The results indicate that different muscle group strengths decline by 5% to 70% with age, depending on which muscle group is tested. These changes have profound effects on whole-body exertion capabilities, which also are shown to depend on specific postures used to perform the exertions.

INTRODUCTION

There is a general recognition in the literature that age changes in muscular strengths are profound. A recent report of the National Research Council (Czaja and Guion, 1990) summarizes strength performance declines with age as follows:

"There is, on average, a decrease in muscle mass with age, which results from a decrease both in the number and the size of muscle fibers. Translated into performance, there are rough estimates that by age 40 average muscle strength is about 95 percent of an earlier maximum in the late 20s; by age 50 it drops to about 85 percent; and by age 65 only 75 percent of the earlier output is still available, with further declines thereafter. However, these are population mean differences and there is a great deal of variability in different muscle groups, in types of muscular performance and individuals."

Viitasalo et al. (1985) compared the static strengths of healthy Finnish men in cohorts of 31-44 years and 71-75 years. They showed average declines of about 30% to 50%, depending on specific muscle groups.

It is the intent of this paper to describe how age affects a selected set of muscle strengths, and how the resulting decreased strengths may affect whole-body lifting capabilities.

METHODS

This study involves two phases. The first phase required the development of a set of static strength data on specific muscle functions for an older population of healthy men and women. Fortunately, data of this type were available from a study of 98 men and women of ages 40 to 80+ years living in the Ann Arbor area, conducted by these investigators (Verbrugge et al. 1991). Table 1 provides some descriptive statistics on these people. In this study 12 different bilateral isometric strengths were measured using a specially designed strength testing chair (Verbrugge et al. 1994). For the purpose of this paper the data were expressed as strength moments for each muscle functions. This provided a set of strength norms for this older population which could then be compared to the population norms now used in the University of Michigan's 2D Static Strength Prediction Program™ (Chaffin and Andersson, 1991). The latter values were collected from several different population studies of younger men and women (18-48 years) employed in manual labor.

In phase two of this investigation the new static strength values were inserted into the existing 2DSSPP™ program. The program was then used to simulate a set of whole-body lifting exertions with both the younger and older aged population norms used for reference. The lifting postures chosen were those available from previous studies (see Chaffin

Ergonomics and Musculoskeletal Disorders

and Andersson, 1991 for summary). Incremental changes in the postures were then made to evaluate how postural compensation may affect the predicted lifting strengths.

RESULTS

Differences in the mean strength values between the older population and the existing norms are shown in Figure 1.

Table 1 Older Population Statistics (Verbrugge et al. 1991)

	Men	Women
Mean Age (yrs)	71.8	74.3
Mean Stature (m)	1.77	1.60
Mean Body Wt. (Kg.)	80.5	62.0
Sample Size	18	80

Total 98 =

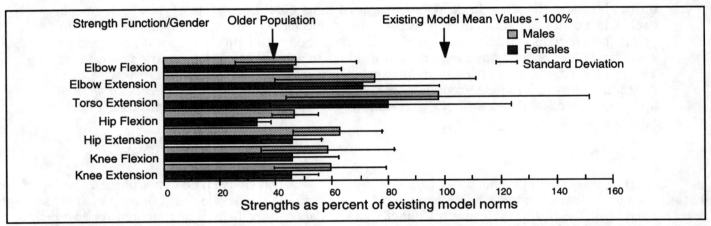

Figure 1: Comparison of older population strengths with existing norms in UM-2DSSPP™

| Arm Lift | Torso Lift | Leg Lift | Floor Lift |

Predicted Lifting Force Capabilities for 50%tile Strengths—(No back compression limit used):

	Arm Lift	Torso Lift	Leg Lift	Floor Lift
Existing Model Male (lbs):	98	112	149	209
Existing Model Female (lbs):	55	62	99	105
Older Population Male (lbs):	42	39	25	81
Older Population Female (lbs):	23	0	0	9

Figure 2: Comparison of Model Predicted 50%tile Lifting Strengths for Existing Norms and Older Population Norms

Four arbitrary but different lifting postures were simulated. The postures and 50%tile lifting capability predictions for the original chosen postures are shown in Figure 2. From inspection it is clear that the lifting capabilities of older men and women is severely compromised by decreases in specific muscle strengths. In this latter regard, knee and hip extension strength declines had the greatest effects for the postures used in the simulations, especially for older females. In fact, it was predicted that 50% of older females would not be able to lift any weight in the torso and leg lift postures.

Clearly the exact postures chosen by individuals attempting a whole-body maximum exertion are critical, and will depend to a large extent on individual muscle strengths at various joints. The reason the older female population was predicted to have no useful lifting capability in the initial torso and leg lift postures illustrated in Figure 2 is because of hip and knee extension strength limits, respec-

tively. By modifying the initial postures to both reduce the joint moment requirements at those joints and increase the moment loads at the joints having higher strength capabilities, increased lifting capabilities were predicted. This is illustrated in Figure 3, wherein lifting capabilities associated with four slightly modified postures are compared to the initial, arbitrary postures for the older female populations.

Though an exhaustive study of strength and postural effects has not been undertaken in this study, the initial results certainly indicate why it is so important to use a model of whole-body exertions to study the complex interactions. This is particularly true when interest in special populations, wherein muscle strengths will not consistently or proportionally vary from one muscle group to the next. Thus a posture which produces high exertion capabilities in one population may not be appropriate for another population.

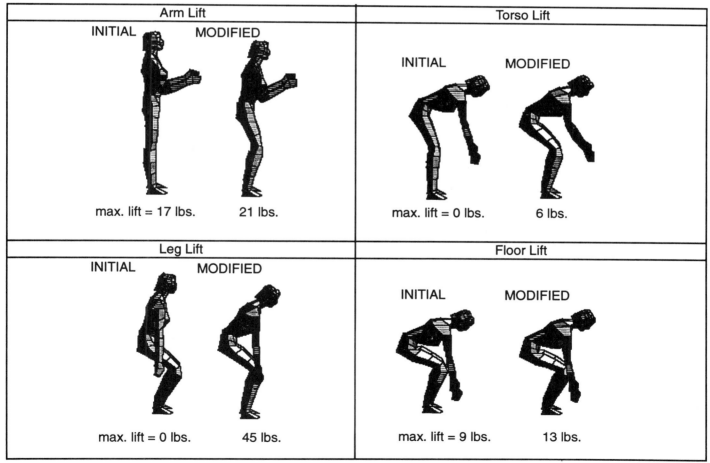

Figure 3. Comparison of Predicted Lifting Strengths in Two Different Postures for Older Female Population.

DISCUSSION

This investigation demonstrates how important specific muscle strengths are in providing whole-body exertion capabilities. Though limited by the availability of normative strength data on older populations, (e.g., shoulder strength declines with age could drastically affect lifting to high shelves) the initial results certainly demonstrate the magnitude of the problem of aging in such physical tasks. Additional tasks are being simulated as part of this investigation.

As reported by several investigators, small changes in postures can have a large effect on predicted, whole-body strength performance (Chaffin and Erig, 1991). Since the declines in muscle strengths with age are not proportional for different muscle groups, it is not clear which types of exertions are most sensitive to postural compensations. Further simulations are planned to explore this aspect of the problem.

Lastly, it is well documented that low-back pain and injury prevalence rates increase in older populations (Biergin-Sorenson, 1982). In the lifting simulations used in this investigation, the higher lifting strengths in the younger populations resulted in much higher spinal disc compression forces than was the case for the older populations. The high spinal compression forces in the young male population may explain why the incident rate of low back pain is high for young men in heavy manual labor (Chaffin and Page, 1993). One could conjecture that reduced back and hip muscle strengths in older populations are protective of the spinal column, even with its reduced compression tolerance in later years (Chaffin and Ashton-Miller, 1991). This conjecture is to be studied more fully, but one must wonder what specific lifting tasks may be particularly hazardous to the low-back of older individuals.

ACKNOWLEDGMENTS

The authors with to acknowledge partial support from the Multipurpose Arthritis Center NIH Grant No. AR20557, and the Whitaker Foundation Predoctoral Fellowship Program.

REFERENCES

Biergin-Sorenson, F. (1982) Low back trouble in a general population of 30-, 40-, 50-, and 60-year old men and women. Study design representativeness, and basic results, *Danish Medical Bulletin*, 29(6), 289-299.

Chaffin, D.B. and Andersson, G.B.J. (1991) *Occupational Biomechanics*, J. Wiley, New York.

Chaffin, D.B. and Ashton-Miller, J.A. (1991) Biomechanical aspects of low-back pain in the older worker. *Experimental Aging Research,* 17(3), 177-187.

Chaffin, D.B. and Erig, M. (1991) Three-dimensional biomechanical static strength prediction model sensitivity to postural and anthropometric inaccuracies, *IIE Transactions*, 23(3):215-227.

Chaffin, D.B. and Page G. (1993) Postural Effects on Biomechanical and Psychophysical Weight-lifting Limits. *Ergonomics,* 37(4):663-676.

Czaja, S.J. and Guion, R.M. (1990) Human factors research needs for an aging population, Panel on Human Factors Research Issues for an aging population,

Verbrugge, L.M., Reoma, J.R., Woolley, C.B., Chaffin, D.B. and Wery, S.D. (1991) The results of the field testing of quantitative musculoskeletal function in older workers. Report to the National Institute of Occupational Institute of Occupational Safety and Health. NIOSH Contract RFQ 90-40,

Verbrugge, L.M., Chaffin, D.B., Woolley, C.B., Sekulski, R.A. (1994) Musculoskeletal functioning in older adults. To be presented at Symposium on Ergonomics and Aging Research (IEA 1994)

Viitasalo, J.T., Era, P., Leskinen, P.A. and Heikkinen, E. (1985) Muscular strength profiles and anthropometry in random samples of men aged 31-35, 51-55 and 71-75 years, *Ergonomics,* 28(1), 1563-1574.

POSTURAL STABILITY WHILE WALKING AND CARRYING LOADS IN VARIOUS POSTURES

M.A. Holbein, Ph.D.[1] and M.S. Redfern, Ph.D.[1,2]
University of Pittsburgh
Departments of 1) Industrial Engineering and 2) Otolaryngology
Pittsburgh, PA

Falls, over-exertion injuries and other potential consequences of balance losses continue to be serious ergonomic concerns. Stability issues are important in the prevention of these injuries, especially when the task is complicated by handling loads. However, stability analyses are not typical components of ergonomic job analyses. This study demonstrated that stability assessments can be effective in recommending load-carrying strategies. In particular, the effects of load positioning and magnitude on stability were investigated. Unladen walking was also tested for comparison. Several stability measures were defined based on the body-and-load's center of mass displacement in the frontal plane. Statistical differences among the load positions and magnitudes were found and are discussed. Results were consistent across measures. Additional work is needed to better define the limits of stability while carrying and to relate these, or other, stability measures to the likelihood of a balance loss.

INTRODUCTION

Postural stability is one of several ergonomic issues relevant to manual material handling tasks. However, it has generally not been explored, especially compared to other issues, such as low back stresses and psychophysical load-handling limits. Injuries may result from a loss of stability whether balance is recovered or a fall results. Stability losses while handling loads may lead to low back injuries, musculoskeletal overexertions, and the consequences of colliding with the load or the surroundings.

Stability while walking has recently received research attention, although these studies have not addressed load-carrying (Redfern and Schumann, in press; MacKinnon and Winter, 1993; Winter, 1987). Also, several physiological and biomechanical studies have investigated carrying loads in various postures, with contradictory results (Neumann et al., 1992; Nottrodt and Manley, 1989; Ghori and Luckwill, 1985; Bobet and Norman, 1984). The purpose of this study was to investigate load-carrying from a postural stability perspective. The main objective was to determine the effects of load positioning on stability while walking and carrying. Effects of load magnitude on stability were also investigated. The stability measures were based on the body-and-load's center of mass (COM) sway

perpendicular to the path of progression, i. e. in the frontal plane.

METHODS

Apparatus

The load was held in a 26x33x26 cm box with padded handles. Fifteen reflective markers were taped to the subject's left and right sides at relevant anthropometric landmarks (Holbein, 1993). A marker was also centered on the load. Four digital video cameras were focused on the center of a level 6 m walkway from different views to record three-dimensional human movement.

Protocol

Fifteen male subjects (mean age 24.7 years, s.d.=4.7 years) were paid to participate. Their natural cadence for unladen walking was found and trials were conducted at this cadence. They were handed the load and instructed to look forward and walk in a straight line for the full length of the walkway. A middle stride between successive right heelstrikes was analyzed. The maximum available walkway was used to insure that steady-state gait was attained prior to the stride of interest and maintained throughout it.

Experimental Design

Five load-carrying postures (Figure 1) were chosen based on frontal plane symmetry and height of the load. The load was held on the dominant side, usually the right, in the one-handed asymmetric postures. Two levels of load magnitude were tested, an empty box and an 11.4 kg load. The experimental design was a full-factorial of the eleven posture and load combinations and was replicated three times. All trials were randomized separately for each subject.

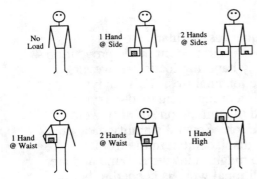

Figure 1: Load-carrying postures.

Dependent Variables

The body's COM was estimated using the video data and anthropometric data (Dempster, 1955) and then combined with the load position to estimate the body-and-load system's COM. (Details of the method were presented by Holbein, 1993.) Dependent variables were defined to describe mediolateral displacements of the body-and-load system's COM (COM ML) from the path of progression (Figure 2). The path of progression was defined as a straight line from mid-ankles at the beginning and the end of the stride. The average COM ML displacement to the dominant and non-dominant sides of the path were calculated separately (cm_d, cm_nd). The maximum deviation was cm_max. The range of COM ML deviation from the path was the sum of the maximum deviations to the left and right (cm_range). Finally, cm_sq was a measure of the average deviation (cm_sq=square root of sum of squared deviations/N).

Statistical Analysis Techniques

Several statistical techniques were used in analyzing the data, with a 0.05 significance level applied throughout. Contrasts were defined to investigate the effect of adding an external load to the body. Each laden posture and each load magnitude were compared to unladen walking. Further analysis of the laden conditions included ANOVA and Tukey's Honestly Significant Difference Tests. Finally, linear functions of the postures were defined to test the symmetry and height characteristics of this variable.

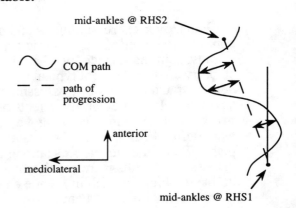

Figure 2: Mediolateral COM displacements with respect to the path of progression. (Compensation was made if subject did not walk directly ahead.)

RESULTS

Unladen walking was compared to carrying a load by defining specific contrasts. Carrying asymmetrically always resulted in significantly larger COM ML responses than walking unladen, with the exception of smaller cm_d values. Thus, carrying a load on one side of the body, compared to not carrying anything, increased COM ML displacements overall and to the unloaded side, and decreased displacements to the loaded side. Table 1 illustrates this for one asymmetric posture (1-Hand-High). Comparisons of the two load magnitudes with unladen walking showed these same results, i. e. larger displacements except smaller cm_d values, whether an empty or loaded box was carried.

Table 1: Mean (and s.d.) COM mediolateral responses (cm): Overall and some specific postures. (overall N=495, unladen n=45, laden postures n=90)

Dep. Measure	Overall	Unladen	1-Hand-High	2-Hands-at-Waist
cm_sq	0.34 (0.14)	0.26 (0.07)	0.48 (0.14)	0.26 (0.07)
cm_nd	2.56 (1.15)	1.91 (0.79)	3.71 (1.10)	1.80 (0.76)
cm_d	0.93 (0.88)	1.53 (0.72)	0.18 (0.41)	1.48 (0.82)
cm_max	4.41 (1.61)	3.50 (1.06)	6.05 (1.54)	3.50 (0.96)
cm_range	5.58 (1.41)	5.33 (1.36)	6.33 (1.39)	5.15 (1.22)

The overall mean COM ML displacement to the non-dominant side (cm_nd), averaged across all subjects, postures and loads was considerably greater than that to the dominant side (cm_d) (Table 1). This trend is evident especially for the one-handed posture, but also for the unladen and two-handed posture shown. Furthermore, the maximum COM ML deviation (cm_max) was to the non-dominant side in about 65% of the trials when walking unladen or carrying with two hands, and almost exclusively to the non-dominant/contra-lateral side for asymmetric postures.

Subsequent analyses focused on differences among carrying conditions, therefore excluding unladen walking. ANOVA showed that the posture and load conditions and their interaction significantly affected all the COM ML measures defined. Further analysis showed that the various postures resulted in COM ML responses from the greatest to smallest magnitudes in the following order: 1-hand-high, 1-hand-at-waist, 1-hand-at-side, 2-hands-at-side, 2-hands-at-waist. The opposite order was found for cm_d. Statistically, holding with 1-hand-high produced greater (smaller for cm_d) COM ML displacements than every other posture, and 1-hand-at-waist resulted in larger responses (smaller for cm_d) than all but 1-hand-high. Finally, 1-hand-at-side produced statistically larger (smaller for cm_d) displacements than 2-hands-at-waist. These pairwise differences were significant especially when holding the heavier load and often found with the lighter load.

Contrasts defined to investigate the load symmetry and height characteristics of the postures were consistent with the above findings. Overall, the asymmetric postures resulted in more COM ML movement (less for cm_d) than their similar two-handed counterparts. These contrasts also showed that COM ML displacements increased (decreased for cm_d) as the load was held higher.

Differences between the two load magnitudes were also investigated. The cm_sq, cm_nd and cm_max responses were statistically larger when carrying the heavier load with 1-hand-high or 1-hand-at-waist. The cm_range did not respond the same as these measures. Increased cm_range was found with the heavier load on the shoulder, compared to the empty box, but smaller ranges resulted when carrying the heavier load in most other postures. Finally, cm_d was negatively affected by increased load magnitudes.

CONCLUSIONS AND DISCUSSION

Larger COM ML displacements may indicate reduced stability. The subjects' goal was to walk in a straight line and the goal of the postural control system is to remain upright while carrying. One measure of the difficulty in achieving these goals may be the amount of mediolateral movement or compensation of the body-and-load system. For example, the contra-lateral COM compensation while carrying with one hand may be an attempt to counteract ipsi-lateral moments of force, caused by the load, acting to collapse the body toward that side. By this definition of stability, asymmetric carrying was less stable than symmetric carrying and unladen walking. Carrying the load higher off the ground also reduced stability.

In terms of the specific measures, rather than of the interpreted definition of stability, larger contra-lateral and overall COM ML displacements were recorded for asymmetric postures with the load held higher. In previous studies, increased muscle activity was recorded for loads placed contra-laterally (Neumann et al., 1992) and higher (Bobet and Norman, 1984). The results of this and the previous studies are consistent since they all indicate that larger compensations or responses are required for specific tasks.

The conclusions presented here regarding load positioning and magnitude effects on stability are very similar to those reported in previous investigations of load-holding by Holbein and Redfern (1993a; 1993b). In these studies, the same postures and loads tested here were studied for their effects on functional stability limits and on responses to support-surface perturbations while standing and holding loads. The similarities in the conclusions across all tasks in these studies demonstrates a potential for developing general material handling guidelines from a postural stability perspective. Future research may identify situations where recommendations based on stability corroborate and contradict those based on other approaches, such as biomechanics and psychophysics.

Based on the analysis and assumptions presented here, recommendations were made for load positioning while carrying which could result in increased stability. However, there were several limitations of this study. First, although the dynamic components of the carrying task likely affect postural stability, they were not analyzed. It is not known whether stability measures incorporating these dynamic components would significantly change the recommendations. It is also recognized that several assumptions were necessary to estimate individuals' whole-body center of mass. Future research is needed to address the sensitivity of these estimates to errors in standard anthropometric data used as input.

ACKNOWLEDGMENTS

The authors acknowledge M. Reza Ghajarnia and Daniel A. Jacob for data collection and processing assistance. M.A. Holbein is currently at the University of Michigan, Center for Ergonomics, Ann Arbor, MI.

REFERENCES

Bobet, J. and Norman, R.W. (1984). Effects of load placement on back muscle activity in load carriage. European Journal of Applied Physiology, 53, 71-75.

Dempster, W.T. (1955). Space Requirements of the Seated Operator, WADC-TR-55-159. Ohio: Aerospace Medical Research Laboratories.

Ghori, G.M.U. and Luckwill, R.G. (1985). Responses of the lower limb to load carrying in walking man. European Journal of Applied Physiology, 54, 145-150.

Holbein, M.A. (1993). Ergonomic Load-Holding and Carrying Strategies Based on the Biomechanics of Human Motion and Postural Stability. Unpublished doctoral dissertation, University of Pittsburgh, Pittsburgh, PA.

Holbein, M.A. and Redfern, M.S. (1993a). Functional stability limits while holding loads in various postures. In Proceedings of the Second IE Research Conference.

Holbein, M.A. and Redfern, M.S. (1993b). Postural stability while holding loads in various postures. In Proceedings of the Human Factors and Ergonomics Society 37th Annual Meeting. Santa Monica, CA: Human Factors Society.

MacKinnon, C.D. and Winter, D.A. (1993). Control of whole-body balance in the frontal plane during human walking. J. of Biomechanics, 26(6), 633-644.

Neumann, D.A., Cook, T.M., Sholty, R.L. and Sobush, D.C. (1992). An electromyographic analysis of hip abductor muscle activity when subjects are carrying loads in one or both hands. Physical Therapy, 72(3), 207-217.

Nottrodt, J.W. and Manley, P. (1989). Acceptable loads and locomotor patterns selected in different carriage methods. Ergonomics, 32(8), 945-957.

Redfern, M.S. and Schumann, T. (In Press). A model of foot placement during gait. Journal of Biomechanics.

Winter, D.A. (1987). Sagittal plane balance and posture in human walking. IEEE Engineering in Medicine and Biology Magazine, (September), 8-11.

RELATIONSHIPS BETWEEN THE NIOSH (1991) LIFTING INDEX, COMPRESSIVE AND SHEAR FORCES ON THE LUMBOSACRAL JOINT, AND LOW BACK INJURY INCIDENCE RATE BASED ON INDUSTRIAL FIELD STUDY

Waldemar Karwowski, Marenda Caldwell, and Paul Gaddie
Center for Industrial Ergonomics, Department of Industrial Engineering
University of Louisville, Louisville, Kentucky 40292
Tel 502 852-7173; Fax 502 852-7397

The main objective of this study was to investigate relationships between the values of NIOSH (1991) Lifting Index calculated for as set of industrial manual handling tasks, the corresponding (estimated) compressive and shear forces on the lumbosacral joint (L5/S1), and the back injury incidence rates based on analysis of the epidemiological field data. A strong positive correlation was observed between the estimated compressive forces on L5/S1 and the lifting index (LI) values, and between the incidence rates (IR) of low back injury and the LI index. Two sets of regression models describing the relationships between the lifting index (LI) and the compressive forces on the L5/S1 were developed. It was shown that the LI=1.0 corresponds to about 1.8 kN and 2.4 N of compression on the L5/S1, for the destination and origin of the lift, respectively. For the lumbar compressive strength values (with safety factor of one standard deviation) of 4.1 kN for males (40 years of age), proposed by Jager and Luttman (1992), the corresponding values of the lifting index are as follows: LI=4.1 (for the lift destination) and 6.4 (for the lift origin). Implications of results for prevention of back injury due to manual lifting were discussed.

BACKGROUND

As discussed by Chaffin and Andersson (1993), many epidemiological studies have shown the importance of the spinal loading on the incidence of low back pain and risk of occupational low back disorders. For example, Herrin et al (1986), concluded that the biomechanical criterion of maximal back compression appears to be a good predictor not only of risk of low back incidents but of overexertion injuries in general. They also showed that the incidence of low back problems was 1.5 times lower for lifting jobs with compressive forces below 4.5 kN than for the job with predicted compressive forces between 4.5 kN and 6.8 kN, and Recently, Leamon (1994, in press) discussed validity of various criteria for the prevention of occupationally induced low back pain disability, and pointed out, that the correlation reported by Herrin et al (1986), between the low back injury incidence rate and the maximum back compression was 0.13, while the correlation between low back injury incidence rate and maximum back compression and all overexertion injuries was 0.17.

Jager and Luttman (1992), investigated the maximum load bearing capacity of the lumbar spine, and concluded that the lumbar compressive strength at an age of 40 years is approximately 6.7 kN for males and 4.7 kN for females. They also suggested the need to reduce these values to 4.1 kN for males and 3.2 kN for females in order to avoid overestimation of the individuals lumbar strength. It should be noted that the predicted compressive forces on the L5/S1 joint of 3.4 kN and 6.4 kN, respectively, correspond to the NIOSH's (1981) Action Limit and Maximum Permissible Limit values.

Chaffin and Park (1973), suggested that the low back pain incidence in repetitive lifting was less than 5% when the predicted compressive force on L5/S1 was below 2.5 kN, but it increased to over 10% when the predictive L5/S1 compression exceeded 4.5 kN. Even though according to the revised NIOSH (1993) position, the limit of 3.4 kN on the L5/S1 may not protect the entire workforce, the biomechanical limit of 3.4 kN was maintained in development of the 1993 Revised Lifting Equation. Furthermore, the NIOSH (1991) document introduced a single value of the Recommended

Weight Limit (RWL), which was designed to be acceptable for the 90% of female workforce population. The concept of the Lifting Index (LI), defined as the ratio between the actual load lifted on the job and the RWL, was also introduced.

Evaluation of the Revised (1991) Lifting Equation should include an assessment of practical implications that the adoption of the recommended weight limit (RWL) will most likely have in preventing low back injury. NIOSH (1993) pointed out that *"In theory, the magnitude of LI may be used as a gauge to estimate the percentage of workforce that is likely to be at risk for developing lifting-related low-back pain. The shape of the risk function, however, is not known"*.

OBJECTIVES

NIOSH (1993), pointed out that at present time,there is uncertainty whether a value of LI=1.0 is a reliable boundary for differentiating between the low and high risk for the working population with respect to the low back injury or pain. NIOSH (1993) also suggested that the prospective studies are needed to identify the compressive force levels at L5/S1 joint that increase the risk of low back injury. The main objective of this study was to investigate the relationships between the calculated NIOSH (1993) Lifting Index values for industrial lifting tasks, the corresponding (estimated) compressive forces on the lumbosacral joint (L5/S1), and the back injury incidence rates based on analysis of the epidemiological field data.

METHODS AND PROCEDURES

The main steps in this study were as follows: 1) review of the extent of musculoskeletal injuries at the plant in terms of their frequency and severity (based on the incidence rates), 2) identification of critical lifting jobs which pose the greatest health hazards to the workers, and 3) 3-d biomechanical analysis of the lifting jobs and estimation of the compressive and shear forces on L5/S1, 4) calculation of the RWl and the LI values for the analyzed jobs, and 5) a comprehensive comparison of the relationships between the lifting index (LI), compression on the L5/S1 for the evaluated tasks, and the corresponding injury rates.

For the purpose of analysis, the plant was divided into departments: packaging, blending, muffins, maintenance, shipping/receiving, sanitation, and management. Examination of the plant's health and safety records revealed that more than sixty percent of the low back injuries (during the 1991-1993 years) were associated with the

Blending and Packaging Departments. A total of 24 lifting tasks were identified and further analyzed.

Injury data was attained from company accident logs, OSHA Form 200 records, and daily production sheets. These records revealed the type of injury, date of injury, department and job title of the injured, as well as any days loss from work due to the injury. Data covered the period between July 1990 through June 1993. The incidence and severity rates were determined for the entire plant, by department, year, and job title. Using company worker's compensation claims, a cost was associated with each department, job title, and body part injury. The body part most injured was the back, followed by the hand/finger(s).

Ergonomic task analysis

Each of the 24 lifting tasks for the six targeted jobs were observed and video taped. The lifting guidelines established by NIOSH (1993) were used to establish the acceptability of the lifting activities associated with these tasks. The compressive and shear forces exerted on the lower back (L5/S1) during lifting tasks were determined using the 3DSSPP (1993) biomechanical model developed at the University of Michigan's Center for Ergonomics.

RESULTS AND DISCUSSION

The weight of loads lifted varied from 1.13 kg to 45.14 kg (2.5 lbs to 100.3 lbs). Many worker postures involved twisting which caused asymmetric lifting of loads. Eighteen of the twenty-four lifts (75%) were asymmetrical either at the origin and/or destination. The frequency of lifts ranged from 0.2 to 6.0 lifts per minute.

The calculated lifting index (LI) values ranged from 0.62 to 13.36. The estimated compression forces on the L5/S1 joint ranged from 1,956 to 9,134 Newtons at the origin of lift, and from 730 to 6,748 Newtons at the destination of the lift. The shear forces ranged from 302 to 852 Newtons at the origin of lift, and from 217 to 624 at the destination of lift. The results showed that only 12.50% of the lifting tasks were acceptable under the 1981 Guide, and 4.17% under the 1993 Guide, if the criterion of LI=1.0 is adopted. Of the twenty-four tasks, the back compression force design limit of 3400 N was exceeded in 62.5% of all tasks, while the upper (permissible) limit of back compression of 6400 N was exceeded in 16.67% of the tasks.

Comparison of task evaluations

In twenty-two of the twenty-four cases, the (1993) RWL was found to be lower than the AL found using the 1981 lifting equation. Only two of the twenty-four lifts resulted with a RWL greater than the AL. According to the NIOSH (1993) design criterion of LI=1.0, all but one lifting tasks was deemed not acceptable, as compared to twenty-one of twenty-four lifts under the NIOSH (1981) guide.

Correlation analysis

A strong positive correlation was observed between the (maximum) compressive forces on L5/S1 and the (maximum) lifting index for the job (r = 0.88, p = 0.05). A high correlation was also found between the estimated (average) compression forces on L5/S1 and the (average) lifting index (LI) values (r = 0.867, p = 0.057).

Significant correlations were also found between the (maximum) incidence rate (IR) of low back injury and the (average) RWL (r=0.97, p=0.004). In addition, the correlation coefficients between the incidence rate (IR) of low back injury for 1992 and the (maximum) LI (r=0.94, p=0.016), and between the incidence rate (IR) of low back injury for 1993 and the LI (r=0.967, p=0.007), were also significant.

The above results allowed to perform the regression analysis and develop two sets of models (for the lift origin and destination) describing the relationships between the lifting index (LI) and the compressive forces on the L5/S1 (see Figures 1 and 2), and another two sets of models for the relationship between the lifting index (LI) and the shear forces on L5/S1 (see Figures 3 and 4). These models explained between 41% and 73% of the variance in the data.

The developed models allowed to relate values of the lifting index (LI), to the specific levels of the (estimated) compression on the L5/S1 joint. It was shown that the LI=1.0 corresponds to about 1.8 kN and 2.4 N of compression on the L5/S1, for the destination and origin of the lift, respectively. These results can be compared to those reported by Chaffin and Park (1973), who suggested that the low back pain incidence in repetitive lifting was less than 5% when the predicted compressive force on L5/S1 was below 2.5 kN.

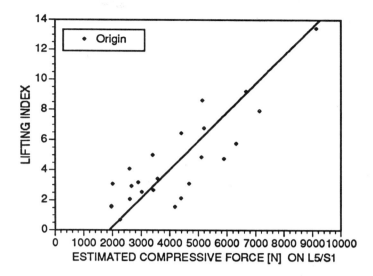

$$LI = 0.00138386 \,(CF) - 1.418 \quad [R^2 = 0.73]$$

Figure 1. Relationship between the compressive forces (CF) on L5/S1 and the NIOSH's Lifting Index at origin of lift.

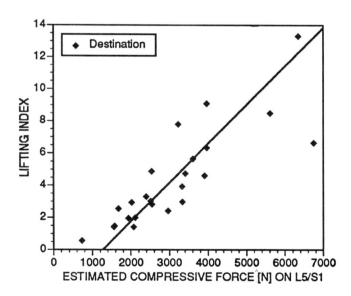

$$LI = 0.00170275 \,(CF) - 0.87125 \quad [R^2 = 0.705\}]$$

Figure 2. Relationship between the compressive forces (CF) on L5/S1 and the NIOSH's Lifting Index at destination of lift.

For the lumbar compressive strength values (with safety factor of one standard deviation) of 4.1 kN for males (40 years of age), proposed by Jager and Luttman (1992), the corresponding values of the lifting index were as follows: LI=4.1 (for lift destination) and 6.4 (for lift origin). Furthermore, it was shown that the LI=3.3 (lift origin) and LI=4.9 (lift destination) values correspond to the (1981) AL of 3.4 kN compression on L5/S1, while the LI=7.4 (lift origin) and LI=10.0 (lift destination) values correspond to the (1981) MPL of 6.4 kN.

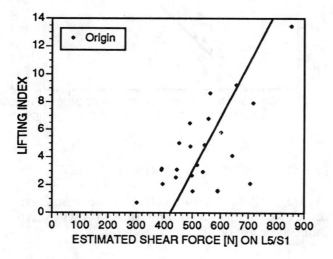

$$LI = 0.0157859 \, (SF) - 4.0753 \quad [R^2 = 0.41]$$

Figure 3. Relationship between the shear forces (SF) on L5/S1 and the NIOSH's Lifting Index at origin of lift.

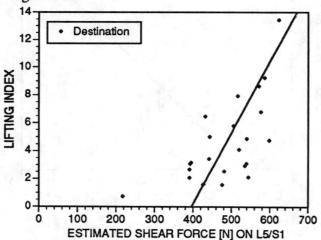

$$LI = 0.02146 \, (SF) - 5.9597 \quad [R^2 = 0.42]$$

Figure 4. Relationship between the shear forces (SF) on L5/S1 and the NIOSH's Lifting Index at destination of lift.

CONCLUSIONS

In conclusion, the results of this field study offer new empirical data that can be used for interpretation and validation of the Revised NIOSH (1991) Lifting Equation.

REFERENCES

CIRIELLO, V. M., SNOOK, S. H., BLICK, A. C. and WILKINSON, P. L., 1990, The effects of task duration on psychophysically-determined maximum acceptable weights and forces, *Ergonomics*, **33** (2), 187-200.

CHAFFIN, D. B., HERRIN G. D. and KEYSERLING, W. M., 1978, Pre-employment strength testing - an updated position, *Journal of Occupational Medicine*, **20**(6), 403-408.

JAGER ,M. and LUTTMAN, A., 1992. The load on the lumbar spine during assymetrical bi-manual materials handling. Ergonomics, 25, 783-805.

LEAMON, T. B., 1994, Research to reality - a critical review of the validity of various criteria for the prevention of occupationally induced low back pain disability, *Ergonomics*, (accepted for publication).

MILLAR, J. D., 1988, Summary of proposed national strategies for the prevention of leading work-related diseases and injuries, part 1, *Amer. J. Ind. Med.*, **13**, 223-240.

MINISTRY OF DEFENSE, London, 1984, Defense standard, *Human Factors for Designers of Equipment, Part 3: Body strength and Stamina*, MD 00-25, United Kingdom.

NIOSH, 1981, *Work Practices Guide for Manual Lifting*, NIOSH Technical Report No. 81-111, U.S. Department of Health and Human Services, National Institute for Occupational Safety and Health, Cincinnati, OH.

WATERS, T., PUTZ-ANDERSON, V., GARG, A. and FINE, L., 1993, Revised NIOSH equation for the design and evaluation of manual lifting tasks. *Ergonomics*, **36**(7), 749- 776.

SPENGLER, D. M. J., BIGOS, S. J., MARTIN, N. A., ZEH, J., FISHER, L. AND NACHEMSON, A., 1986, Back injuries in industry: a retrospective study, *Spine*, **11**, 241-256.

THE EFFECTS OF TASK CONDITIONS ON TRUNK MUSCULAR FATIGUE DURING DYNAMIC LIFTING

Sang H. Kim, Min K. Chung and Wook G. Lee

Department of Industrial Engineering
Pohang University of Science and Technology, Pohang, Korea

To investigate the effects of MMH task conditions on the activity and fatigue of the trunk musculature, EMG signals from eight major trunk muscles were analyzed during 120 minutes of repetitive dynamic lifting tasks. Two independent task variables were the work strategy of weight and lifting frequency combination and the body posture. The dependent variables were the amplitude of EMG signals and the amount of median power frequency (MPF) decrease over time for the eight trunk muscles. The results of the study indicated that the recruitment and the level of force exertions of the trunk muscles during manual lifting tasks are a function of the direction and the weight of the external load. The different activities of the muscles due to given task conditions also influence the fatiguing process of the individual muscle. The muscles in the dorsal part of trunk were activated during the symmetric task conditions, while the muscles on the contralateral side to the workload were more strongly activated during the asymmetric task conditions. The decreasing trends of MPF were found in some activated muscles, and they were more pronounced for the asymmetric posture than for the symmetric posture. It was also seen that the muscles became fatigued faster for light load-high frequency conditions than for heavy load-low frequency conditions.

INTRODUCTION

Although many jobs are becoming mechanized and automated today, workers are still required to perform manual materials handling (MMH) tasks in many occupational situations. Approximately one third of all industrial jobs in the United States involves some form of MMH such as lifting, lowering, holding, carrying, pushing or pulling (NIOSH, 1981; Cook and Neumann, 1987). Since MMH tasks are performed by using the muscle force of workers, it is important in ergonomic job design to find out the effects of task conditions on muscular activity and fatigue. In this sense, analysis of electromyographic (EMG) signals is generally accepted as an essential methodology in recent MMH studies.

For all the explicitness shown when muscular fatigue is attendant on the repetitive exertion of muscular strength, few studies have considered it. Since muscular fatigue degrades the overall performance of workers, it is critical for MMH job design to evaluate the level of muscular fatigue during a given task condition. In general, the utilization of EMG signals in biomechanical studies of the back have concentrated on the amplitude of EMG activity elicited by certain trunk muscles as an indication of how hard the muscles work during specific loading tasks (Sommerich and Marras, 1992). Most previous studies, which used EMG signals as a measure of fatigue, were limited within the bound of isometric sustained contrac-

tions of some localized muscles for short durations (Arendt-Nielson et al., 1989; Hansson et al., 1992; Van Dieën et al., 1993; Caffier et al. ,1993).

To investigate the effects of MMH task conditions on the activity and fatigue of the trunk musculature, EMG signals from eight major trunk muscles during 120 minutes of dynamic lifting tasks were analyzed in this study.

METHOD

Subjects

Nine healthy male subjects with no history of low back disorders participated in the study. Their mean age, weight and height were 25.2 years (23-28 years), 172.2 cm (165-179 cm) and 63.8 kg (55-71 kg), respectively.

Experimental Design

EMG signals from eight trunk muscles were measured during 120 minutes of repetitive dynamic manual lifting tasks. The eight muscles included the left and right erector spinae (ERSL and ERSR), latissimus dorsi (LATL and LATR), external obliques (EXOL and EXOR), and rectus abdomini (RABL and RABR). These are the primary muscles crossing the transverse plane at the L3/L4 level (Schultz and Andersson, 1981).

Two independent task variables considered in this study were the work strategy of weight of load and frequency of lifting combination and the body posture. Four different task conditions summarized in Table 1 were studied. The dependent variables were the amplitude of EMG activities and the amount of median power frequency (MPF) decrease over time for the eight trunk muscles.

Table 1. Task conditions in this study
(Duration: 120min.)

Category	Weight (%MVIS[*])	Frequency (times/min.)	Body Posture (deg.)
Task S1	10%	6	Symmetric (0)
Task S2	20%	3	
Task AS1	10%	6	Asymmetric (90)
Task AS2	20%	3	

[*] The weight of the load was determined subjectively on the basis of the maximum voluntary isometric strength obtained by the standard static strength testing procedure (Caldwell et al. 1974). Note that the amount of total workload is maintained at the same level for all task conditions.

Procedures

The subjects were given a physical adjustment session of thirty minutes for warming-up and muscle conditioning before the main practice. After the physical adjustment session, EMG signals of the eight trunk muscles were collected for data normalization while the subjects exerted maximum voluntary isometric strengths (MVIS) at each body posture. The subject was asked to lift the load from the floor to the worktable of 72 cm in height and then lower it consecutively in one repetition of a task. The position of the load was in front of the subject for symmetric conditions and on the right hand side of the subject parallel to his mass center for asymmetric conditions.

Eight pairs of surface electrodes were used to collect the EMG signals. Data collection was conducted for the first two repetitions in every 10 minutes. The sampling duration was 7 seconds which was enough to include resting periods of the eight muscles before and after performing the task. The raw EMG signals were prefiltered using an analog Butterworth filter (band-pass 30-360 Hz) and then converted analog-to-digital at a sampling frequency of 1024 Hz.

Data Processing

Root mean square (RMS) amplitude of the raw EMG signals sampled from each muscle within each task condition were obtained subjectively at every 10 msec with window size of 100 msec. It is identical to sample smoothed RMS value at 100 Hz. Average RMS value during activation period for each muscle was calculated and then normalized relative to its maximal and minimal values observed respectively during MVIS trials and resting period.

The power spectral analysis of the EMG signals obtained at every sampling trial (0-12) for each muscle was performed by means of 512 points Fast Fourier Transform (FFT) to investigate the changes of MPF's over time. Thirteen points of the MPF's of each muscle within each task condition were normalized to their initial value obtained at time 0.

RESULTS

RMS EMG Amplitude

To evaluate whether the EMG amplitudes of the eight trunk muscles were dependent on the task conditions, multivariate analysis of variance (MANOVA) procedures were conducted on the RMS data. As shown in Table 2, all muscle activities were significantly affected by posture and frequency/load combinations of the task. Figures 1 through 3 show the mean levels of RMS EMG amplitude of the muscles across the subjects for each task condition.

For the symmetric task conditions (Task S1 and S2), the erector spinae and the latissimus dorsi muscles in the dorsal (posterior) part of trunk were found to be active. As shown in Figures 1 and 2, the mean levels of

Table 2. Statistical results from the MANOVA and ANOVA procedures of the RMS data

Effect	Parameter	MANOVA	ERSL	ERSR	LATL	LATR	EXOL	EXOR	RABL	RABR
Body Posture	F	51.3303	69.45	86.23	27.72	161.10	NT	NT	NT	NT
	df	8, 925	1, 932	1, 932	1, 932	1, 932				
	p	< 0.001	< 0.001	< 0.001	< 0.001	< 0.001				
Work Strategy	F	473.1494	73.48	54.68	162.79	204.42	147.59	45.20	NT	NT
	df	8, 925	1, 932	1, 932	1, 932	1, 932	1, 932	1, 932		
	p	< 0.001	< 0.001	< 0.001	< 0.001	< 0.001	< 0.001	< 0.001		
Body Posture * Work Strategy	F	43.9032	0.45	0.15	73.05	8.46	NT	NT	NT	NT
	df	8, 925	1, 932	1, 932	1, 932	1, 932				
	p	< 0.001	NS	NS	< 0.001	< 0.005				

"NT" indicates that ANOVA procedure was not conducted due to inactivity of the muscle at that condition.
"NS" indicates that the result of the ANOVA procedure is statistically non-significant.

RMS EMG amplitude of the ERSL and ERSR were much greater than that of the LATL and LATR. There were no significant differences in the amplitude between the ERSL and ERSR and between the LATL and LATR. The results indicated that the erector spinae were the prime movers of the workload and that good symmetry was maintained between the left and right sides of the muscles during the tasks.

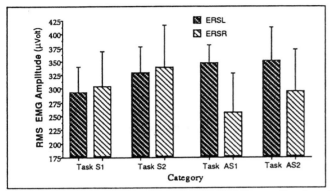

Figure 1. Root mean square (RMS) EMG amplitude of the erector spinae muscles for four task conditions. Mean values and +1 std.'s for nine subjects are shown.

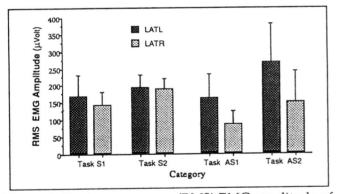

Figure 2. Root mean square (RMS) EMG amplitude of the latissimus dorsi muscles for four task conditions. Mean values and +1 std.'s for nine subjects are shown.

For the asymmetric conditions (Task AS1 and AS2), all muscles except for the RABL and RABR were activated (Figures 2 and 3). The mean amplitudes of the EXOL and EXOR were less than those of the dorsal muscles. There were significant differences in the amplitude between the left and right sides of the muscles activated during the asymmetric lifting tasks. The contralateral muscles (the left side in this study) were more strongly activated than those of the ipsilateral muscles.

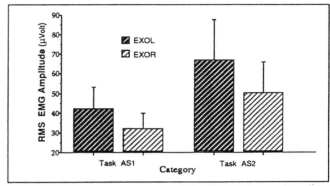

Figure 3. Root mean square (RMS) EMG amplitude of the external oblique muscles for four task conditions. Mean values and +1 std.'s for nine subjects are shown.

%NEMG Changes ,

MANOVA procedures were conducted to investigate the effects of the task conditions on the relative activity of each muscle. The relative activity for a given muscle was defined as %NEMG which is the muscle's proportion to the total NEMG obtained by summing up all NEMG values of the active muscles under the specific task condition. As shown in Table 3, all muscle activities were significantly affected by posture and frequency/load combinations of the task.

As seen in Figure 4, the %NEMG of the ERSL was greater than that of other muscles for all task conditions. In the asymmetric task conditions, the EXOL muscle was determined to serve a fairly large amount of relative activations in spite of its small RMS amplitude. It is noteworthy that the erector spinae muscles showed little increase in the %NEMG despite the weight reduction of external load within the same body posture.

Table 3. Statistical results from the MANOVA and ANOVA procedures of the %NEMG data

Effect	Parameter	MANOVA	ERSL	ERSR	LATL	LATR	EXOL	EXOR	RABL	RABR
Body Posture	F	873.6249	57.32	461.43	0.000	158.13	NT	NT	NT	NT
	df	6 , 823	1 , 828	1 , 828	1 , 828	1 , 828				
	p	< 0.001	< 0.001	< 0.001	NS	< 0.001				
Work Strategy	F	34.3401	31.73	14.55	9.23	19.02	39.37	157.05	NT	NT
	df	6 , 823	1 , 828	1 , 828	1 , 828	1 , 828	1 , 828	1 , 828		
	p	< 0.001	< 0.001	< 0.001	< 0.003	< 0.001	< 0.001	< 0.001		
Body Posture * Work Strategy	F	39.4243	36.46	1.78	17.60	8.46	NT	NT	NT	NT
	df	6 , 823	1 , 828	1 , 828	1 , 828	1 , 828				
	p	< 0.001	< 0.001	NS	< 0.001	NS				

"NT" indicates that ANOVA procedure was not conducted due to inactivity of the muscle at that condition.
"NS" indicates that the result of the ANOVA procedure is statistically non-significant.

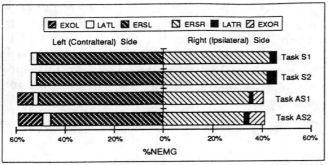

Figure 4. %NEMG of the active muscles for four task conditions, which represent the relative activation levels of the individual muscles.

MPF Decrease

To investigate the differences in decreasing trends of the MPF of each muscle with respect to each task condition, simple linear regression analysis was conducted on the normalized MPF data. Table 4 shows the slope of declination in the fitted regression lines for the muscles which showed significant decreasing trends in their MPF's.

Table 4. Slopes of the fitted regression lines for the muscles which showed statistically significant decreasing trends.

Category	Task S1	Task S2	Task AS1	Task AS2
ERSL	- 0.2243[*]	—	- 0.6270[**]	- 0.4177[**]
ERSR	- 0.3808[**]	—	- 0.2982[*]	- 0.1274[*]
LATL	—	—	- 0.2647[*]	—

[*] significant at $p \leq 0.05$ [**] significant at $p \leq 0.01$

Significant decreasing trends over time in the MPF's were found in some strongly activated muscles during the task, and the amount of decrease was most notable for the ERSL among the eight trunk muscles. The result indicates that the primary muscles which devote relatively high contributions with respect to their maximum ability for counteracting the external load become fatigued faster than other muscles, as would be anticipated.

Comparing the two body postures, the decreasing trends of MPF's were more pronounced for the asymmetric posture than for the symmetric posture. For the task AS1, a significant decrease was found for all muscles of the contralateral side (Figure 5). Within the same body posture, the slopes of MPF decrease of the muscles for the task S1 and AS1 were found to be steeper than those of the task S2 and AS2, respectively. The result indicates that the muscles became fatigued faster for light load-high frequency conditions than for heavy load-low frequency conditions. Figure 6 shows the different MPF decreasing trends of the ERSL muscle according to the task conditions.

DISCUSSION

The preceding results provide a strong evidence that the recruitment and the level of force exertions of the trunk muscles during manual lifting tasks are a function of the direction and the weight of the external load. The different activities of the muscles due to given task conditions also have an influence on the fatiguing process of each individual muscle.

The muscles being recruited during a given task are determined by the body posture following the applied load direction. The muscles on the contralateral side of the external load always play primary roles. The asymmetry of the task has a big influence on the activity of the muscles during lifting. The biased load distributions during asymmetric lifting create a disadvantage for the contralateral muscles and escalate fatigue accumulating in the muscles. It is evident from the finding that the declination slopes of the MPF of active muscles in the asymmetric tasks were always greater than those of the symmetric tasks (Table 4). Within the same body posture, the level of force exertions of the activated muscles is dependent on the weight of the external load, but the amount of change with respect to the weight reduction varies muscle by muscle. It suggests that certain muscles (erector spinae muscles

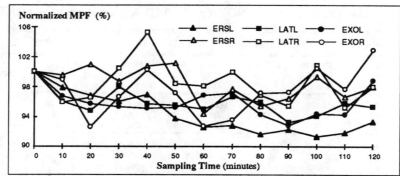

Figure 5. Decreasing trends in the normalized mean power frequency (MPF) of the EMG signals from six active muscles for Task AS1 (10% MVIS × 6 times/min.).

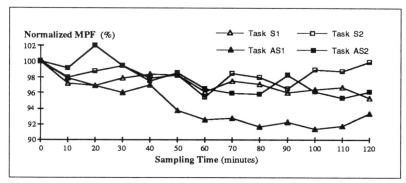

Figure 6. Decreasing trends in the normalized mean power frequency (MPF) of the EMG signals from the left erector spinae muscle (ERSL) for four task condition.

in the lifting tasks) are insensitive to the reduction in the weight of external load since they complement the reduced activation level of other muscles. It is anticipated that those muscles become fatigued faster.

It is important to note that the muscles became fatigued faster for high frequency-light load combinations than for low frequency-heavy load combinations. Much effort has been spent in the industry to prevent musculoskeletal disorders by reducing the weight of the load (Caffier et al., 1993). Reducing the weight with the same lifting frequency would be too conservative to apply to real workplace situations, however. If the total amount of workload assigned to a worker should be maintained at the same level, reducing the weight of the load causes the increase of lifting frequency. Since muscular fatigue degraded the overall performance of workers, a job design based on the underestimated workload of initial healthy state might lead to the low back disorders. The trade-off between the weight of the load and lifting frequency should be considered for ergonomic design of the workplace. If the increase of lifting frequency is the precursor for reducing the weight of the load, it can be claimed that the weight of the external load should be maximized to the level not exceeding the biomechanical safety limit.

It is well known that the power spectral analysis of surface EMG signals is an effective approach to studying localized muscular fatigue during the isometric sustained contractions (Stulen and De Luca, 1982; Gilmore and De Luca, 1985; Seroussi et al., 1989). The same methodology was used to investigate muscular fatigue during intermittent dynamic lifting tasks in this study and the results were quite satisfactory. Due to the large fluctuations in the decreasing trends of MPF which reflect individual difference between subjects, the level of muscular fatigue may not be strictly proportional to the face value of the slope in the MPF decrease for the EMG signals of that muscle. The steepness of the MPF decreasing trends, however, indicates the physiological demand of the task and fatigue level of the muscle in terms of ordinal scale.

REFERENCES

Arendt-Nielson, L., Mills, K. R. and Forster, A. 1989, Changes in muscle fiber conduction velocity, mean power frequency, and mean EMG voltage during prolonged submaximal contractions, Muscle & Nerve, 12, 493-497.

Caffier, G., Heinecke, D. and Hinterthan, R.. 1993, Surface EMG and load level during long-lasting static contractions of low intensity, International Journal of Industrial Ergonomics, 12, 77-83.

Caldwell, L. E., Chaffin, D. B., Dukes-Dobos, F. N. et al. 1974, A proposed standard procedure for Static Muscle Strength Testing, American Industrial Hygiene Association Journal, 35, 201-206.

Cook, T. M. and Neumann, D. A. 1987, The effects of load placement on the EMG activity of the low back muscle during load carrying by men and women, Ergonomics, 30, 1413-1423.

Gilmore, L. D. and De Luca, C. J. 1985, Muscle fatigue monitor (MFM): second generation, IEEE Transaction on Biomedical Engineering, 32, 75-78.

Hansson, G. A., Stromberg, U., Larsson, B., Ohlsson, K., Balogh, I. and Moritz, U. 1992, Electromyo- graphic fatigue in neck/shoulder muscles and endurance in women with repetitive work, Ergonomics, 35, 1341-1352.

NIOSH 1981, Work Practice Guide for Manual Lifting, NIOSH Technical Report No. 81-122, US Depart-ment of Health and Human Services, National Institute for Occupational Safety and Health, Cincinnati, OH.

Schultz, A. B. and Andersson, G. B. J. 1981, Analysis of loads on the lumbar spine, Spine, 6, 76-82.

Seroussi ,R., Krag, M. H., Wilder, P. and Pope, M. H. 1989, The design and use of a microcomputerized real-time muscle fatigue monitor based on the median frequency shift in the electromyographic signal, IEEE Transaction on Biomedical Engineering, 36, 515-523.

Sommerich, C. M. and Marras, W. S. 1992, Temporal patterns of trunk muscle activity throughout a dynamic, asymmetric lifting motion, Human Factors, 34, 215-230.

Stulen, F. B. and De Luca, C. J. 1982, Muscle fatigue monitor: a noninvasive device observing localized muscular fatigue, IEEE Transaction on Biomedical Engineering, 29, 760-768.

Van Dieën, J. H., Toussaint, H. M. and Van De Van, A. 1993, Spectral analysis of erector spinae EMG during intermittent isometric fatiguing exercise, Ergonomics, 36, 407-414.

AN INVESTIGATION OF THE VARIABILITY IN HUMAN PERFORMANCE DURING MANUAL MATERIAL HANDLING ACTIVITIES

Gary A. Mirka
and
Ann Baker

Department of Industrial Engineering
North Carolina State University
Raleigh, North Carolina

The goal of this study was to quantify the variability of the three-dimensional kinematic and kinetic parameters describing the motion of the torso during the performance of sagittally symmetric lifting tasks. Subjects performed eight repetitions of simple lifting tasks described by three levels of coupling (poor, fair and good) and seven levels of load (4.5, 9, 13.5, 18, 22.5, 27 and 31.5 kg). The three-dimensional, time dependent position, velocity and acceleration of the lumbar spine were monitored using the Lumbar Motion Monitor. These measures were then input into a dynamic biomechanical model which calculated torque about the L5/S1 joint in the sagittal plane. The results of the kinematic analysis showed significant variability in the magnitude of the peak velocity and acceleration in the sagittal plane and also showed significant motion in the transverse and coronal planes. The kinetic analysis showed an increase in the variability of the peak dynamic torque with greater levels of load but no coupling effect.

INTRODUCTION

Forces exerted on the spine during manual material handling (MMH) tasks can be classified into one of two categories: external and internal forces. To understand spine loading during MMH tasks, it is important to understand all aspects of these two categories of forces. One of the facets which has yet to be fully understood is the stochastic nature of these forces. The multidimensional, indeterminate nature of most biomechanical systems indicates that there are many ways a person can perform a movement or exertion- both from an internal and external perspective. This implies that a more appropriate method of analyzing the biomechanical system is to use a stochastic modelling approach. This stochastic modelling approach can be employed in two different ways: modelling variability in human performance (external) and modelling variability in the biomechanical stresses (internal).

With regard to the variability of the biomechanical stresses, one recent study (Mirka and Marras, 1993) has shown that there is a significant amount of variability in the internal stresses on the spine during lifting motions. In this study, subjects performed controlled trunk extension exertions. As the subjects performed these simulated lifting exertions, the electromyographic activities of ten trunk muscles were collected. The results of this study showed that there was significant variance in the muscle coactivation patterns employed during simple bending motions. These data were further examined using an EMG-driven biomechanical model and it was shown that the peak forces in the spine were 10% greater than the mean in compression, 40% greater than the mean in anterior shear and 50% greater than the mean in lateral shear. These results could be traced to the variable activation of the trunk musculature, specifically the erector spinae muscles. It was further shown that this erector spinae variability affected, through complex coactivation patterns, the shearing forces experienced by the spine. This study illustrated the potential variability of the internal stressors of the spine.

Based on some recent industrial surveillance studies, a measure of the variability of human performance during MMH activities may also be important in describing the etiology of low back injuries. Herrin et al (1986) performed detailed biomechanical analyses of 55 industrial jobs which entailed a total of 2934 MMH tasks. These authors showed that, as opposed to the average or aggregate requirements of a job, it was the most stressful tasks within that job which were found to be most predictive of overexertion injury.

Another industrial study (Marras et al, 1993) used a motion monitoring device to capture the continuous kinematic parameters that describe three-dimensional trunk motion during MMH tasks. Their results showed that the peak kinematic variables (such as peak velocity in the coronal plane) are more predictive of the risk of injury than are the average kinematic variables (such as average velocity in the coronal plane) for a particular task. These two studies indicate that by focussing our attention on the average stresses encountered during a task, valuable information pertaining to risk may be lost. Therefore, by quantifying the variability of human performance during manual materials handling tasks, valuable insight into workplace risk can be gained. The objectives of the present study were to quantify the variability of the kinematic and kinetic parameters describing human performance of sagittally symmetric lifting activities and to investigate the effects of different workplace variables on the magnitude of this variability.

METHOD

Subjects

Seven male college graduate and undergraduate students served as subjects in this experiment. None had a history of low back impairment/disorder. Some static anthropometry is shown below.

Table 1. Subject Anthropometry

	Mean	Std
Age (yrs)	25	2.98
Mass (kg)	82.9	6.31
Stature (cm)	179.0	7.64
Spine Length (cm)	57.4	1.92
Shoulder Height (cm)	150.2	6.99
Trunk Circumference (cm)	90.1	6.56

Experimental Design

Independent variables. The independent variables in this study were the magnitude of the load and the quality of the coupling interface between the subject and the box being lifted. The levels of load magnitude were: 4.5, 9, 13.5 18, 22.5, 27, and 31.5 kg. The coupling levels were good, fair and poor as described by the revised lifting guidelines (Putz-Anderson and Waters, 1991). All lifts were performed in sagittally symmetric postures.

Dependent variables. The dependent variables were the kinematic parameters describing the position and motion of the lumbar trunk in three-dimensional space. Subsequent calculations using a dynamic biomechanical model of the torso allowed for the calculation of another dependent measure: total torque in the sagittal plane about the L5/S1 joint.

Design. Each combination of independent variables was repeated 8 times (4 lifts per minute for two minutes) per subject. The presentation of coupling conditions was randomized within load levels while the load levels began at 4.5 kg and then increased in 4.5 kg increments.

Apparatus

The motion of the lumbar trunk was monitored using the Lumbar Motion Monitor (LMM). This device was secured to the subject's back and measured the angular position, velocity and acceleration of the lumbar spine in the sagittal, coronal and transverse planes. The trunk position data was collected at a rate of 60Hz. For a more complete description of the LMM, see Marras et al (1992).

Subjects were asked to lift a 35x35x30cm wooden box that weighed 4.5 kg. Cast iron weights of 4.5 kg were added to create the designated load levels. In the good coupling condition, the subjects used cut-out handles on the box. In the fair coupling condition, the subjects were instructed to lift the box from underneath, allowing the fingers to be flexed at 90°. In the fair condition, the box was placed on a stool to control for the distance the subjects had to stoop in order to reach the box. In the poor coupling condition the subjects were instructed to the lift the box with a compression type hold on the sides of the box.

Procedure

Subjects began the experimental session with a brief warm-up and collection of anthropometric variables. They were then fitted with the LMM and were encouraged to move through a complete 3-D range of motion to assure themselves that the LMM would not inhibit normal trunk motion. Baseline trunk position readings were collected as the subjects stood in a relaxed vertical posture (0°) and in a sagittally symmetric 90° forward bend posture. These calibration data values would later be used in normalization of the data during processing.

Once these preliminary trials were complete, the subject moved to the lifting area. The weight of the first trial was 4.5 kg, and the coupling was set using the randomization scheme. The lift rate was set at 4 lifts per minute, and the duration was two minutes for each experimental condition, thus rendering the eight repetitions. The subjects were asked to lift the box "using the maximum comfortable speed that you would normally lift an object of this weight" while keeping their elbows and knees at a constant degree of flexion throughout the range of lifting motion. Most subjects chose to lift with straight arms and a slight angle of flexion of the knees. After completing this first set of eight lifts, the subjects rested for three minutes while the box was prepared for the next trial. This process continued until all of the trials within the capability of the subject were completed.

Biomechanical Model

The biomechanical model calculated the instantaneous dynamic external torque about the lumbosacral joint. The body was partitioned into a 5-link system consisting of two upper extremity links, two lower extremity links, and a composite head/neck/trunk link beginning at the lumbosacral joint. Using regression equations from Dempster (1955), and Pheasant (1986) as well as the static anthropometry of the subject population, estimates were made of 1) the trunk mass, 2) the distance between center of mass of trunk and L5/S1, 3) the mass of the arms, 4) the distance between the gleno-humeral joint center of rotation and L5/S1.

The time dependent sagittal position data was first normalized with respect to the 0° and 90° calibration values collected at the beginning of the experiment. This normalized LMM data and the anthropometric values were then used as inputs to the dynamic biomechanical model which output the time dependent external torque in the sagittal plane about the L5/S1 intervertebral joint. The equation used to calculate the sagittal moment about L5/S1 is shown below:

$$M(L5/S1) = MT*g*R1*\sin\theta + MAB*g*R2*\sin\theta + MAB*R2^2*(\sin\theta)^2*\alpha + I\alpha \quad (1)$$

Variable Definitions
MT = mass of trunk (kg)
g = gravitational constant (9.81 m/sec/sec)
R1 = distance from L5/S1 to COG of trunk (m)
θ = sagittal angle of the trunk (upright=0 deg)
MAB = mass of the arms plus mass of box (kg)
R2 = distance from L5/S1 to shoulder joint (m)
α = angular acceleration of trunk (rad/sec/sec)
I = mass moment of inertia of the trunk (kg m^2)

This equation was used to calculate the moment about L5/S1 in the sagittal plane at each instant in time at a frequency of sixty data points per second.

Data Processing

The first step in data processing was to normalize the data so that each trial began when the subject's hands came in contact with the box. In each plane of the body, the range of motion, peak velocity and peak acceleration were then obtained for the remainder of the lifting motion. Finally, the kinematic data from the LMM were input into the dynamic biomechanical model so that the time dependent torque about the lumbosacral joint could be plotted.

RESULTS AND DISCUSSION

Looking at the data qualitatively, Figure 1 illustrates the time dependent response of the acceleration in the sagittal plane. In this figure, time t=0 corresponds that point during the lift when the subject's hands first touched the box. This figure shows graphically the variability in the magnitude of the peak acceleration as well as the variability in the time history of acceleration.

The results of a more quantitative analysis of trunk kinematics showed significant amount of motion in all three planes. Comparing the three levels of kinematic data, the results of this study show that the higher derivatives of motion become relatively more variable. The average coefficient of variation (CV) for the range of motion in the sagittal plane was 2.5% while the CV for peak velocity was 8.1% and the CV for peak acceleration was 10.4%.

From an ergonomic point of view the greatest impact of this kinematic variability is in its effect on trunk kinetics. These effects were shown when the trunk motion data was input into the dynamic biomechanical model. The results of this analysis are shown in Table 2. In an attempt to understand the effect of workplace factors on the magnitude of this variability, differences in the variance of the peak torques were tested using a Bartlett test. These results showed a significant effect due to the load weight, with greater weight levels showing a higher level of variability. This significant weight effect is graphically illustrated in Figure 2. Further analysis showed that coupling by itself had no significant effect on the magnitude of the variance, but at near maximal levels of load, poor coupling generated the greatest variability. Numerically, consideration of the range of potential torques about L5/S1 shows that the

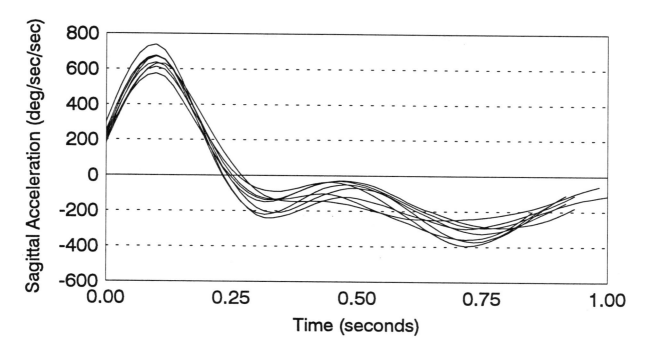

Figure 1. Time dependent sagittal acceleration. Load=22.5 kg, Coupling='Good'

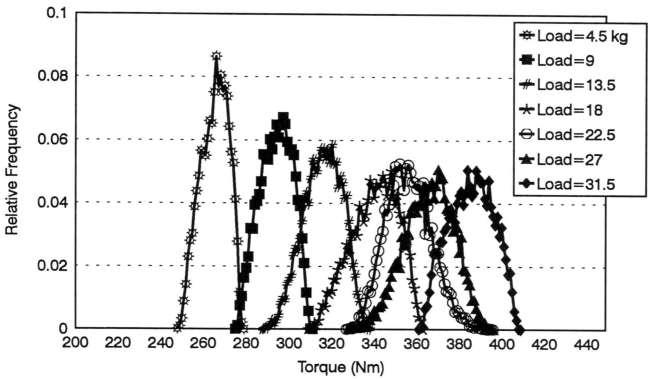

Figure 2. Distributions of peak sagittal torque as a function of load magnitude.

torques at 2 standard deviations above the mean are between 5% and 15% higher than those at the mean.

Table 2. Mean and standard deviation of torque in the sagittal plane (in Nm)

Coupling Quality

Load	Good Mean	Std	Fair Mean	Std	Poor Mean	Std
4.5	268.9	7.4	270.8	7.6	265.8	6.3
9.0	299.5	9.1	294.3	7.4	292.2	7.3
13.5	317.8	10.5	316.6	8.6	310.2	9.5
18.0	340.4	12.7	340.6	11.1	321.5	10.9
22.5	353.5	13.9	356.8	10.3	319.4	18.1
27.0	375.2	17.5	366.2	11.3		
31.5	390.1	12.3	386.3	10.0		

Kinematic results from the transverse and coronal planes showed that there was also a significant amount of dynamic activity in these "off-planes" even though the lifts were performed in sagittally symmetric postures. These results are shown in Table 3. These kinematic parameters did not, however, show significant trends as a function of load or quality of coupling.

Table 3. Mean and standard deviation of kinematic parameters in the coronal and transverse planes

	Mean	Std
Coronal Range of Motion (deg)	4.6	1.7
Transverse Range of Motion	3.1	1.7
Max Coronal Velocity (deg/sec)	13.6	4.6
Max Transverse Velocity	7.8	3.1
Max Coronal Acceleration (deg/sec^2)	63.2	22.6
Max Transverse Acceleration	40.4	13.7

CONCLUSIONS

This study has shown that there is significant variability in human performance- from both a kinematic and kinetic perspective. As ergonomists we need to be concerned with, or at least be aware of, the worst case scenario. This study has shown that by considering only the average performance profile, the peak torque is underestimated by between 5% and 15% and points to a need for the stochastic modelling of biomechanical systems to be considered.

References

Dempster, W.T. 1955, Space Requirements of the Seated Operator, Wright Air Development Center TR-55-159, Wright-Patterson Air Force Base, Dayton, OH (AD 87 892).

Herrin, G. Jaraiedi, M. and Anderson, C. "Prediction of Overexertion Injuries Using Biomechanical and Psychophysical Models," *American Industrial Hygiene Association Journal*, 47, 322-330 (1986).

Marras, W., Fathallah, F., Miller, R., Davis, S. and Mirka, G. "Accuracy of a Three-Dimensional Lumbar Motion Monitor for Recording Dynamic Trunk Motion Characteristics," *International Journal of Industrial Ergonomics*, 9, 75-87 (1992).

Marras, W., Lavender, S., Leurgans, S., Rajulu, S., Alread, G. Fathallah, F. and Ferguson, S. "The Role of Dynamic Three-Dimensional Trunk Motion in Occupationally Related Low Back Disorders: The Effects of Workplace Factors, Trunk Position and Trunk Motion Characteristics on Risk of Injury," *Spine*, 18, 617-628 (1993).

Mirka, G.A. and Marras, W.S. "A Stochastic Model of Trunk Muscle Coactivation During Trunk Bending," *Spine*, 18, 1396-1409 (1993).

Pheasant, S. 1986, Bodyspace: Anthropometry, Ergonomics and Design, (Taylor & Francis, London), 133-134.

Putz-Anderson, V. and Waters, T.R. 1991, Revision in NIOSH Guide to Manual Lifting, Paper presented at national conference entitled "A National Strategy for Occupational Musculoskeletal injury prevention -- Implementation issues and research needs." University of Michigan, Ann Arbor, Michigan.

Waters, T.L., Putz-Anderson, V., Garg, A. and Fine, L.J. "Revised NIOSH Equation for the Design and Evaluation of Manual Lifting Tasks," *Ergonomics*, 36, 749-776 (1993).

ACKNOWLEDGMENT

This publication was partially supported by grant number KO1 OH00135-01 from NIOSH. Its contents are solely the responsibility of the authors and do not necessarily reflect the official views of NIOSH.

A STUDY OF THE INTERACTION BETWEEN LOAD AND COUPLING DURING LIFTING

Gary Mirka, Ann Baker, Angela Harrison, Dan Kelaher and Joseph Davis
North Carolina State University
Raleigh, North Carolina

The National Institute for Occupational Safety and Health (NIOSH) has increased the applicability of its lifting equation to a wider range of jobs by relaxing some of the simplifying assumptions of the original equation. Specifically, NIOSH has added an asymmetry factor and a coupling factor in the revised lifting equation. Two of the remaining simplifications, however, are that (1) interactions between terms are not considered and (2) the biomechanical analysis still utilizes a static modelling approach in its calculations. The purpose of the present research was to investigate the interaction between coupling and load magnitude under dynamic lifting conditions. Subjects lifted a box under various combinations of coupling and load. The dependent variables in this study were the peak external moment about the lumbosacral joint (as calculated by a dynamic biomechanical model) and the peak vertical ground reaction forces. The results show that at low levels of load there was little difference in peak torque across the different coupling conditions. However, when loads greater than 13.5 kg were combined with poor coupling, there was a fundamental change in the dynamics of the lifting motion. The results of this study indicate that the role of coupling under dynamic lifting conditions has both a perceptual and biomechanical effect which should be considered when designing manual materials handling tasks.

INTRODUCTION

In an effort to broaden the applicability of the NIOSH lifting equation to a wider variety of industrial tasks, Waters et al (1993) developed the revised NIOSH equation. In this equation, two new workplace factors, asymmetry and coupling, have been added so that the actual lifting capacity in non-optimal lifting conditions can be better evaluated. Although this new equation did allow for more realistic modelling of the workplace, two limitations of the revised equation are that 1) a static biomechanical modelling approach is still used and 2) the factors are evaluated individually without considering the interactive effects of the lifting variables. The focus of this study was to evaluate these two issues by investigating the complex relationship between coupling quality and load magnitude under dynamic lifting conditions.

Some recent research has shown the importance of coupling during manual materials handling tasks. It has been reported that proper design and placement of handles can reduce the physiological and perceived stress in manual materials handling tasks (Drury and Pizatella, 1983; Smith and Jiang, 1984; Drury et al, 1989). Drury et al (1989) showed that, from a psychophysical and physiological viewpoint, lifting boxes with handles provided lower rated perceived exertions (RPEs), lower ratings of body-part discomfort frequency and lower heart rates than those boxes without handles. Smith and Jiang (1984) reported that handles reduced subjective levels of fatigue of the hands in their study of bag lifting. These studies show the positive effects handles have on the perceptions of the people performing lifting tasks.

Contrary to the perceptually beneficial aspects of good coupling via handles are two studies which showed greater biomechanical lumbar stress with better coupling (Freivalds et al., 1984; Garg et al., 1982). Freivalds et al. (1984) examined biomechanical aspects of lifting during a psychophysical lifting test. These authors compared a 38 cm square box with handles to one without handles (corresponding to 'good' and 'fair' coupling according to the revised lifting guide). They found faster rise times and higher peak forces on lifts of boxes with handles than of boxes without handles even though in some lifts subjects lifted larger loads in the boxes without handles. Similar findings were shown by Garg et al. 1982 in which peak moments and compressive forces were consistently lower for boxes without handles as compared to boxes with handles. It is suggested from these studies that the subjects recognized the hazards of lifting boxes with no handles and were more cautious in their approach, thereby altering the normal dynamics of the lift. A limitation of these studies is that they did not investigate how this perceived risk responded to different levels of load. That is, does the observed change in lifting dynamics occur under all loading conditions or does the magnitude of the load affect this response? The specific objective of the present research is to quantify the interactive biomechanical effects of coupling and load under dynamic lifting conditions.

METHODS

Subjects

Seven male subjects from the university student population participated in this study. The mean (and standard deviation) of age, mass, stature, and spine length (L5-C1) were: 25 (2.98) years, 82.9 (6.31) kg, 179.0 (7.64) cm, 57.4 (1.92) cm. Prior to the study, each subject indicated that he had no history of back problems.

Apparatus

The Lumbar Motion Monitor (LMM) (Chattanooga Group, Inc., Chattanooga, TN), was used to quantify three-dimensional trunk motion (angular position, angular velocity and angular acceleration in the sagittal, coronal and transverse planes) for input into a dynamic biomechanical model. In addition to the LMM data, three-dimensional ground reaction forces were collected using a Bertec force platform (Model 4060A, Bertec Corporation, Worthington, OH)

A 35x35x30cm high 1/2" plywood box was constructed for this study. The box had 20x8cm cut-out handles located 25cm above the base. These values were within the container size criteria specified in the revised lifting guidelines (Putz-Andersson and Waters, 1991).

The 'good' coupling required lifting the box using the handles located 25 cm above the base. (See Figure 1.) The 'fair' coupling involved lifting the box from the bottom while it was resting on a 25 cm high pedestal. For the 'poor' coupling lift, a compression lift was performed with the hands also centered at approximately 25 cm above the box base. The starting position of the box relative to the ankle joint was constant for each subject across all conditions. The mass of the box itself was 4.5 kg, and the load was increased by placing cast iron plates in the box.

Experimental Design

Independent variables. The independent variables in this study included the levels of load being lifted and the quality of coupling between the lifter and

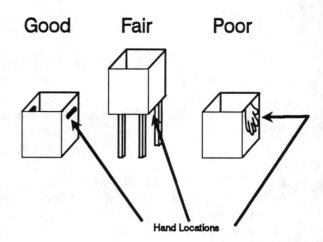

Figure 1. Location of hands during lifting tasks under the good, fair and poor coupling conditions

the box. Seven levels of load were used in this study (4.5, 9, 13.5, 18, 22.5, 27, 31.5 kg) and three levels of coupling were used (good, fair and poor as defined in the revised NIOSH equation). Each of these combinations was repeated 8 times with fifteen seconds between consecutive lifts.

Dependent variables. The dependent variables in this study were the peak external torque about L5/S1 developed during the dynamic lifting motion and the peak vertical ground reaction force as measured by the force platform.

Procedure

After collection of anthropometric measurements and a brief warm-up, subjects were fitted with the LMM. A familiarization period was allowed for each subject to become accustomed to having the LMM on his back. Each subject was encouraged to move through a complete three-dimensional range of motion to assure himself that the apparatus would not inhibit normal motion. Baseline trunk position readings were collected as the subjects stood in a relaxed vertical posture (0°) and in a sagittally symmetric 90° forward flexed posture. These calibration data values were later used to normalize the data.

The lift rate was set at 4 lifts per minute, and the duration was two minutes for each experimental condition, thus rendering the eight repetitions. The levels of coupling were randomized within the load levels. The subjects were asked to lift the box "using the maximum comfortable speed that you would normally lift an object of this weight" while keeping their elbows and knees at a minimal degree of flexion (< 15°) throughout the range of lifting motion. Most subjects chose to lift with straight elbows and a slight angle of flexion of the knees. After completion of one set of lifts, the subjects rested for three minutes while the box was prepared for the next trial. This process continued until all of the trials within the capability of the subject were completed. Under the poor coupling conditions, load levels above 22.5 kg were not completed by the majority of the subjects and were therefore eliminated from the analysis.

Biomechanical Model

The biomechanical model computed the instantaneous external torque about the lumbosacral joint center. The body was partitioned into a 5-link system consisting of two upper extremity links, two lower extremity links, and a composite head/neck/trunk link beginning at the lumbosacral joint. Using data and regression equations from Dempster (1955), and Pheasant (1986) as well as the static anthropometry of the subject population, estimates were made of the 1) trunk mass, 2) distance between center of mass of trunk and L5/S1, 3) the mass of the arms, and 4) the distance between the gleno-humeral joint center of rotation and L5/S1. The equation used to calculate the sagittal moment about L5/S1 is shown below:

$$M(L5/S1) = MT*g*R1*\sin\theta + MAB*g*R2*\sin\theta + MAB*R2^2*(\sin\theta)^2*\alpha + I\alpha \quad (1)$$

Variable Definitions
MT = mass of trunk (kg)
g = gravitational constant (9.81 m/sec/sec)
R1 = distance from L5/S1 to COG of trunk (m)
θ = sagittal angle of the trunk (upright=0 deg)
MAB = mass of the arms plus mass of box (kg)
R2 = distance from L5/S1 to shoulder joint (m)
α = angular acceleration of trunk (rad/sec/sec)
I = mass moment of inertia of the trunk (kg m^2)

RESULTS

The experimental data was analyzed using the Analysis of Variance (ANOVA) package of Statistical Analysis System (SAS Institute Inc, Cary, NC). This analysis indicated that when the peak torque (as computed by the biomechanical model) is considered, statistical significance exists for the coupling load interaction ($p<.0001$) and for main effects of load ($p<.0001$) and coupling ($p<.001$). However, when these results are displayed graphically (Figure 2) it is evident that the coupling factor is overshadowed by the interactive effect and therefore should not be considered as a significant main effect. These results further show

that under the poor coupling condition, weights above 13.5 kg caused a shift in the dynamic lifting strategy employed by the subject and increasing the load beyond 18 kg actually caused a decrease in peak torque. An examination of the acceleration data confirmed the hypothesis of decreased acceleration at these higher load levels only for poor coupling.

The peak vertical ground reaction forces as recorded by the force platform showed a slightly different trend. The statistical analysis showed that the load main effect was significant ($p<.0001$) and the interaction between load and coupling was also significant ($p<.01$), but coupling alone was not significant. Figure 3 illustrates this response.

DISCUSSION

A strength of the NIOSH approach has been that static workplace dimensions are relatively easy to measure and therefore the lifting equations can easily be implemented in industry to help understand potential for injury. Unfortunately there has been sparse literature (Liles and Mahajan, 1985) showing these equations to be effective in the reduction of low-back injuries. One of the often cited reasons for the apparent ineffectiveness is that the model does not account for the additional biomechanical effects of dynamic lifting.

As models and standards begin to reflect the addition of dynamic effects, they must also begin to reflect the human performance aspect of lifting. Modelling the total biomechanical stress associated with a dynamic lifting task is no longer limited to simple workplace dimensional information but now can be affected by the lifter himself. When static models of the trunk during lifting were considered adequate to represent lumbar stress, the human performance component could be easily overlooked. Models must address human performance issues because the human has the ability to alter the dynamics of the system by choosing an appropriate velocity/acceleration to be used to perform the task. It is thought that this appropriate velocity/acceleration is derived from a cognitive integration of perceived risk, fatigue, and production requirements as well as certain sociological factors. From an ergonomic point of view, one of the interesting issues is the element of perceived risk and how it affects the biomechanics of lifting.

Recent work by Thompson and Chaffin (1993) indicates that the ability of the human to perceive biomechanical stress in the lumbar region is extremely limited. In this study, the authors show that there was no correlation between the Borg rating (rating of perceived

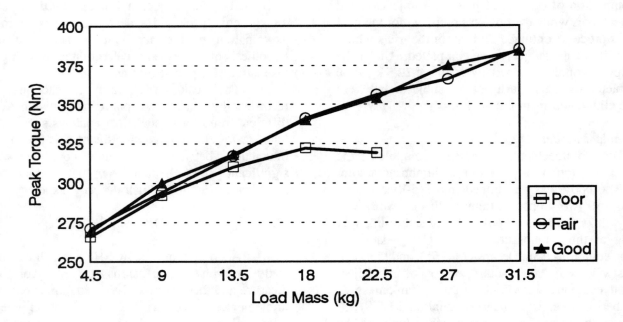

Figure 2. Effects of coupling and load on the peak external moment about L5/S1

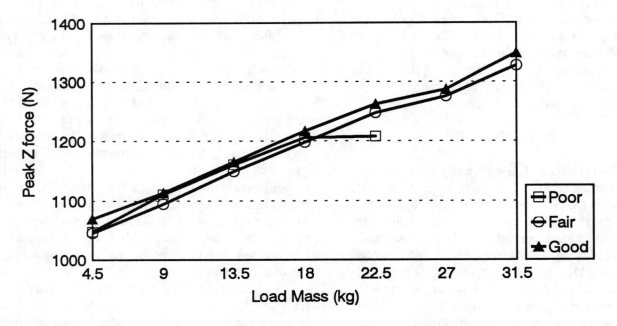

Figure 3. Effects of coupling and load on the peak vertical ground reaction forces

exertion) and the peak L5/S1 compression forces during lifting. These results indicate that during psychophysical lifting tasks, the limiting factor is not the lumbar stress but some other factor such as the stress encountered at some other link in the body. By comparison, the structures of the upper extremity have a much greater level of innervation than those in the spine. This becomes important because although people may have some psychophysical preference for boxes with handles this preference may originate in the perceived exertions of the upper extremities (hands, elbows, shoulders etc.) and may not reflect the stress on the low-back.

The results of the work by Thompson and Chaffin (1993) are supported by the present research in that there was no apparent alteration in the subject's perceived risk at higher load levels for the good and fair coupling conditions. That is, with increasing load they did not reduce their rate of acceleration and thereby the total external moment. It was only when the effect of poor coupling began to limit the dynamics of the lift that the subjects responded by reducing their rates of acceleration. This result points to a recognition by the subject of perceived level of upper extremity stress and associated risk of dropping the box, but not a perceived stress of the low-back.

CONCLUSIONS

As dynamic models of lifting are developed, quantitative information regarding the dynamics of trunk motion during manual materials handling tasks needs to be obtained. One approach to this problem could be to require industry to use some type of motion analysis system to quantify required trunk dynamics for each job. Another approach may be to perform laboratory experiments with the goal of trying to find predictable patterns of motion (position, velocity and acceleration) as a function of static workplace variables.

The present study is the first in a series to try to document some of these quantitative responses. The specific results of this study are that there appears to be no effect of increasing load (up to 31.5 kg) on the dynamics of lifting under the good and fair coupling conditions. However, under the poor coupling conditions, a break point occurred at about 13.5 kg of load. At this point the dynamics of the torso were significantly altered, perhaps to accommodate the upper extremity stress. This research should not be viewed as a call for lifting tasks to be designed without handles, but rather as an illustration of the importance of the interaction between perception and human performance and the effect of this interaction on the biomechanics of the lumbar spine.

References

Dempster, W.T. 1955, Space Requirements of the Seated Operator, Wright Air Development Center TR-55-159, Wright-Patterson Air Force Base, Dayton, OH (AD 87 892).

Drury, C.G. and Pizatella, T. 1983, Hand placement in manual materials handling, *Human Factors*, **25**, 551-562.

Drury, C.G., Deeb, J.M., Hartman, B., Woolley, S., Drury, C.E. and Gallagher, S. 1989, Symmetric and asymmetric manual materials handling: Part I: Physiology and psychophysics, *Ergonomics*, **32**, 467-489.

Freivalds, A., Chaffin, D.B., Garg, A. and Lee, K.S. 1984, A dynamic biomechanical evaluation of lifting maximum acceptable loads, *Journal of Biomechanics*, **17**, 251-161.

Garg, A., Chaffin, D.B. and Freivalds, A. 1982, Biomechanical stresses from manual load lifting: a static vs dynamic evaluation, *IIE Transactions*, **14**, 272-281.

Liles, D.H. and Mahajan, P. 1985, Using NIOSH Lifting Guide Decreases Risk of Back Injuries, *Occupational Health and Safety*, 57-60.

NIOSH, 1981, Work Practices Guide for Manual Lifting. Department of Health and Human Services, National Institute for Occupational Safety and Health (NIOSH), Publication No. 81-122.

Pheasant, S. 1986, Bodyspace: Anthropometry, Ergonomics and Design, (Taylor & Francis, London), 133-134.

Putz-Anderson, V. and Waters, T.R. 1991, Revision in NIOSH guide to manual lifting, Paper presented at national conference entitled "A National Strategy for Occupational Musculoskeletal injury prevention -- Implementation issues and research needs." University of Michigan, Ann Arbor, Michigan.

Smith, J.L. and Jiang, B.C. 1984, A manual materials handling study of bag lifting, *American Industrial Hygiene Association Journal*, **45**, 505-508.

Thompson, D.D. and Chaffin, D.B. 1993, Can Biomechanically Determined Stress be Perceived?, *Proceedings of the Human Factors and Ergonomics Society 37th Annual Meeting*, pp. 789-792.

Waters, T.L., Putz-Anderson, V., Garg, A. and Fine, L.J. 1993, Revised NIOSH equation for the design and evaluation of manual lifting tasks, *Ergonomics*, **36**, 749-776.

A BIOMECHANICAL ASSESSMENT OF AXIAL TWISTING EXERTIONS

Kevin P. Granata and William S. Marras
Biodynamics Laboratory
The Ohio State University

ABSTRACT

Axial twisting of the torso has been identified as a significant risk factor for occupationally related low-back disorders. The purpose of this investigation was to examine the influence of dynamic twisting parameters upon spinal load. Measured trunk moments and muscle activities were employed in a biomechanical model to determine loads on the lumbar spine. Spinal loads were examined as a function of dynamic torsional exertions under various conditions of force, velocity, position, and direction. Results demonstrate significant flexion-extension and lateral moments were generated during the twisting exertions. Muscle coactivity was significantly greater than equivalent levels measured during sagittal lifting exertions. Relative spinal compression during dynamic twisting exertions was twice that of static exertions. Spine loading also varied as a function of whether the trunk was twisted to the left or right, and the direction of applied torsion, i.e. clockwise versus counter-clockwise. The results may help explain, biomechanically, why epidemiological findings have repeatedly identified twisting as a risk factor for low-back disorder

INTRODUCTION

Axial twisting of the torso has been identified as a significant risk factor for occupationally related low back disorders (Kelsey et.al., 1984; Punnet et.al., 1991, Marras et.al., 1993). Studies of trunk muscle activity have described significant muscle co-contraction during twisting exertions (Carlsoo, 1961; Zetterberg et.al., 1987; Basmajian and DeLuca, 1993). Pope et. al. (1986) reported a high degree of coactivation of the antagonist muscles as well as the erector spinae and rectus abdominis muscles during torsional exertions. To realistically estimate spinal loads during twisting exertions, one must include the influence of muscle co-contraction in the biomechanical analyses (Granata and Marras, 1995). Previous EMG-assisted models were able to account for the influences of measured muscle coactivity, but succeeded in predicting the torsional moments only by permitting muscle forces to far exceed realistic levels. Since the previous models were unable to accurately assess muscle forces, the resulting estimates of spinal loads must be re-evaluated.

Recent developments have improved EMG-assisted modeling theories resulting in analyses which accurately simulate dynamic torsional moments (Marras and Granata, 1995). The developments involve appropriate interpretation of the force generating capacity from muscle cross-sectional areas as a function of spinal level. Since the recent analyses were able to accurately predict isometric and isokinetic twisting moments, associated levels of spinal loading were likely to be more realistic. With an accurate and realistic biomechanical model, one may examine the influence of dynamic task parameters upon the relative magnitudes of loads experienced by the spine during twisting exertions.

The objectives of this study were to document kinetic coupling of trunk moments and gain an appreciation of how spinal loads are influenced by task parameters during dynamic twisting exertions. The goal was to describe the biomechanical association between dynamic twisting exertions and the risk of low-back disorders.

METHODS

Twelve male subjects (weight 76.4±8.4 kg., height 177.0±16.4 cm.) performed isometric and isokinetic, torsional exertions at 100% and 50% of their maximum voluntary contraction (MVC) effort. Torsional exertions were applied in both clockwise and counter-clockwise directions. Isometric twisting exertions were achieved with the trunk rotated axially 20° to the right (20R), 20° to the left (20L), and sagittally symmetric. Isokinetic exertions were performed at twisting velocities of 10 and 20 deg./sec. through a range of motion ±24° about a symmetric posture. Isokinetic motions were performed in both clockwise and counter-clockwise directions. Subject positions and motions were controlled by a Kin/Com isokinetic dynamometer, while trunk motions were measured by a Lumbar Motion Monitor. Applied trunk moments and forces were determined from mathematical translation of measured force plate data. The experimental trunk position, and motion characteristics were selected to represent trunk twisting motions observed in industry (Marras et.al., 1993).

Myoelectric activity of trunk muscles were collected using bipolar surface electrodes over the right and left pairs of the latissimus dorsi, erector spinae, rectus abdominis, external oblique and the internal oblique muscles as reported in Mirka and Marras (1993). EMG signals were processed and normalized using activity levels collected during MVC exertions.

Dynamic spinal loads were predicted from a validated, EMG-assisted model (Marras and Granata, 1995). The biomechanical model computed muscle forces by solving the equation of dynamic equilibrium, while distributing the forces among the muscles as dictated by the relative EMG values. The force in each muscle was determined from normalized EMG activity, muscle cross-sectional area, modulation factors describing the muscular force-length and force-velocity relations, a common gain value describing the muscle stress capacity, and a three-dimensional muscle unit-vector. Muscle generated trunk moments were determined from the sum of the individual muscle moments computed from the vector products of the muscle force vectors and respective moment arm vectors relative to L5/S1. Three-dimensional loads on the lumbar spine were determined by vector summation of the ten muscle equivalents. Statistical analyses were performed to examine the significance of postural and dynamic parameters upon the spinal loads.

RESULTS

Although subjects were asked to produce only torsional moments, significant forces about the other body axes were developed (Figure 1). During torsional exertions, coupled moments were generated in the sagittal plane equivalent to 22% of the extension maximum and in the coronal plane equivalent to an average of 34% of the lateral maximum.

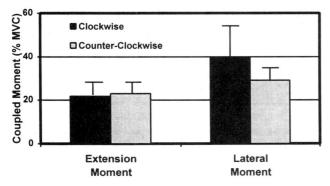

Figure 1. Significant coupling moments, i.e. 22% extension MVC and an average of 34% lateral MVC, were generated during torsional exertions.

Maximum torsional moments varied significantly (p<.01) as a function of twisting direction, angle and velocity. Greater peak isometric moments were generated in the clockwise direction (96 Nm) than the counterclockwise direction (83 Nm). Significantly

greater isometric torque was generated when the trunk was pre-rotated 20° away from the direction of applied torsion (107 Nm) compared to the sagittally symmetric condition (95 Nm). Thus, when subjects applied a clockwise torsional moment, the magnitude was greater when the trunk was pre-rotated to the left than when sagittally symmetric (Figure 2a). Torsional moments generated during maximum isometric exertions (90 Nm) were significantly greater than the peak moments generated during dynamic exertions (76 Nm) (Figure 2b). Hence, these results show torsional moment is a function of twisting direction, angle, and velocity.

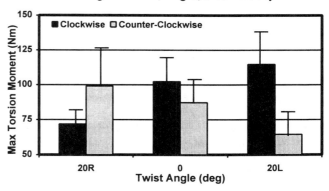

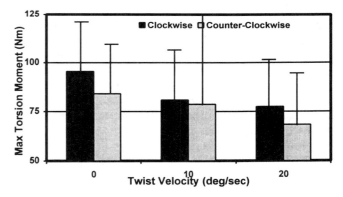

Figure 2. Maximum applied torsional moment was influenced by the direction of the twisting effort, trunk pre-rotation angle (a), and twisting velocity (b).

Muscle coactivity was described as the average activity level of the co-contracting muscles relative to the prime mover, i.e. most active muscle. The average coactivity (Figure 3) during twisting exertions (42%) was significantly (p<.01) greater than during sagittal lifting (extension) exertions (26%) determined from the data of Granata, (1993). However, there were no significant differences in average coactivity associated with twisting angle or direction. Measured EMG activity of the vertically oriented muscles were influenced by task parameters including exertion level and torsional direction. One would expect such changes in the obliquely oriented muscles but not necessarily in the vertically oriented muscles. The left latissimus dorsi, left

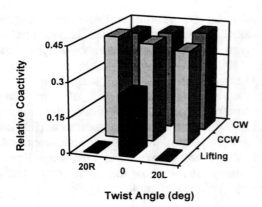

Figure 3. Average muscle coactivity during twisting exertions (42%) was significantly greater than during lifting exertions (28%). There was no statistical difference between coactivity measured during torsional exertions as a function of twisting angle.

internal oblique, and right external oblique muscles were generally more active during counter-clockwise exertions than the right latissimus dorsi, right internal oblique and left external oblique muscles during clockwise exertions. Thus, agonist muscles were generally at a higher level of activity in the counter-clockwise direction of force application. This may due to the behavior of the left external oblique muscle, which was significantly more active when performing as an antagonist, i.e. counter-clockwise, than when acting as an agonist, i.e. clockwise. Finally, the analyses indicated that exertions performed at increased velocities required greater EMG per unit torsional moment, as expected from the force-velocity relations described in the literature (Hill, 1938).

Relative spinal loads, expressed in terms of spinal force per unit of torsional moment, changed significantly (p<.01) as a function of exertion level, direction and velocity. It is particularly noteworthy that spinal compression at 10 deg./sec. was twice that of the average isometric value (Figure 4c). A pairwise comparison of relative spine load between the 20R and 20L conditions indicated that a significantly elevated compression load was present in the 20R condition (14.3 N/Mz) compared to the loading during the 20L condition (8.6 N/Mz). Thus, the magnitude of spinal load depended upon whether the trunk was twisted to the right or the left (Figure 4b). Finally, the analyses indicated that counter-clockwise exertions resulted in about twice the lateral shear loading (10.2 N/Mz) compared to clockwise loading (5.4 N/Mz) (Figure 4a).

DISCUSSION

This effort has shown that the torso is limited in its ability to generate twisting torque, particularly under dynamic conditions. During MVC exertions, the average torsional moments produced by our subjects were on the order of 52 Nm, with moment peaks of 90 Nm. Maximum isometric torsional moments measured during sagittally symmetric exertions (95 Nm) were similar to the moments reported by others (91 Nm) (McGill, 1991; Parnianpour et.al., 1991; McGill, 1992). The magnitude of the twisting moments represent approximately 50% of the extension moment that subjects were able to produce, as reported by Parnianpour et.al. (1988). Hence, the body's ability to produce twisting moment is far more limited than its ability to generate lifting moment.

Increased maximum torsion at pre-rotated angles may be attributed to the relation between muscle length and contractile strength of the agonist muscles.

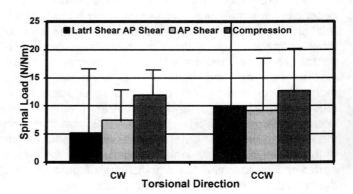

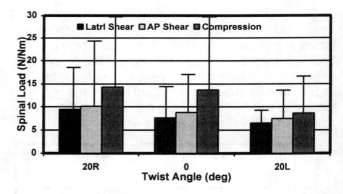

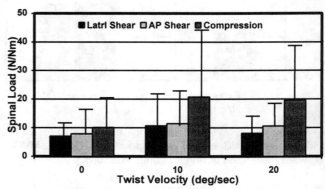

Figure 4. Spinal Compression and shear per unit torsional moment are significantly influenced by torsional direction (a), twist angle (b) and twisting velocity (c).

McGill (1992) agrees muscle length may contribute to this effect but concludes passive tissues may play a more significant role.

The magnitude of counter-clockwise torsional moments were typically less than 80% of the clockwise values. McGill (1991) reported similar discrepancies between directions of applied moment. The directional influence may be associated with increased coactive antagonism in the counter-clockwise direction, most notably in the left external abdominal oblique. Conversely, the variation in average coactivity as a percent of the most active muscle was not statistically influenced by torsional direction. Increased activity in the rectus abdomini and erector spinae It is necessary to recognize that most of our subjects (10 of 12) were right handed. Hand preference may influence the muscular strength and control associated with the clockwise versus counter-clockwise torsional exertions and the related maximum applied moments.

Velocity was a critical factor in a subject's ability to apply a twisting moment. Maximum torque production during dynamic exertions was 43 to 50% lower than isometric values, even at low levels of velocity. This affect was expected as per the physiologic force-velocity relation of skeletal muscle (Hill, 1938), but the magnitude of the change was suprising.

Although instructed to generate torsional exertions, the applied moments were not purely torsional. Along with the torsional moments, subjects generated simultaneous extension (average 22% extension MVC) and lateral moments (average 34% lateral MVC). These coupled moments may indicate torsional exertions are biomechanically more complex than sagittal plane lifting exertions.

When one considers the true capacity to apply torsional moments, reduced torque capacity associated with twisting velocity, and the effect of torque application direction, it is not difficult to understand why twisting is often cited in the epidemiological literature as a significant factor in occupationally related low-back disorders. These analyses indicate that task requirements may readily exceed the capacity of the trunk in torsional exertions. This is particularly true when dynamic motion and counter-clockwise motions are involved.

Measurements of muscle activity indicate that the right latissimus dorsi, right internal oblique (measured posteriorly) and the left external oblique contribute to clockwise torsional exertions, as expected. The contralateral muscles contribute to counter-clockwise exertions. The bi-lateral behavior of EMG activities of the erector spinae muscles indicate they also contribute to torsional exertions. Similar to the measurements of Pope et.al. (1986), significant antagonistic co-contraction was also evident. While acting as an antagonists to the torsional moment, the

right external oblique generated 28% of its MVC myoelectric value. Similar behavior was observed in the left external oblique. The magnitude of the antagonistic activities in the external obliques, combined with significant activities in the rectus abdomini help to explain the coupled flexion moments, and indicate the need for trunk stability during torsional exertions.

Biomechanical analyses indicated that compressive forces were significantly greater during dynamic twisting exertions than during isometric exertions (Figure 4c). This velocity trend agrees with the dynamic relations associated with lifting tasks (Freivalds et.al., 1984; McGill and Norman, 1985; Granata and Marras, 1993; Granata and Marras, 1995). Examination of the spinal loads also demonstrates increased shear loading accompanied the compressive forces. When compressive loads on the spine are augmented by shear and torsional loads, the risk of disc injury is increased (Shirazi-Adl et.al, 1986; Shirazi-Adl, 1989). Thus, twisting may create a situation where the combination of forces acting upon the spine places one at a higher risk of exceeding the spinal tolerance to these forces.

These findings agree with the assessment of Marras et. al. (1993) who documented trunk motions associated with both high and low risk industrial tasks. Significantly greater velocities were observed in the high risk jobs. The average velocity of the high risk tasks were similar to our 10 deg./sec. velocity condition. Thus, the trends associated with the spine loads predicted by the model correlate well with epidemiological findings.

One must consider several potential limitations when considering the results of this study. The dynamic geometry of the spinal curvature has been ignored in this research. Although spinal curvature may influence the coordinate rotation of the spinal forces, i.e. the distribution of the load between shear and compression, the magnitude of the load would not be affected. Further research could determine loads on individual elements of the lumbar spine by combining results generated by this model with a geometric model of the lumbar spine. Although the length-modulation factor employed in the biomechanical model accounts for some passive muscle forces, passive disc and ligament forces have been ignored in this analysis. The passive loads may play a role in spinal loading, especially during tasks requiring significant twisting angles.

CONCLUSIONS

The capacity to generate torsional moments was significantly reduced by twisting direction, angle and velocity. During torsional exertions, coupled moments in the sagittal and lateral planes were also developed. Biomechanical analyses indicated that spinal loads increased with exertion load, velocity and twist angle. These results support epidemiological findings that indicate the risk of low-back pain may be related to exertion load, velocity, and twisting angle. Future efforts must explore how the musculo-skeletal system changes in more realistic, multi-dimensional, coupled postures and exertions.

ACKNOWLEDGMENT

Funding for this study has been provided by the Ohio Bureau of Worker's Compensation, Division of Safety and Hygiene.

REFERENCES

Basmajian JV, DeLuca CJ. (1983) <u>Muscles Alive, Their Functions Revealed by Electromyography</u> 5th ed., Baltimore, Wiliams and Wilkes.

Carlsoo S. (1961) The static muscle load in different work positions: An electromyographic study. *Ergonomics*, 4: 193-198

Freivalds A, Chaffin DB, Garg A, Lee KS. (1984) A dynamic biomechanical evaluation of lifting maximum acceptable loads. *J. Biomechanics*, 17 (4): 251-262

Granata KP, Marras WS. (1993) An EMG-assisted model of loads on the lumbar spine during asymmetric trunk extensions. *J. Biomechanics*, 26 (12): 1429-1438

Granata K.P. and Marras WS. (1995) The influence of trunk muscle coactivity upon dynamic spinal loads, *Spine*, 20 (8): 913-919.

Hill AV. (1938) The heat of shortening and the dynamic constants of muscle. *Proc. R. Soc. Biol.*, 126: 136-195

Kelsey KL, Githens PB, White AA III, et.al. (1984) An epidemiologic study of lifting and twisting on the job and risk for acute prolapsed lumbar intervertebral disc. *J. Ortho. Res.*, 2 (1): 61-66

Marras WS. and Granata KP. (1995) A Biomechanical Assessment and Model of Axial Twisting in the Thoraco-Lumbar Spine, *Spine*, (in press).

Marras WS, Lavender SA, Leurgans SE, et.al. (1993) The role of dynamic three-dimensional trunk motion in occupationally-related low back disorders: The effects of workplace factors, trunk position and trunk motion characteristics on risk of injury. *Spine*, 18 (5), 617-628.

McGill SM. (1991) Electromyographic activity of the abdominal and low back musculature during the generation of isometric and dynamic axial trunk torque: Implications for lumbar mechanics. *J. Orthop. Res.*, 9 (1): 91-103.

McGill SM. (1992) The influence of lordosis in axial trunk torque and trunk muscle myoelectric activity. *Spine*, 17 (10): 1187-1193.

McGill SM, Norman RW. (1995) Dynamically and statically determined low back moments during lifting. *J. Biomechanics*, 8 (12): 877-885.

Mirka GA and Marras WS. (1993) A stochastic model of trunk muscle coactivation during trunk bending. *Spine*, 18 (11): 1396-1409.

Parnianpour M, Nordin M, Khanovitz N, Frankel V. (1988) The triaxial coupling of torque generation of trunk muscles during isometric exertions and the effect of fatiguing isoinertial movements on the motor output and movement patterns. *Spine*, 13 (9): 982-992.

Parnianpour M, Campello M, Sheikhzadeh A. (1991) The effect of posture on triaxial trunk strength in different directions: Its biomechanical consideration with respect to incidence of low-back problems in construction industry. *Intl. J. Ind. Ergon.*, 8: 279-287.

Pope MH, Andersson GBJ, Broman H, Svensson M, Zetterburg C. (1986) Electromyographic studies of the lumbar trunk musculature during the development of axial torques. *J. Orthop. Res.*, 4 (3): 288-297.

Punnet L, Fine LJ, Keyserling WM, Herrin GD, Chaffin DB. (1991) Back disorders and non-neutral trunk postures of automobile assembly workers. *Scand. J. Work Envir. Health.*, 17: 337-346.

Shirazi-Adl A, Ahmed AM, Shrivastava SC. (1986) Mechanical response of the lumbar motion segment in axial torque alone and in combination with compression. *Spine*, 11 (9): 914-927.

Shirazi-Adl A. (1989) Stress in fibers of a lumbar disc, analysis of the role of lifting in producing disc prolapse. *Spine*, 14 (1): 96-103.

Zetterberg C, Andersson GB, Schultz AB. (1987) The activity of individual trunk muscles during heavy physical loading. *Spine*, 12 (10): 1035-1040.

STABILITY LIMITS IN EXTREME POSTURES:
EFFECTS OF LOAD POSITIONING AND FOOT PLACEMENT

M.A. Holbein, Ph.D. and D.B. Chaffin, Ph.D.
Center for Ergonomics, University of Michigan
Ann Arbor, Michigan

Although injuries related to postural stability are prevalent, ergonomic job analyses have traditionally not addressed stability issues. In this research, functional stability limits are quantified for persons standing in extreme postures under various external load and foot positioning conditions. Six subjects were tested while standing unladen and while holding a 5.2 kg load. The foot positions, or bases of support (BOS), were varied in width of the stance and sagittal separation of the feet. They were asked to lean and displace their center of gravity (COG) as far as possible in eight directions to the sides and front of the body. Stability measures based on these COG displacements were calculated. All controlled variables significantly affected the stability measures. When standing unladen, subjects extended their COG to within 99% of their BOS limit. Movement was much more restricted when leaning while handling a load (89%), especially holding it with one hand on the shoulder (84%). On average, increased separation of the feet in a particular direction resulted in larger COG displacements in that direction. The results are discussed relative to their effects on balance and stability modeling.

INTRODUCTION

Loss of balance can result in falling, low back and other musculoskeletal overexertions, and collisions with the surroundings. However, stability analyses are not typical components of ergonomic job analyses. Biomechanical, physiological and psychophysical evaluations are more common.

There is no standard definition of postural stability in the literature, however, a theory of mechanics is often adopted (Davis, 1983; Winter, 1987; Gollhofer et al., 1989; Holbein, 1993). A stable posture is described as the center of gravity (COG) being within the base of support (BOS). The BOS is the area within the outline of both feet in contact with the supporting surface. In a static situation, the vertical ground reaction force vector must act at the same horizontal location as the COG in order to maintain equilibrium and can only act within the BOS. The theoretical maximum stability area is, therefore, defined by the BOS, for example from the great toes to heels anteroposteriorly. However, the *functional* stability area may be smaller than this theoretical area due to limiting muscle strengths, internal postural control, control of any external load, or other factors. The functional stability region (FSR) may be defined as the area within which the COG can be contained by a person (Holbein, 1993).

The purpose of this research was to assess the FSR of persons holding loads while standing in extreme postures, i. e. while leaning as far as possible in several directions. The effects of load position, foot position (or BOS) and direction of the lean were investigated. Isometric strength tests were also conducted to determine if muscle strength is a significant factor limiting functional stability regions.

METHODS

Apparatus

The load was held in a 26x33x26 cm box with padded handles. The load was 5.2 kg and securely centered in the bottom of the box. Fifteen passive reflective markers were attached to the subject at relevant anthropometric landmarks and on the load. The MacReflex motion analysis system (Qualisys AB, Sweden) with two infrared cameras was used to obtain three-dimensional marker positions. An isometric strength testing chair was used to measure maximum voluntary force production at major joints.

Protocol

Four male and two female young, healthy right-handed subjects were paid to complete two data collection sessions, approximately three to four hours each, on non-consecutive days. Seven standard strength tests were performed using the protocol described by Stobbe (1982).

For the leaning trials, the bases of support for each subject and the desired lean directions were marked on the floor. Subjects were instructed that their goal was to move the COG of themselves and the load as far as possible in the lean direction given, while being able to hold steady in the extreme posture for a few seconds. Significantly altering the position of the load relative to the body was not permitted. Subjects were permitted to lift any portion of their feet but had to maintain some floor contact with both feet and keep both feet pointed forward at all times.

Subjects placed their feet and were handed the load. Starting from upright, they moved slowly in the first direction given. Final postures were self-selected, i. e. rotation about any joint was permitted. They verbally indicated when they were at their extreme and steady. They held the posture (approximately 3 sec total) while their postural data were recorded for one second (collected at 12 Hz) and until given the next lean direction. Subjects returned to upright between each lean and relaxed.

Experimental Design

Six load conditions (POSTURE) with varying load height and frontal plane symmetry and an unladen condition were tested (Figure 1). Four BOSs were defined based on individual subjects' shoulder width and shod foot length (Figure 2a). Eight lean directions (LEAN), primarily anterior, were investigated (Figure 2b).

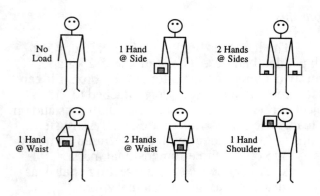

Figure 1: Postures tested.

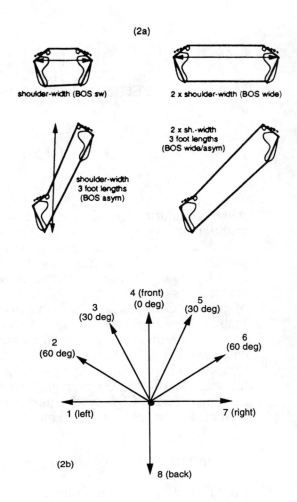

Figure 2: BOSs and lean directions tested.

The experimental design was a full factorial of the loading postures, BOSs and lean directions and was replicated twice for each subject. Therefore, a total of 2,304 trials were conducted (6 postures * 4 BOSs * 8 leans * 2 repeats * 6 subjects = 2,304). Conditions were blocked on posture and BOS combinations. Blocks were then randomized for each subject and lean directions were randomized within each block.

Data Analysis

The BOS limit was defined using toe marker positions while the feet were in full contact with the floor and foot measurements locating the 5th metatarsal heads and the backs of the heels (Figure 2a). Twelve frames of video data (1 sec) at each extreme posture were averaged to determine reflective marker coordinates. Fewer frames were averaged if subjects did not remain steady for the full second. The COG of the body-and-load was estimated using the video data and anthropometric data cited by Chaffin and Andersson (1991). Finally, lean directions were reassigned or trials discarded when actual COG directions were not close to the desired directions.

Dependent measures of stability, or functional stability limits (FSL), were defined from the COG displacements in each lean direction. FSL% is the distance from the center of the BOS (CBOS) to the COG expressed as a percent of the distance from the CBOS to the BOS limit in that direction, i. e. the theoretical maximum. SWAY is the angle between the COG's vertical projection and upright, with the vertex at the center of the BOS (see insert, Figure 4b). Unlike FSL%, SWAY is not referenced to the BOS limit.

In the statistical analysis, ANOVA was used to determine if the independent variables affected the dependent stability measures. Contrasts were used to determine which levels of the main effects were significantly different. Results of the numerous contrasts tests were used to explain and support the trends in the main effect means across its levels, rather than to enumerate all significant pairwise differences.

RESULTS

The main effects of Subject, Posture, BOS and Lean all significantly affected both dependent measures (Table 1). Two-way interactions were most often significant as well. However, only the Subject*Posture and BOS*Lean interactions considerably affected trends in the means and were analyzed further.

Table 1: P-values (* indicates p<0.0001)

	FSL%	SWAY angle
R^2	0.76	0.74
Subject	*	*
Posture	*	*
BOS	*	*
Lean	*	*
Subj * Posture	*	*
Subj * BOS	*	*
Posture * BOS	0.56	0.39
Subj * Lean	*	*
Posture * Lean	0.01	0.09
BOS * Lean	*	*

The FSL% and SWAY dependent measures experienced similar effects of Posture (Figure 3). Overall, the unladen posture resulted in larger FSL% (mean = 99%) and SWAY (mean = 17 deg) than the laden postures (means = 89% and 14 deg). Holding with 1 hand on the shoulder resulted in the smallest

stability responses (means = 84% and 13 deg) compared to holding the load in other positions.

BOS and Lean direction effects were different for the FSL% compared to the SWAY measures. Staggering the feet in the sagittal plane (BOS asym.) resulted in significantly larger FSL% responses (mean = 103%) than other foot positions (Figure 4a). With the feet in a wide stance (BOS wide and BOS wide/asym.), relatively small FSL% values were found (means = 86% and 85%, respectively). SWAY angle measures were smallest with the shoulder-width stance (BOS sw, mean = 12 deg) and significantly larger with any other foot placement (means = 16 to 17 deg) (Figure 4b).

Leaning directly forward (Lean 4) resulted in the largest FSL% measures (mean = 117%) (Figure 5a). The mean response decreased as the lean directions became increasingly asymmetric to the sagittal plane, especially when leaning toward the right side (Leans 5 to 7). When subjects leaned backward, much smaller FSL% responses resulted compared to leans in other directions (mean = 47%). The SWAY angles were also smallest when leaning backward (mean = 7 deg) but the largest SWAY was found with the slightly right-side asymmetric Lean 5 (mean = 21 deg) (Figure 5b). The most extreme right leans (5 and 6) and the most extreme left leans (1 and 2) also resulted in somewhat large SWAY measures (means = 14 deg to 15 deg).

DISCUSSION

Stability limits have traditionally been assumed to be at the theoretical base of support limit. This work indicates that people generally either do not or cannot reach their theoretical stability limit. Restricted COG displacements cover an average of only 91% of the theoretical maximum.

The FSL% measure was greater than 100% under some conditions, however, indicating that the subject's COG sometimes surpassed their theoretical limit. However, the actual distance that the COG extended beyond the BOS was typically small. (Of the cases where FSL% was greater than 100%, the distance of the COG beyond the BOS was less than three inches in 90% of the cases and less than one inch in 47% of them.) Several factors likely contributed to the possible overestimation of FSL%. First, the BOS model used was a simplified six-sided polynomial outlining the feet, and probably underestimated the actual BOS area (see Figure 2a). This is especially true for the toe-to-toe segment of the shoulder-width, asymmetric BOS, where most of the large FSL% values were found. In addition, the distance from the center of the BOS to the BOS limit is small for this condition, causing the FSL% calculation to be especially sensitive when this

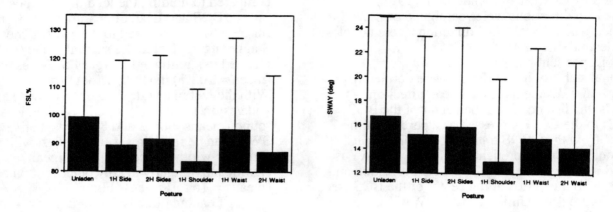

Figure 3: FSL% (a) and SWAY (b) Means (& std.dev.) by Posture
Averaged over All Subjects, BOSs & Lean Directions (n=304 to 345)

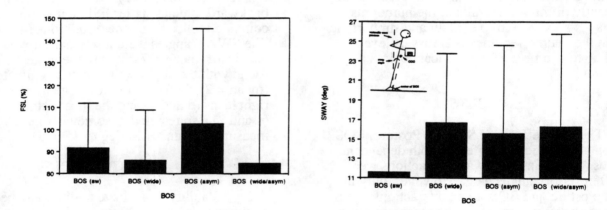

Figure 4: FSL% (a) and SWAY (b) Means (& std.dev.) by Base of Support
Averaged over All Subjects, Postures & Lean Directions (n=429 to 539)

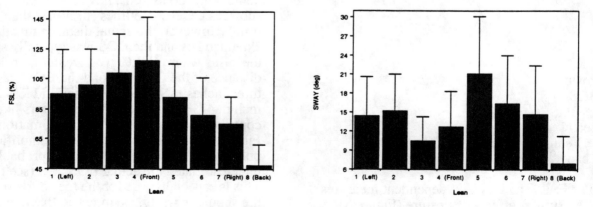

Figure 5: FSL% (a) and SWAY (b) Means (& std.dev.) by Lean Direction
Averaged over All Subjects, Postures, & BOSs (n=137 to 359)

segment is used. For any BOS and BOS segment, FSL% may also have been overestimated since subjects were permitted to partially lift and rotate their feet, allowing additional displacement of the COG without also adjusting the BOS model. Finally, estimations involved in locating the subjects' COG could also have affected FSL%. These results indicate that rectangular or simple polynomial representations of the BOS used in this and previous studies should be carefully interpreted.

Differences in the stability measures across experimental conditions were large and all controlled variables were highly significant in explaining these differences. Combining levels of some variables to generalize results may be useful . For example, categorizing the bases of support based on width and symmetry, combining results from all lean directions to estimate a two-dimensional stability envelope, and checking for variations in whole-body postures and COG height across subjects may provide further insight into the effects of these measures on stability.

Not surprisingly, subject variability was high. Results of the strength tests did not indicate that muscle strength greatly affects stability as was defined in this work. Although the smallest stability responses were often from the weakest person, the second weakest person often had the largest responses. Examining correlations between specific joint strengths, whole-body postures assumed, and the resulting maximum COG displacement may provide further insight into the effects of strength on functional stability limits.

Stability issues have been investigated in previous studies. Holbein (1993) tested persons holding loads and standing with the feet shoulder-width apart and reported functional stability limits of 50% to 60% of maximum. However, subjects were permitted to rotate only about the ankles and limited lean directions were tested. This current work defines stability limits in directions off the sagittal and frontal planes and allows more realistic whole-body postures. Davis (1983) measured sway angles as were measured here and found similar results, but functional versus theoretical stability limits were not addressed. Kerk (1992) and Grieve (1979a,b) used stability constraints to model hand force exertion capabilities given fixed whole body postures. This current work investigates voluntary stability limits, which may be reached using any desired body posture, and compares them to theoretical constraints.

REFERENCES

Chaffin, D.B. and Andersson, G.B.J., Occupational Biomechanics (2nd edition; New York: Wiley Interscience, 1991).

Davis, P.R., "Human Factors Contributing to Slips, Trips and Falls," Ergonomics, Vol. 26, No. 1 (1983), pp. 51-59.

Gollhofer, A., Horstmann, G.A., Berger, W. and Dietz, V., "Compensation of Translational and Rotational Perturbations in Human Posture: Stabilization of the Centre of Gravity," Neuroscience Newsletters, Vol. 105, (1989), pp. 73-78.

Grieve, D.W., "The Postural Stability Diagram (PSD): Personal Constraints on the Static Exertion of Force", Ergonomics, Vol. 22, No. 10 (1979a), pp. 1155-1164.

Grieve, D.W., "Environmental Constraints on the Static Exertion of Force: PSD Analysis in Task Design", Ergonomics, Vol. 22, No. 10 (1979b), pp. 1165-1175.

Holbein, M.A., "Ergonomic Load-holding and Carrying Strategies Based on the Biomechanics of Human Motion and Postural Stability" (unpublished Ph.D. Dissertation, Industrial Engineering Department, The University of Pittsburgh, 1993).

Kerk, C.J., "Development and Evaluation of a Static Hand Force Exertion Capability Model Using Strength, Stability and Coefficient f Friction" (unpublished Ph.D. Dissertation, Industrial and Operations Engineering, The University of Michigan, 1992)

Stobbe, T.J., "The Development of a Practical Strength Testing Program for Industry" (unpublished Ph.D. Dissertation, Industrial and Operations Engineering, The University of Michigan, 1982).

Winter, D.A., "Sagittal Plane Balance and Posture in Human Walking," IEEE Engineering in Medicine and Biology Magazine, (September, 1987), pp. 8-11.

SIMULATION OF THE 1991 REVISED NIOSH MANUAL LIFTING EQUATION

Waldemar Karwowski and Paul Gaddie
Center for Industrial Ergonomics
Department of Industrial Engineering
University of Louisville
Louisville, KY 40292, USA

Digital computer simulation of the 1991 Revised NIOSH Lifting Equation was performed using SLAM II in order to examine the behavior of this equation under a variety of realistic industrial lifting tasks. The results showed that over all conditions studied (represented by 100,000 randomly generated lifting task scenarios), the recommended weight limit (RWL) values for the 99.5% of all tasks were equal to or lower than 12.5 kg or 27.5 lbs. With respect to lifting time exposure, the RWL values for the 99.5% of cases were equal to or lower than 13 kg (or 28.6 lbs) for up to one hour of lifting, 12.5 kg (or 26.4 lbs) for less than 2 hours of exposure, and 10.5 kg (or 23.1 lbs) for lifting over an 8-hour shift. From a practical point of view, the results of this study define the threshold RWL values (TRWL), that can be used by practitioners for the purpose of immediate risk assessment of manual lifting tasks performed in industry.

INTRODUCTION

Low-back pain and injuries due to manual lifting are recognized as one of the leading causes of work-related illnesses and injuries in industry. In order to reduce the risk of such injuries, simple and easy to apply design guidelines for manual lifting tasks should be adapted. The National Institute of Occupational Safety and Health (NIOSH) proposed the recommended weight limit (RWL) as the guide for evaluating manual tasks in industry (Waters et al, 1993). The 1991 Revised Lifting Equation expands beyond the previous guidelines, and can be applied to a larger percentage of lifting tasks (Waters et al, 1993). The RWL is designed to protect 90% of the mixed (male/female) industrial working population against low back pain.

One way to investigate the practical implications of the 1991 Lifting Equation for industry is to determine the likely results of the equation when applying a realistic and practical range of values for the risk factors (Karwowski, 1992). This can be done using the modern computer simulation techniques. The purpose of this study was to develop such a computer simulation approach in order to examine the behavior of the 1991 Revised Lifting Equation under a broad range of conditions. For this purpose, probability distributions for all the relevant risk factors were defined, and a digital simulation of the revised equation was performed.

THE 1991 REVISED NIOSH LIFTING EQUATION

Table 1 shows definitions of the relevant terms utilized by the 1991 Revised Lifting Equation. The recommended weight limit (RWL) is the product of the load constant and six multipliers which account for seven risk factors, i.e.: RWL (Kg) = LC * HM * VM * DM * AM * FM * CM. The multipliers are defined in terms of the related risk factors, including the horizontal location, vertical location, vertical travel distance, frequency of lift, and asymmetry angle. The multipliers for frequency and coupling are defined using relevant tables. In addition to lifting frequency, the work duration and vertical distance factors are used to compute the frequency multiplier. Table 2 shows the coupling multiplier.

SIMULATION MODELING APPROACH

A computer simulation model was written in SLAM II, a Simulation Language for Alternative Modeling, and run on a VAX 8600 Cluster mainframe computer. SLAM II, developed and supported by Pritsker & Associates (Pritsker, 1986), provides a unified modeling framework, including a set of graphical symbols representing nodes and branches, and supports discrete, continuous and combined simulations.

Table 1. Terms of the NIOSH 1991 equation.

Multiplier	Formula [cm]
Load constant	LC = 23 kg
Horizontal	HM = 25 / H
Vertical	VM = 1 - (0.003 | V -75 |)
Distance	DM = 0.82 + 4.5 / D
Asymmetry	AM = 1 - 0.0032 A
Frequency	FM = from Table*
Coupling	CM = see Table 2

*(see Waters et al, 1993)

H -   The horizontal distance of the hands from the midpoint of the ankles, measured at the origin and destination of the lift (cm).

V -   The vertical distance of the hands from the floor, measured at the origin and destination of the lift (cm).

D -   The vertical travel distance between the origin and destination of the lift (cm).

A -   The angle of asymmetry--angular displacement of the load from the sagittal plane, measured at the origin and destination of the lift (degrees).

F -   Average frequency of lift (lifts/minute).

C -   Load coupling, the degree that appropriate handles, devices or lifting surfaces are present to assist lifting and reduce the possibility of dropping the load.   A qualitative measure--good, fair, and poor.

Table 2. The coupling multiplier.

Couplings	V < 75 cm	V ≥ 75 cm
Good	1.00	1.00
Fair	0.95	1.00
Poor	0.90	0.90

Using the process orientation employing a network structure with nodes and branches, the NIOSH 1991 Lifting Equation was modeled as the product of the six factor multipliers represented as attributes of an entity flowing through the network. The developed SLAM simulation is a sequence of operations which draws random values from each of the relevant factor distributions and tables, and then calculates the corresponding RWL values.

An objective of the simulation was to examine the behavior of the 1991 Revised NIOSH Lifting Equation with realistic distributions for the lifting risk factors, and to determine the resulting distribution of the RWL values. The results of simulation provided distribution graphs and descriptive statistics for all the considered lifting factors, corresponding multipliers, and the RWL values.

THE LIFTING FACTORS AND MULTIPLIERS

Factor Distributions

Seven risk factors are used in the 1991 Equation to define the lifting task environment As much as possible, the probability distributions for these factors were chosen to be representative of the real industrial workplace (Ciriello et al, 1990; Brokaw, 1992, Karwowski and Brokaw, 1992; Marras et al, 1995).   Except for the vertical travel distance factor, coupling and asymmetry multipliers, all factors were defined using either normal or lognormal distributions.

For all the factors defined as having lognormal distributions, the following procedure was used to adjust for the required range of real values whenever necessary.   After the given distribution was generated, it was then adjusted for the range, i.e. truncated for values below the lower limit and above the upper limit. Specifically, the values outside the range were removed, and the new values that would fall within the desired range were generated. This process normalized the derived distributions to the area of 1.0.

Horizontal Factor

The horizontal factor (H) represents the distance of the hands from the midpoint of the ankles, measured in centimeters at the origin and the destination of lift. This horizontal distance was assumed to be the maximum of the two positions, and that no intermediate position was greater than this maximum. The range of horizontal distance as specified by NIOSH (Waters et al, 1993) is from 25 cm to 63 cm, and is used to define the horizontal multiplier (HM). A lognormal distribution for H (mean value of 44 cm, standard deviation of 6 cm) was used to generate random values within the required range. These values were chosen because H = 44 cm is the midpoint of the range from 25 cm to 63 cm, and the standard deviation of 6 cm allows almost all the values generated to fall within this range. It should be noted that the lognormal distribution gives greater weight to the values which are located near the mean.

Vertical Height Factor

The vertical height factor (V), indicates the vertical distance of the hands from the floor, with the range from the "floor position" (of 0 cm at a minimum) to the upper limit of vertical reach of 175 cm (Waters et al, 1993). This factor was defined using a lognormal distribution (mean =100 cm, std = 25 cm). It should be noted that the vertical height factor is also used as an independent variable in the definition of vertical multiplier (VM), the coupling multiplier (CM), and the frequency multiplier (FM).

Distance Multiplier

The vertical travel distance factor (D), i.e. the difference between the origin and destination of the lift, is used by NIOSH (Waters et al, 1993) to define the distance multiplier (DM). For the purpose of this simulation, the vertical travel distance (D) was assumed uniform over the interval of (0, [175-V]). Since the vertical distance traveled is the difference between the origin and the destination heights, this distance can be no greater than the one defined by the origin of lift and the maximum vertical reach limit. A variable called the maximum vertical distance (MAVED) was created based on the vertical factor (V). The travel distance factor (D) was then drawn from a uniform distribution ranging from 0 to MAVED. The MAVED variable was then modified in order to reset all values of D which were less than 25 cm to 25 cm. This procedure guaranteed that the DM was set at 1.0 for such D values.

Asymmetry Factor

The asymmetry factor is defined for the [0 - 135 degrees] range of the upper body twisting angle. This factor was modeled using a lognormal distribution (45, 15), i.e. the mean of 45° and a standard deviation of 15°. This mean and standard deviation provide a natural range from 0° to 90° of twist for the underlying distribution. The mean of 45 degrees was chosen based on the conservative assumption that in most of the industrial lifting tasks the upper body twisting does not exceed 90 degrees.

The standard deviation was chosen to ensure that almost all the lifting posture asymmetry values will fall within the NIOSH- specified range limit.

Work Duration Factor

The work duration factor is used to define the frequency multiplier (FM). For the purpose of this simulation, the work duration was conservatively modelled as a normal variable with the mean of 2.0 hours and standard deviation of 2.0 hours. This distribution was then adjusted for the range, i.e. truncated for values of work duration of zero or less, and those which were greater than 8 hours. For the purpose of simulation, the values outside the range of work duration were removed, and the new values that would fall within the desired range were generated. This process normalized the derived distribution to the area of 1.0. These random values were then used in the selection of the frequency multiplier from the NIOSH table for FM (Waters et al, 1993).

Frequency Factor

The frequency variable F (lifts/minute), for the range of lifting frequency from 0.2 lifts/min to 15 lifts/min, is used in the definition of the frequency multiplier (FM). The frequency multipliers (FMs) were selected from the NIOSH table based on randomization of three factors: 1) the frequency of lift, 2) work duration, and 3) vertical height (V). Due to the discrete nature of the values for frequency of lift, the probability distribution function for this factor was generated using normal distribution. The parameters for this distribution were defined as follows: a mean of 4.7 lifts/min, and a standard deviation 1.75 lifts/min. These values were derived based on the information provided by results of industrial surveys of manual lifting tasks. For example, Ciriello et al. (1990) reported that 94% of industrial lifting tasks are performed at frequencies of 4.3 lifts/min or slower. Brokaw (1992) analyzed 31 industrial lifting tasks and reported that in about 80% of these tasks the observed frequency was lower than 5 lifts/min. For the purpose of this simulation, however, it was decided that better representation of the higher lifting frequencies is warranted by the fact that such frequencies of lifting tasks are common today in the service industry, and by the attention given to such tasks by NIOSH (1993), as reflected by their inclusion in the frequency multiplier table.

The randomly generated values of the frequency of lift which were less than or equal to zero were eliminated, and the truncated normal distribution was re-normalized to equate the area under the curve to 1.0. In other words, the resulting distribution was re-normalized, and used to generate the probability density function values for entries from the frequency of lift table.

Coupling Multiplier

Coupling describes the quality of coupling between the hand and the load. This quality has three linguistic values, i.e. *good, fair* and *poor*. The discrete values for the coupling multiplier (CM) were chosen based on the results of an industrial survey of thirty-one manual lifting tasks performed in three different companies reported by Brokaw (1992). This survey showed that slightly less than 10% of the observed lifting tasks had *good* couplings, about 64 % had *fair* couplings, while about 26% had *poor* couplings. Due to the limited sample size of the above study, it was decided to allow for more liberal coupling assumption for the simulation purposes. Therefore, the probability of *good* hand-to-container coupling was set at 0.3, for *fair* coupling at 0.5, and the probability for *poor* coupling was set at 0.2.

RESULTS AND DISCUSSION

The SLAM II computer simulation was run for a total of 100,000 trials, i.e. the 100,000 randomly selected scenarios which realistically define the industrial tasks in terms of the NIOSH (1991) Revised Lifting Equation. Descriptive statistical data were collected for all the input (lifting) factors, the respective multipliers, and the resulting recommended weight limits. The input factor distributions were examined in order to verify the intended distributions.

The results of simulation regarding the values of RWL are summarized in Figure 1. Overall, for all the lifting conditions studied, the distribution of recommended weight limit values had a mean of 7.22 kg and a standard deviation of 2.09 kg (with a minimum of 0 kg and maximum of 17.2 kg). From a practical point of view, it should be noted that in 95% of all cases, the RWL was at or below the value of 10.5 kg or about 23.1 lbs. Furthermore, in 99.5% of all cases the RWL value was at or below 12.5 kg or 27.5 lbs. That implies that when the Lifting Index (LI) is set to 1.0 for task design or evaluation purposes, only half of one percent of the (simulated) industrial lifting tasks would have the recommended weight limits greater than 12.5 kg.

With respect to differences in the RWL values due to lifting task duration, in the 99.5% of the examined cases (i.e. for 99.5% of the simulated lifting tasks), the recommended weight limit values were equal to or were lower than 13.0 kg (or 28.6 lbs) for up to one hour of lifting task exposure, 12.5 kg (or 27.5 lbs) for less than 2 hours of exposure, and 10.5 kg (or 23.1 lbs) for lifting over an 8-hour shift. From a practical point of view, these values define simple and straightforward lifting limits, i.e. the threshold RWL values (TRWL) that can be used by practitioners for the purpose of gross risk assessment of manual lifting tasks performed in industry.

The above results indicate that under most of the examined lifting conditions (99.5% of the simulated cases), one can reasonably expect that an implementation of the 1991 NOSH Revised Lifting Equation at the level of Lifting Index of 1.0, which is designed to protect 90% of the mixed industrial working population, would necessitate redesign of manual lifting tasks according to the TRWL values reported above.

CONCLUSIONS

The 1991 Revised NIOSH Manual Lifting Equation is designed to ensure that the RWL will not exceed the acceptable lifting capability of 99% of the male workers and 75% of the female workers. According to Waters et al, (1993), this amounts to protecting about 90% of the industrial workers if there is a 50/50 split between males and females. From a practical perspective, the results of this simulation study define the threshold RWL values (TRWL), which are useful for immediate risk assessment of manual lifting tasks performed in industry. Such threshold values, if exceeded, could provide an indication of the need for a more thorough examination of the identified tasks, as well as evaluation of physical capacity of the exposed workers.

REFERENCES

Brokaw, N. 1992, *Implications of the Revised NIOSH Lifting Guide of 1991: A Field Study.* Unpublished M.S. Thesis, Department of Industrial Engineering, University of Louisville, Louisville, Kentucky.

Ciriello, V. M., Snook, S. H., Blick, A.C., and Wilkinson, P. L. 1990, The effects of task duration on psychophysically-determined maximum acceptable weights and forces. *Ergonomics,* **33**, 187-200.

Karwowski, W. and Brokaw, N. 1992, Implications of the Proposed Revisions in a Draft of the Revised NIOSH Lifting Guide (1991) for Job Redesign: A Field Study, *Proceedings of the 36th Annual Meeting of the Human Factors Society,* Atlanta, GA, 1992, pp. 659-663.

Karwowski, W. 1992, Comments on the assumption of multiplicity of risk factors in the draft revisions to NIOSH Lifting Guide, *Advances in Industrial Ergonomics and Safety IV,* 905-910.

Marras, W. S., Lavender, S. A., Leurgans, S. E., Fathallah, F. A., Ferguson, S. A., Allread, W. G., and Rajulu, S. L. 1995, Biomechanical risk factors for occupationally related low back disorders. *Ergonomics,* 38, 377-410 .

Pritsker, A. A. B. 1986, *Introduction to Simulation and SLAM II,* Third Edition (Wiley, New York).

Waters, T. R., Putz-Anderson, V., Garg, A., and Fine, L.J. 1993, Revised NIOSH equation for the design and evaluation of manual lifting tasks, *Ergonomics,* **36,** 749-776.

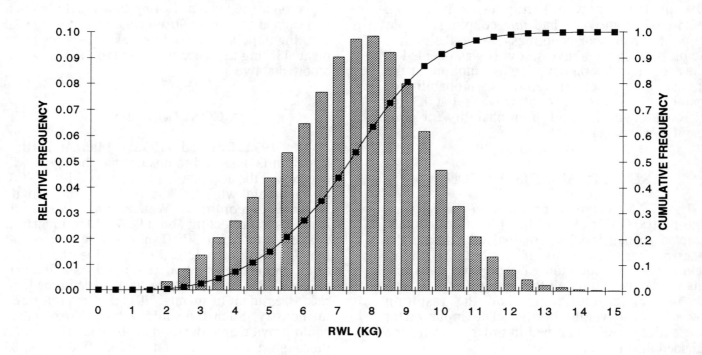

Figure 1. Distribution of the simulated RWL values.

THE EFFECTS OF LIFTING FREQUENCY ON THE DYNAMICS OF LIFTING

Gary A. Mirka
and
Daniel P. Kelaher

Ergonomics Laboratory
Department of Industrial Engineering
North Carolina State University
Raleigh, North Carolina

The goal of this study was to quantify the effects of different lifting frequencies (3, 6 and 9 lifts/minute) at different lifting heights (30 and 60 cm) on the kinematics of the lumbar region. Each of these lifting tasks was performed for twenty minutes. The time dependent traces of the both the mean and standard deviation of sagittal acceleration showed subject dependent trends over time. Averaged across time, the results of this study reveal that there is a non-linear increase in the sagittal acceleration with greater frequency of lifting.

INTRODUCTION

The frequency multiplier found in the NIOSH equation (Waters et al, 1993) is based on data from psychophysical studies (Snook and Ciriello, 1991) and physiological studies (Garg, 1976, Garg et al, 1978). The psychophysical approach was used to develop the multiplier for frequencies up to four lifts/minute while the physiological results were used to find the multipliers for frequencies greater than four lifts/minute. These studies were useful in predicting preferred workload of an individual (psychophysical) or the cardiovascular cost of lifting (physiological), but they did not discuss the biomechanical implications of variable lifting frequency. Given that frequency of lift does not fit well into a static description of a work environment, trunk kinematics during manual lifting tasks need to be considered in order to get a more complete picture of the risk associated with frequency of lift in free dynamic MMH tasks. The goal of the current study was to better understand the effects of lifting frequency on these trunk kinematic parameters.

METHODS

Subjects

Five male and five female college students volunteered for this study. The subjects had no history of a low back disorder. Written informed consent was given by all of the subjects at the beginning of the experiment.

Equipment

Motion analysis and heart rate monitoring equipment were utilized for this experiment. A telemetry-based Lumbar Motion Monitor was placed on the back of the individual to measure the angular position, velocity, and acceleration of the lumbar spine in three-dimensional space. The heart rate of the subjects was measured using an Accurex II Heart Rate Monitor.

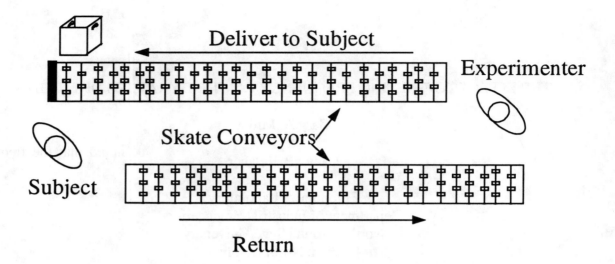

Figure 1. Overhead view of experimental environment

Experimental Setup and Task

Wooden boxes with handles ("good" coupling as defined by the 1991 Lifting Equation) were delivered to the subjects via gravity-fed passive skate-wheel conveyers at approximately 5° slope. The mass of the boxes were 11.4 kg for men and 7.4 kg for the women. The conveyors were placed parallel to each other 2 feet apart and were offset by 2 feet. This resulted in a 90° rotation requirement for transferring the boxes from the first to the second conveyer (Figure 1). Subjects worked for twenty minutes and then were given a five minute rest period.

The boxes were placed on the conveyer at the prescribed frequencies to simulate equally spaced packages from an automatic conveyer system. Each box was sent to the subject with the handles in line with conveyor to standardize the initial lifting posture. The subjects were then asked to lift the box after it reached the box stop and place it on the top of the return conveyer with the handles perpendicular to the midline of the conveyer. The subjects were given no instructions as to the lifting style to be used. Research has shown that this free-style lift allows for the greatest psychophysical lifting capacity (Garg and Saxena, 1979) while providing data which would be most representative of industrial data.

Experimental Design

The independent variables for this study were frequency of lift and the height of the handles as the box was lifted. The starting heights were 30 and 60 cm from the floor, while the height of the handles while placing the box on the return conveyer was 70 cm. The three lifting frequencies chosen for this study were 3, 6, and 9 lifts/minute. These frequencies were chosen for comparison with the data from previous research (Garg and Saxena, 1979) and to cover the range of frequencies typically seen in industry (Marras et al, 1993). The duration of each experimental condition was twenty minutes. Data was collected on each of the lifts for the 3 and 6 lifts/minute condition while computer limitations only allowed for data collection on every other lift for the 9 lifts per minute condition

The dependent variables were the kinematic parameters describing the motion of the lumbar spine (range of motion, velocity, and acceleration) in the three planes (sagittal, coronal, and transverse). This kinematic data was collected at a rate of 60 samples per second. Heart rate was collected as the instantaneous value at one minute intervals. Sagittal acceleration is the only dependent variable which will be discussed in the current paper.

Data Analysis

The peak value of sagittal acceleration was extracted from each of the data files. This peak value was computed as the average of the peak value and one data point on each side of this peak value. This was done to reduce data processing induced variability. This resulted in one peak value per trial. The time dependent standard deviation was then calculated from seven consecutive trials.

RESULTS

The results of this study are shown graphically in Figures 2-5. Figure 2 shows the response of sagittal acceleration to the different levels of starting height and lift frequency. A statistical analysis revealed a significant effect for both starting height and lift frequency ($p<.001$). Figure 3 shows an example of a time dependent response of the peak sagittal acceleration. The data in this part of the analysis was highly variable from subject to subject. In fact when these were averaged across subjects no significant trends resulted. It was only when the data was plotted by subject did the analysis reveal these interesting results. Finally, Figures 4-5 shown how the variability in the peak sagittal acceleration changed as a function of time. Again these trends were highly subject dependent.

DISCUSSION

The results of the present study suggest that in addition to the increase in some of the physiologic parameters which have been shown previously, there are some critical biomechanical parameters which are affected by increased frequency of lifting.

It should be noted that the frequency levels chosen in this experiment were not so high as to require continuous work from the subject. If the frequency levels chosen had been at these excessive levels, it would not have been surprising for the frequency levels to be highly correlated with some of these kinematic parameters because the subjects' trunk dynamics would have to increase to keep up with the incoming work. The results of this study have shown, however, that under the conditions which do not require continuous work there still was an increase in most kinematic parameters with increased lifting frequency. Even under the highest work rate (9 lifts per minute) there still was between 3 and 4 seconds of rest wherein the subject was standing erect and waiting for the next box to arrive. This result indicates that instead of a required increase in trunk kinematics, as would be the case under conditions of high lift rates (>15/minute), there appears to some other mechanism at work.

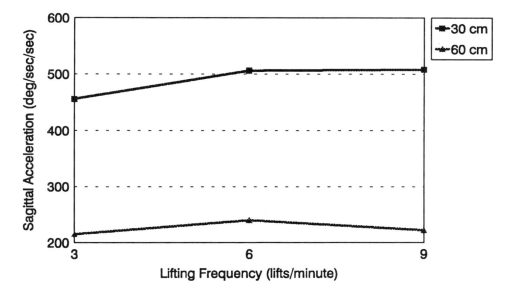

Figure 2. Response of sagittal acceleration to different levels of lifting frequency and starting heights. Note non-linear trend in these graphs.

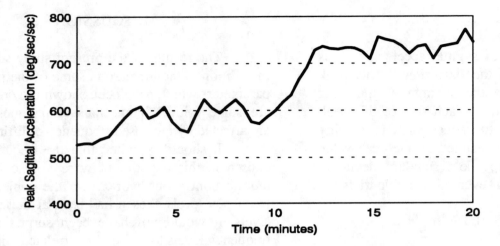

Figure 3. Time dependent response of the peak sagittal acceleration. Condition: Frequency = 9
       lifts/minute, starting height = 30 cm.

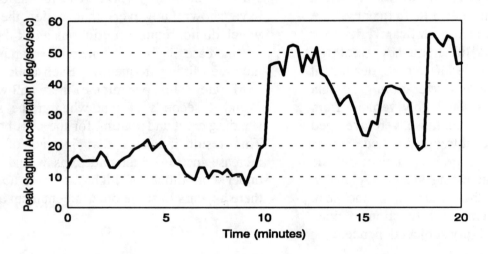

Figure 4. Time dependent response of the standard deviation of the peak sagittal acceleration.
       Condition: Frequency = 9 lifts/minute, starting height = 30 cm.

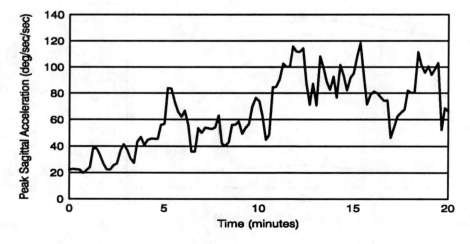

Figure 5. Time dependent response of the standard deviation of the peak sagittal acceleration.
       Condition: Frequency = 9 lifts/minute, starting height = 30 cm.

There has quite a bit of work done in the area of the effects of frequency during lifting. Much of the work has been in the area of psychophysics and cardiovascular physiology. Of particular relevance to the current study is a paper by Garg and Saxena (1979) which described a psychophysical study wherein subjects performed lifts at constant frequencies (3, 6, 9, and 12 lifts/minute) and were asked to find their maximum acceptable weight of lift. They found an interesting energy minimization at a frequency of 9 lifts/minute. The metabolic cost per unit work curve (i.e., Kcal/Kg*m or HR/Kg*m ratio) was parabolic with the minimum point occurring at the frequency of 9 lifts/minute (Garg and Saxena, 1979; Garg and Banaag, 1988). These studies were very useful in predicting the weights and workloads that subjects would choose as a function of the lifting frequency, but they did not discuss any change in the lifting biomechanics, a limitation which has been overcome somewhat with the current study. It is interesting to note that there does seem to be a sort of leveling off of the sagittal acceleration, a results consistent with that of the studies mentioned above.

The time dependent data shown in Figures 3-5 showed some very interesting subject dependent trends. Of particular interest is data presented in Figures 3 and 4. At about the ten minute mark there is an abrupt change in the variability of the subjects performance and an increase in the peak sagittal acceleration. Reviewing written notes taken during subject data collection it was noted that it was around this point that the subject changed from a squat lift to a stoop lift. This would explain the trends shown in these figures. Figure 5 shows a gradual increase in the variability of the data. This may be attributed to the gradual onset of fatigue. Not shown in this paper are data which show fairly high levels of variability early in the twenty minute period which tended to level off as the subject progressed through the experiment. This might indicate a short adjustment period as the subject became acclimated to the lifting task. Taken as a whole, the time dependent data suggests that it may be important for ergonomists to consider the changes over time of the human performance aspect of manual material handling. These parameters could become markers describing variables such as fatigue or warm-up periods which could be useful in designing and monitoring workplace design and evaluation.

ACKNOWLEDGMENTS

This publication was partially supported by grant number KO1 OH00135-02 from NIOSH. Its contents are solely the responsibility of the authors and do not necessarily the official views of NIOSH.

REFERENCES

Garg, A. 1976, "A Metabolic Rate Prediction Model for Manual Materials Handling Jobs," PhD Dissertation, University of Michigan.

Garg, A. and Banaag, J. "Maximum acceptable weights, heart rates, and RPEs for one hour's repetitive asymmetric lifting," *Ergonomics*, 31: 77-96 (1988).

Garg, A. and Saxena, U. "Effects of lifting frequency and technique on physical fatigue with special reference to psychophysical methodology and metabolic rate," *American Industrial Hygiene Association Journal*, 40: 894-903 (1979).

Marras, W., Lavender, S., Leurgans, S., Rajulu, S., Alread, G., Fathallah, F. and Ferguson, S. "The Role of Dynamic Three-Dimensional Trunk Motion in Occupationally Related Low Back Disorders: The Effects of Workplace Factors, Trunk Position and Trunk Motion Characteristics," *Spine*, 18: 617-628 (1993).

Snook, S.H. and Ciriello, V.M. "The Design of Manual Handling Tasks: Revised Tables of Maximum Acceptable Weights and Forces," *Ergonomics*, 34, 1197-1213 (1991).

Waters, T.L., Putz-Anderson, V., Garg, A. and Fine, L.J. "Revised NIOSH Equation for the Design and Evaluation of Manual Lifting Tasks," *Ergonomics*, 36, 749-776 (1993).

A BIOMECHANICAL INVESTIGATION OF THE ASYMMETRIC MULTIPLIER IN THE REVISED NIOSH LIFTING EQUATION

Maury A. Nussbaum, Don B. Chaffin, and George B. Page
Center for Ergonomics, The University of Michigan
Ann Arbor, Michigan, USA

There is growing evidence, from epidemiological and biomechanical sources, that lifting performed in asymmetric postures is a risk factor for the development of a musculoskeletal injury. In the recent update of the NIOSH Lifting Guide, a linear Asymmetric Multiplier was added to account for this type of risk. The present study addresses the form of this Multiplier through analysis of several asymmetric lifting tasks. Both spinal loading and a derived metric of muscle injury risk were calculated as a function of asymmetry angle. The results suggest that there is a non-linear increase in injury risk with respect to asymmetry. Only moderate increases in risk were predicted for asymmetry of 0°-30°, and sharply increasing risk as asymmetry reaches 90°, implying that ergonomic intervention should be concentrated on tasks with the highest asymmetries.

INTRODUCTION

Evidence from several sources has suggested that asymmetric lifting is an important risk factor in the development of musculoskeletal injury. The work of Punnett et al. (1991) is an example of the epidemiological indications of risk associated with an asymmetric work load during lifting. In an industrial case-referent study, they found that the occurrence of self-reported back disorders was related to the presence of severe trunk asymmetry. Specifically, an odds ratio of 5.9 was determined when there existed severe (>20°) trunk rotation or lateral bending during exertions. It is probable that physical exertions, performed while the torso is at or near any of the motion extremes, place excessive loads on some components of the spinal column and either cause or predispose the system to mechanical trauma.

In contrast to earlier work involving symmetric lifting, biomechanical analysis of torso asymmetry requires more complex 3-D experimental techniques and models. Some of these models have been used to document altered spinal loading patterns associated with asymmetry (Bean et al. 1988), while other have examined the force contributions of the various torso muscles (Marras and Mirka, 1990; McGill, 1991). To date, despite several suggestive results, there has been no conclusive mechanical demonstration of the cause and effect relationship between asymmetry and spinal injury, nor any quantification of dose-response characteristics.

Given the established risk of asymmetric lifting, and the need to reduce musculoskeletal injury in the workplace, NIOSH (Waters et al. 1993) revised its popular Lifting Guide through the inclusion of an Asymmetric Multiplier (AM). The AM is a linear function of the asymmetry angle, defined as load rotation in the horizontal plane (Figure 1). The justification for the form of the AM comes primarily from three studies demonstrating decreases (on the order of 30%) in maximum acceptable loads and lifting strengths with 90° of postural asymmetry. The AM was created as a linear interpolation between 0° (no decrement) and 90° (30% decrement), with further extrapolation to 135° of asymmetry.

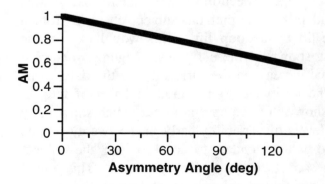

Figure 1. The Asymmetric Multiplier as a function of asymmetry angle (Waters et al., 1993), where 0° (load in front) and 90° (load at side) are relative to the midsagittal plane. The AM is not defined for angles > 135°.

The goal of the present work is to re-examine the logic of the AM, specifically its assumption of a linearly increasing risk with asymmetry angle, using biomechanical criteria. The studies cited as the foundation for the AM are used to obtain asymmetric lifting postures. These postures are then examined using contemporary biomechanical models of the low back. Several results were produced and used to suggest an appropriate form for future guidelines to control the incidence of work-related injury and to aid in the design of material handling jobs. This allows for the addition of a biomechanical basis for the Asymmetric Multiplier and estimation of whether the current form is physically protective.

METHODS

The lifting postures described in the work of Garg and Badger (1986) were duplicated in biomechanical simulations performed in the present work. All postures involved lifting boxes from floor level, with handles placed 10cm from the floor. The boxes were lifted 0°, 30°, 60°, and 90° from the mid-sagittal plane, with all lifts performed on the right side of the body. Three box sizes were studied, each of which spanned 51cm in the frontal plane and was 25cm high. The small, medium, and large boxes varied in their sagittal plane dimension: 25cm, 38cm, and 51cm, respectively. Whereas the subjects in the Garg and Badger (1986) experiment lifted freestyle, and were observed to employ a range of stoop and squat techniques, simulations were performed assuming a semi-squat lifting technique. For simplicity, subject anthropometry was fixed as a 50th%-ile male, and hand force vectors were assumed to be 50N in magnitude and vertically oriented.

A series of biomechanical models were employed to evaluate the musculoskeletal consequences of the various asymmetric lifts. Resultant moments about the mid-lumbar spine (L3/L4) and gross torso postures were determined using a kinematic model (3DSSPP™). A detailed geometric representation of the torso (Nussbaum and Chaffin, 1995) was used to estimate muscle moment arms and lines-of-action, muscle length-tension effects, and passive forces and moments developed by muscles and the spinal column. Reactive forces developed by the torso musculature, and associated spinal compression and shear, were calculated using a non-linear optimization algorithm (Hughes et al., 1994). The optimization-based model uses the sum of the cubed power of muscle stresses (force/area) as an objective along with equilibrium constraints about the L3/L4 motion segments and physiological constraints on the upper and lower bounds of individual muscle forces. The 10 muscles included in the geometric and optimization models are the following bilateral pairs: Erector Spinae (ES), Rectus Abdominus (RA), External Oblique (EO), Internal Oblique (IO), and Latissimus Dorsi (LD).

It has been suggested that injury to myofibril structures may occur as a result of muscles contracting while in an elongated state (Gareis et al. 1992). More specifically, it appears that some muscles, when elongated, are still capable of generating an active component of force while the passive component is also elevated. The implication is that active contraction and elongation can lead to intra-muscular force that is in excess of a maximum force that would otherwise be determined when the muscle is closer to resting length. Working from this hypothesis, a relationship between muscle contraction force, muscle length, and the relative injury potential was determined. Muscle injury risk, MIR, is a multiplicative function of normalized length and normalized total muscle force. Explicitly ignoring contraction dynamics and repetition, the relationship can be described as:

$$MIR_i = \frac{l_i}{l_{0i}}\left(\frac{F_i + P_i}{F\max_i}\right)$$

where i indexes the muscle, l and l_o are the muscle length and resting length, F and P are the active and passive muscle force components, and $Fmax$ is the maximum muscle force estimated at muscle resting length. All muscles were assumed to have resting lengths with the torso in an upright posture.

The musculoskeletal effects of postural asymmetry were evaluated using the various measures described above. For each posture and box size, the dependent measures of spinal compression, spinal shear components, and muscle injury risk factors were determined. The dependent measures were evaluated with respect to box size and asymmetry angle using ANOVA (critical $\alpha=0.05$).

RESULTS

The maximum hand load was defined as the largest force (total in both hands) that resulted in a spinal compression of less than the NIOSH Action Limit of 3400N. The maximum hand load (Figure 2) did not show a significant relationship with either box size (p=0.08) or asymmetry angle (p=0.64). As expected, when the small box was lifted at 0° (in front of the body), more load could be handled than using a larger box. This effect was canceled, however, with asymmetry. The larger box was more stressful when lifted at the side (i.e., lower box weight produced 3400N spinal compression), than when at 30° or 60° asymmetry.

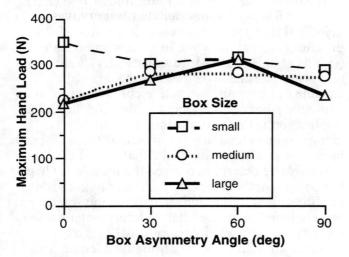

Figure 2. Maximum hand loads (total in both hands) that resulted in spinal compression <3400N versus asymmetry angle and box size.

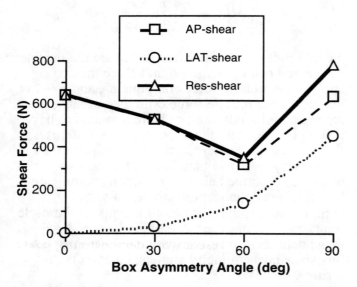

Figure 3. Spinal shear forces (absolute magnitude) as a function of asymmetry angle and averaged across box size. Resultant shear (Res-shear) is the norm of the AP and LAT components. Box weight = 100N.

The AP and Lateral shear force components, and their resultant, showed differing dependencies on the asymmetry angle (Figure 3). All three values were affected by asymmetry angle (p=0.001), but there were no significant effects of box size (p=0.9). AP-shear and the resultant were largest at the extremes of 0° and 90°, and appeared to be minimized at 60°. Lateral shear, in contrast, showed large increases for asymmetry angles >30°.

The derived measure of muscle injury risk (MIR), was significantly affected by specific muscles (p=0.001) and asymmetry angle (p=0.03), but not box size (p=0.9). The significant muscle effect is explained by the observation that only minimal antagonistic activity (<5% MVC) was predicted, leading to near zero force and associated MIR for the LRA, RRA, LIO, and REO muscles. Ignoring these inactive muscles, the average MIR across the three box sizes was not significantly different between 0°, 30°, and 60°, but all three were significantly larger at 90°. Averaging across all muscles yields the observation that the estimated risk of injury increases non-linearly, and nearly monotonically with increasing asymmetry (Figure 4). This result is examined in more detail for the left ES (Figure 5), which is a major agonist over the range of lifting conditions being studied. With increasing asymmetry, the left ES lengthens and its total force (F-total) increases. While the active component (F-active) decreases, the elongation causes a large increase in passive force (F-passive). The combination of high total force and muscle elongation, leads to the high derived risk value for this muscle at 90° of asymmetry.

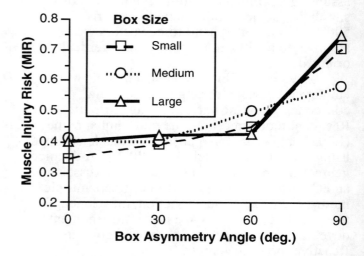

Figure 4. Derived muscle injury 'risk' (see Methods) as a function of asymmetry angle and box size. Box weight = 100N.

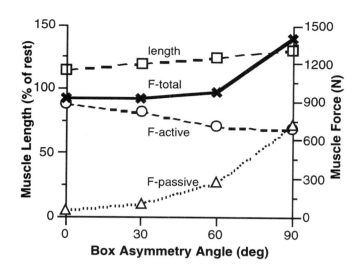

Figure 5. Predicted length and force characteristics of the left Erector Spinae muscle with changing asymmetry angle. Box Size = small; Box weight = 100N.

DISCUSSION

The use of an improved model of the spine and torso musculature appears to indicate that lifting boxes of various sizes poses additional risk when the initial load placement asymmetry is greater than 60°. In general, motion segment shear forces and muscle injury risk factors increased with asymmetry above 60°. The present study did not show as great an effect of asymmetry on the predicted compression force as previous studies, which demonstrated an increase in compression (or decrease in maximum hand load) with asymmetry (Bean et al. 1988; Chaffin and Page, 1993). Instead, there was a complex trade-off between different muscles recruited to maintain equilibrium which yielded a relatively constant maximum hand load versus asymmetry angle. It is suggested that the more realistic patterns of muscle recruitment predicted using the non-linear optimization scheme, and the incorporation of passive sources of spinal moments, explains the differences between the present study and the two sources cited above. The same trade-off between recruited muscles, along with postural changes in muscle lines-of-action and length, led to the result that there was not a simple relationship between spinal shear and asymmetry. The resultant shear vector changed both in magnitude and direction, while the A-P and Lateral components showed unique behaviors. It is difficult to evaluate the risks associated with these predicted patterns of shear forces given the absence of experimental data

indicating the tolerance of motion segments undergoing shear loading.

A measure of musculoskeletal risk was derived from the hypothesis that muscle injury risk is elevated in circumstances wherein a large muscle force (through either active or passive mechanisms) is generated by a muscle that is in a substantially lengthened state. A clear relationship between the derived 'risk' and asymmetry was predicted. This risk, which was not significantly affected by box size in the sagittal plane, was non-linear over the range 0°-90°. From 0° to 60° the risk rose only moderately, yet rose with increasing slope as the asymmetry approached 90°. This relationship is in contrast to the linear form of the Asymmetry Multiplier presented by NIOSH (Waters et al. 1993). Indeed, the results as a whole suggest that musculoskeletal risk increases only minimally with asymmetry of 0° to 30°.

The biomechanical predictions described here for simulated lifting are dependent on the accuracy of the models employed. We estimate that the largest source of error occurs during estimation of a unique set of lumbar muscle activities for an instance of static loading. The non-linear optimization method used has been demonstrated to be robust and consistent with empirical (EMG) data during symmetric postures and loads, as well as asymmetric loads in non-extreme postures (Hughes et al., 1994; Nussbaum et al., 1992). The veracity of the model in extreme postures had not been verified, thus the results for asymmetry angles near 90° must be interpreted with caution.

A metric was introduced which describes the relative risk of muscle injury, specifically during contraction of an elongated muscle. Justification for this measure stems from the hypothesis that muscle fibers may be excessively strained if the active component of force, in combination with high levels of passive force due to stretching, exceed a nominal level of maximum muscle contraction. While there is no empirical or epidemiological evidence available to support this hypothesis, as the present work demonstrates, it is possible to quantify the level of this new metric for a range of loading scenarios, thus allowing for future experimentation to test the stated hypothesis.

The linear relationship given for the Asymmetric Multiplier may place undue emphasis on reducing asymmetry to 0°. The present study, on the other hand, implies that the benefits of ergonomic intervention and redesign may fall short of the costs when the asymmetry is already moderate, and that efforts would best be concentrated on those jobs with the highest levels of asymmetry. Given the sharp increase in musculoskeletal risk predicted for loads with 60° to 90° asymmetry, the justification for a linear extrapolation from 90° to 135° asymmetry is suspect. It may be more reasonable to have a substantially larger penalty for loads whose initial placement requires postural asymmetry greater than 90°. A caveat must be noted, in that the NIOSH limit is based on strength limits. When the load is positioned to one side, the contralateral arm must reach across the body, imposing large shoulder moment loads that may lead to the reported decrements in acceptable loads and strength. Indeed, shoulder strength limits may serve to protect the back in these cases. The present work, and the conclusions given above, have focused solely on the low back. Future efforts must determine the interrelationship between low back risk and maximum acceptable loads and strength. In addition, dynamic lifting in asymmetric postures may impose further risk of muscle injury, especially in the lowering (eccentric) phase of task motion.

ACKNOWLEDGMENT

This work was supported by NIH Grant 1R01-AR39599 in cooperation with Drs. Steven A. Lavender and Gunnar B.J. Andersson.

REFERENCES

Bean JC, Chaffin DB, Schultz AB (1988) Biomechanical model calculation of muscle contraction forces: a double linear programming method. *J Biomech* 21:59-66

Chaffin DB, Page GB (1993) Postural effects on biomechanical and psychophysical weight-lifting limits. *Ergonomics* 37:663-676.

Gareis H, Solomonow M, Baratta R, Best R, D'Ambrosia R (1992) The isometric length-force models of nine different skeletal muscles. *J Biomech* 25:903-916

Garg A, Badger D (1986) Maximum acceptable weights and maximum voluntary isometric strengths for asymmetric lifting. *Ergonomics* 29:879-892

Hughes RE, Chaffin DB, Lavender SA, Andersson GBJ (1994) Evaluation of muscle force prediction models of the lumbar trunk using surface electromyography. *J Orth Res* 12:689-698

Marras WS and Mirka GA (1990) Muscle activities during asymmetric trunk angular accelerations. *J Orth Res* 8:824-832.

McGill SM (1991) Kinetic potential of the lumbar trunk musculature about three orthogonal orthopaedic axes in extreme postures. *Spine* 16:809-815

Nussbaum, MA, Lavender, SA, Chaffin, DB, and Andersson, GBJ (1992) Optimization-based torso muscle force prediction: asymmetric loading in two flexed postures. *Proc of the Second North American Conference on Biomechanics*, Chicago, pp. 525-526.

Nussbaum, MA and Chaffin DB (1995) Development and evaluation of a geometric model of the human torso. *Clin Biom* In Press.

Punnett L, Fine LJ, Keyserling WM, Herrin GD, Chaffin DB (1991) Back disorders and nonneutral trunk postures of automobile assembly workers. *Scan J Work Environ Health* 17:337-346

Waters TR, Putz-Anderson V, Garg A, Fine LJ (1993) Revised NIOSH equation for the design and evaluation of manual lifting tasks. *Ergonomics* 36:749-776.

THE GENERALIZABILITY OF PSYCHOPHYSICAL RATINGS IN PREDICTING THE PERCEPTION OF LIFTING DIFFICULTY

Marc L. Resnick
Florida International University
Miami, FL 33199

Psychophysical rating scales have been used as a parameter in lifting guidelines for workers in industrial settings, used to identify musculoskeletal disorders in the workforce, and used as a surveillance tool to identify workforce discomfort. These scales can be an inexpensive and easy-to-use tool for evaluating a large variety of exertions, especially those which are difficult to evaluate using current biomechanical and physiological models because of complex dynamic or asymmetric movements. In order for these scales to be used quantitatively, however, it is essential that they consistently represent the same level of perception across different subjects and tasks. Twenty subjects rated a variety of elbow flexion exertions on the Borg CR-10 scale under two task formats. The intra-subject, inter-subject, and inter-task variations were measured. Results indicate that the Borg CR-10 scale provides a consistent measure of psychophysical perceptions across a variety of task parameters.

INTRODUCTION

Psychophysical measures of various kinds have been used to investigate acceptable exertion levels (Snook and Irvine 1966, Borg 1970, Ayoub et al 1978) and physical discomfort (Boussenna et al. 1982, Bhatnager 1985). Studies have investigated exertions as diverse as lifting, cycling, pushing, running on a treadmill, and various types of industrial work. In the past fifteen years, psychophysics has also been included as a parameter in lifting guidelines for workers in industrial settings. Buckle et al. (1984) have used psychophysical ratings to identify musculoskeletal disorders in the workforce, and Saldana et al. (1994) have developed a surveillance tool based on psychophysics to identify workforce discomfort.

One type of psychophysical measure, rating scales, has recently been studied by a large number of investigators (Karwowski 1991, Resnick 1993, Thompson 1993, Oberg et al. 1994, and others). Resnick (1994) concluded that rating scales can be an inexpensive and easy-to-use tool for evaluating a large variety of exertions, especially those which are difficult to evaluate using current biomechanical and physiological models. However, comparing results from studies which use different scales, tasks or subjects is tenuous due to a lack of evidence regarding the generalizability of ratings.

One of the most commonly used rating scales, the Category Ratio (CR-10) scale, was developed by Borg (1982). Initial research indicated that each unit on the scale represents one tenth of the maximum muscular strength of the exerted muscle in an aerobic task (Eastman Kodak, 1983). No relationship has been established between ratings and anaerobic exertions. Ratings have also been used to measure the physical discomfort of a variety of body parts. Whether a rating of an aerobic exertion can be compared to the rating of an anaerobic exertion or physical discomfort is not clear. Furthermore, whether different subjects will rate similar levels of discomfort equally on the scale has not been established.

Objective

A study which investigates the generalizability of ratings across the many uses to which they have been applied would provide needed credibility for the widespread applicability of ratings. Specifically, can a subject's rating of a particular task with a given set of muscles be compared to another subject in a different task using the same or different muscles? This study has the following objectives:

1. To investigate the variability of ratings within a single subject/task combination to measure the repeatability of psychophysical ratings.
2. To investigate the variability of ratings between subjects in the same task to measure the extent to which results can be applied to worker populations.
3. To investigate the variability of ratings between tasks to determine whether psychophysical ratings reflect the same underlying physiological mechanisms, such as maximum muscular strength.

METHOD

Subjects

Twenty subjects, thirteen male and seven female, participated in the experiment. Subjects reported no musculoskeletal disorders and were healthy at the time of testing. Subjects were paid for their participation. Informed consent was obtained prior to each session.

Apparatus

Subjects' maximum static lifting strengths and voluntary exertions were measured using a tri-axial force transducer attached to a stationary base by a steel chain. The force transducer is a three channel device attached through an amplifier to an analog to digital converter in a 80386 compatible PC running Workbench for PC software. The transducer was calibrated to 0, 100, and 225 N loads.

The Borg CR-10 rating scale was used as the psychophysical scale to anchor subjects' ratings. Standard Olympic steel weights were used for lifting trials. The weights were placed in a closed pine container so subjects were not aware of the load magnitude. The handle on the box was a 1" diameter steel rod extending horizontally out 6" from each side of the container.

Procedure

The body weight and height of each subject was measured and recorded. All required tasks were explained in detail to the subject at the beginning of each session. Subjects' maximum elbow flexion strengths were measured by having them perform a maximum lift on a handle fixed at elbow height connected to the tri-axial force transducer. The maximum exertion was repeated three times, with two minutes of rest between each repetition. The average of the three trials was considered the subject's maximum voluntary exertion (MVC).

Phase 1. Subjects were instructed to statically lift a randomly presented load at elbow height. Loads were set at 20, 40, 60 and 80% of the subjects' measured maximum lifting strength. Subjects were constrained to sagitally symmetric postures. They were instructed to lift the load for three seconds. Immediately following the lift, subjects rated the exertion on the Borg CR-10 scale and the load was changed. One minute of rest was provided to the subject upon completion of each lift. Subjects were allowed to request additional rest if they felt fatigued, though this never occurred. Each subject performed a total of twenty lifts, corresponding to five repetitions of each load. Each exertion was presented in random order.

Phase 2. Subjects exerted a static lifting force on the triaxial force transducer which corresponded to their perception of each of the five verbal anchors from the Borg CR-10 scale (very weak, weak, moderate, strong, very strong). Postures were identical to those used in Phase 1. The handle was located at elbow height and sagitally symmetric postures were required. The magnitude of the force for each exertion level was recorded by the data acquisition software. The software began recording as soon as the exertion exceeded 10 N. Subjects were instructed to continue the force for three seconds. One minute of rest was provided to the subject upon

completion of each exertion. Again, additional rest would have been provided if requested, though it never was. Each subject performed a total of twenty five lifts, corresponding to five repetitions of each exertion level. Each exertion was presented in random order.

Experimental Design

Subjects performed the phases in random order. The order of the trials were completely randomized within each phase. The entire experimental procedure lasted between two and three hours.

RESULTS

Repeatability of Borg CR-10 ratings

The standard deviation of each set of five repetitions was calculated for each subject/load combination from Phase 1. The standard deviations ranged from 0.0 to 1.5, with most falling below 1.0. Because the average ratings ranged from 0.5 to 9, it is more useful to represent the standard deviation as a percentage of the mean rating for that condition. Two-thirds of the conditions had a standard deviation of 10% of the mean or less (Figure 1). The only cases where this ratio was large was for the very weak exertions where the mean rating was '1'.

Generalizability between subjects

Borg CR-10 ratings as a function of the percent of the subject's strength exerted are shown in Figure 2 for each subject in Phase 1. The normalized values are used because of the predictive advantage obtained from this transformation (see Resnick, 1995 for a complete explanation of the normalization process and its benefits). The r-squared of 0.86 shows that 86% of the variance in the rating is explained simply by the normalized load. Figure 3 shows this same relationship for Phase 2. Again, a significant portion (81%) of the variance is explained by the normalized load. Thus the amount of a subject's strength used in an exertion explains most of the variance in Borg CR-10 ratings. Including other variables such as time of day, subjects' age, gender, and motivation, etc. are not needed to determine fairly accurate estimates of the perception of exertion difficulty.

Generalizability between tasks

Comparisons were also made between the two phases. Figure 4 shows the relationship between the Borg CR-10 ratings and the loads lifted as a percent of the subjects' strengths in both phases for both male and female subjects. Because subjects' strengths did not include any factor for the mass of the arms, the intercept of this line was set to zero. The data from both phases fall on the same regression line (rating = 0.10 * load, r^2= 0.81). This indicates that for a heterogeneous population (including both genders), all subjects interpreted Borg CR-10 ratings in the same way. Furthermore, this

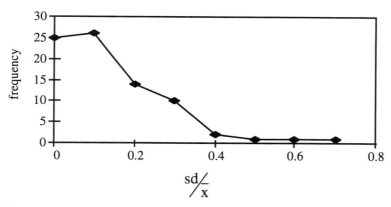

Figure 1. Distribution of Standard Deviations of the Borg CR-10 ratings
for each subject/load combination

$$rating = 0.10 * load \quad r^2 = 0.86$$

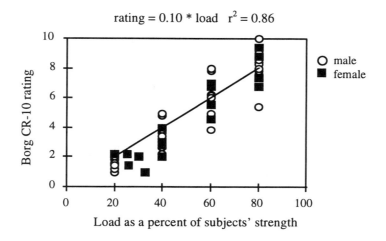

Figure 2. Borg CR-10 rating versus load as a percent of subject's strengths for all subjects
(each symbol represents the data from one subject)

$$rating = 0.10 * load \quad r^2 = 0.81$$

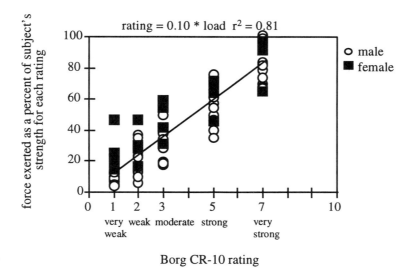

Figure 3. Percent of a subject's peak strength exerted for each Borg CR-10 rating in Phase 2

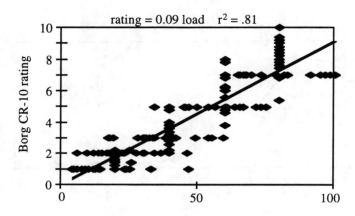

Figure 4. Borg CR-10 ratings for each load normalized as a percent of the
subject's strength in both phases of the study

strongly suggests that the Borg CR-10 scale was interpreted in the same way in both phases, independent of the task specifics. In other words, the relationship between ratings and lifting force when subjects rate given loads (phase 1) is the same as the relationship when subjects exert forces corresponding to standard verbal anchors (phase 2). This result supports the generalizability of psychophysical ratings. If the ratings were not reflexive, the meanings of psychophysical ratings on the Borg CR-10 scale would not be consistent across experimental designs, considerably limiting the utility of comparing ratings determined in one study to those from another study.

DISCUSSION

In order for psychophysical ratings to be an effective tool to measure exertion difficulty, it is necessary that they represent a repeatable and generalizable measure of the perceptions produced by the exertion. If ratings need to be scaled for a variety of individual characteristics such as personality or motivation, their application would be severely limited. Ratings on a scale should be determined by task parameters with minimal variance due to personal factors.

The first objective of this study was to determine whether individuals will rate the same exertion consistently. In most cases, the standard deviation of the rating was below 1. This indicates that subjects had consistent perceptions of each exertion.

The second objective was to determine whether the perceptions of a worker population could be quantified reliably. The results of this study show that only one individual characteristic, peak strength, needs to be measured in order to explain most of the variability in ratings on the Borg CR-10 scale. Documenting a worker's strength would add the administrative requirement of strength testing each worker before they could be assigned to a particular job. This practice has been recommended for years (Chaffin, Herrin, Keyserling and Foulke 1977), and has several benefits. The guidelines would be more effective in matching workers with

jobs, thus the overall injury rate will be reduced by protecting weaker workers from injury. The testing would also provide a baseline on record for any future workers' compensation investigations or rehabilitation efforts. Finally, if this data could be anonymously compiled, anthropometric strength data would be continuously updated for the working population.

The final objective of the study was to determine whether the factors which contribute to psychophysical perceptions in one task are also the factors which contribute in other tasks. In both phases of the study, the subjects' strengths explained most of the variance in the ratings, indicating that strength limits are a global contributor to perceptions of exertion, at least for this type of task.

When workers' peak strengths are known, Borg CR-10 ratings can be a repeatable and generalizable measure of exertion difficulty. They can be used to determine job acceptability in an inexpensive and quick fashion by staff who are not sophisticated experts in ergonomics. Development of exertion guidelines using rating scales can significantly broaden the usage of ergonomic criteria in the design of jobs and the selection of workers to perform the jobs.

An additional result which bears note is the fact that the slopes of the regression lines in both phases are 0.1, the same result reported by Eastman Kodak (1983). Surprisingly, it seems that subjects are consistent in using increments on the Borg CR-10 to reflect ten percent of their capability on whatever measure is being tested, whether it be percent of static strength as in this study, or percent of maximum aerobic muscle strength as in the previous study. This is significant because subjects in both studies were not aware of their levels of exertion, thus the internal calibration of the subjects is quite high.

It must be noted that the subject population used in this study was limited. Though subjects represented both genders and a variety of ethnicities, the range of ages was limited to 18 - 31. Whether older workers would still show this consistent behavior remains to be seen. However, adding an age factor to the guidelines would not make them substantially more complicated. A further limitation of this study was the use of static, single joint exertions. Whether whole body exertions also follow this trend is unknown. Other factors besides strength may play a significant role in determining the perceptions of exertion in other types of task. Further study is necessary to determine whether psychophysical rating scales can be used for other types of exertions.

REFERENCES

Aghazadeh F., Lee K.S., and Waikar A. 1988, Utilization of the direct estimation method to determine the capacity to lift loads, In *Proceedings of the Human Factors Society 32nd Annual Meeting.* 687-689.

Ayoub M. M., Dryden R. D., McDaniel J.W., Knipfer R.E., and Aghazadeh F. 1978, Modeling of lifting capacity as a function of operator and task variables, *Safety in Manual Materials Handling* NIOSH: Cincinnati, OH.

Bhatnager V., Drury C.G., and Schiro S.G. 1985, Posture, postural discomfort, and performance, *Human Factors*, **27**, 189-199.

Borg G. 1970, Perceived Exertion as an Indicator of Somatic Stress, *Scandinavian Journal of Rehabilitative Medicine*, 2(3), 92-98.

Borg G. 1982, Psychophysical Bases of Perceived Exertion, *Medicine and Science in Sports and Exercise*, **14**(5), 377-381.

Bousenna M., Corlett E.N., and Pheasant S.T. 1982, The relation between discomfort and postural loading at the joints, *Ergonomics*, **25**, 315-322.

Buckle P.W., Stubbs D.A., and Baty D. 1984, Musculoskeletal disorders (and discomfort) and associated work factors, in N. Corlett, J. Wilson and I Manenica (eds) *The Ergonomics of Working Postures* (Taylor and Francis, London), 19-30.

Chaffin, D.B., Herrin G.D., Keyserling W.M. and Foulke J. Pre-employment strength testing in selecting workers for materials handling jobs, US Dept of Health and Human Services, National Institute of Occupational Safety and Health, CDC Report #99-74-62, 1977, Cincinnati, OH.

Chen F., Aghazadeh F., and Lee K.S. 1992, Utilization of the Direct Estimation Method to Predict the Maximum Acceptable Weight of Symmetrical and Asymmetrical Lift, In: *Proceedings of the Human Factors Society 36th Annual Meeting*, 679-683.

Karwowski W. 1991, Psychophysical Acceptability and Perception of Load Heaviness by Females, *Ergonomics*, **34**(4), 487-496.

National Institute for Occupational Safety and Health, 1981, "Work Practices Guide for Manual Lifting" Technical Report 81-122, NIOSH, Cincinnati.

Oberg T. 1994, Subjective and objective evaluation of shoulder muscle fatigue, *Ergonomics,* **37**(8), 1323-1333.

Resnick M. L. 1993, Biomechanics, Kinematics, Psychophysics and Motor Control in the Application of Material Handling Devices (MHDs), Ph.D. Dissertation, University of Michigan, Ann Arbor, MI.

Resnick M.L. 1995, Normalizing psychophysical exertion guidelines using individual strengths, In *Proceedings of the Industrial Engineering Research Conference.*

Resnick M. L., and Chaffin D. B. 1994, The effects of materials handling device (MHD) implementation on stresses at the low back, Submitted to *International Journal of Industrial Ergonomics.*

Saldana N., Herrin G.D., Armstrong T.J., and Franzblau A. 1994, A computerized method for assessment of musculoskeletal discomfort in the workforce: a tool for surveillance, *Ergonomics, 37(6), 1097-1112.*

Snook S.H., and Irvine C. H. 1967, Maximum Acceptable Weight of Lift, *American Industrial Hygiene Association Journal*, **28**, p322.

Thompson D. 1993, The Perception of Physical Stress as a Measure of Biomechanical Tolerance, Ph.D. Dissertation, University of Michigan, Ann Arbor, MI.

PREDICTION OF TWO-PERSON TEAM LIFTING CAPACITY

Valerie J. Rice, Marilyn A. Sharp, Bradley C. Nindl,
and Randall K. Bills
Occupational Physiology Division
U.S. Army Research Institute of Environmental Medicine
Natick, MA 01760-5007

Predictive models for team lifting capacity are important for task and equipment design, as well as worker selection and placement. The purpose of this study was to develop a prediction equation for single gender and mixed gender two-person team lifting from the floor to knuckle height. Men (n=23) and women (n=17) were combined into teams of two men (n=26), two women (n=24), and one man with one woman (n=25). Independent variables included incremental dynamic lift, 38 cm upright pull, dead lift, fat free mass, and body mass. A least squares linear regression was used. In addition, an equation was developed from deadlift strengths only. The lightest individual deadlift and the sum of the individual deadlifts were the best predictors of team lifting capacity ($R^2 = 0.90$, SEE = 16). The results indicate that two-person team lifts to knuckle height are determined by the weaker team member.

INTRODUCTION

Two-person teams are recommended for lifting heavy or bulky loads. Aircraft maintenance and repair, railroad, mining, and patient care tasks frequently require two-person lifts. Models that can predict the lifting capacity of two-person teams are needed for effective employee selection and placement, and task design.

Military Standard 1472D suggests doubling the recommended load for a single person lift when lifting in two-person teams (Naval Publications and Forms Center, 1989). However, no references or data are provided to substantiate this recommendation.

Karwowski (1988) examined lifting capacity of men and women in single gender two-person teams, and developed a prediction equation for two-person lifting. Although limited by a small sample size, it is one of the first attempts to develop a method of predicting the maximum permissible lifting capacity of two-person teams. The most significant factor was the maximum lifting capacity of the heavier (body weight) team member.

Fox (1982) examined male two-person repetitive lifting (4 lifts/min) from the floor to a height of 80 cm. A model was developed using 20 of 28 team combinations, with eight combinations randomly selected to validate the model. The only significant variable was the lightest individual maximum acceptable load and the model had little predictive value.

The purposes of this study were to develop a prediction equation for single gender and mixed gender two-person team lifting, and to compare our equation with previously published prediction equations.

METHOD

Participants were 23 male and 17 female soldiers. Table 1 contains the physical characteristics of the subjects. All subjects were medically screened and signed an informed consent form following a detailed briefing. Participants were randomly assigned to teams of two men (n=26), two women (n=24), and one man with one woman (n=25).

Independent variables included fat free mass, body mass, and three strength measures. Body composition was assessed using hydrostatic weighing (Goldman and Buskirk, 1961). One repetition maximum strength tests (1 RM) were determined for a dead lift, a machine lift using a vertical weight stack

Table 1. Physical characteristics of subjects (mean ± SD).

	Men	Women
n	23	17
age (yr)	20.3 ± 1.7	26.7 ± 6.4[*]
height (cm)	177.9 ± 6.4	163.3 ± 4.2[**]
weight (kg)	76.3 ± 12.2	61.1 ± 7.8[**]
body fat (%)	16.8 ± 6.2	26.1 ± 5.5[**]
fat-free mass (kg)	62.9 ± 8.0	44.7 ± 4.7[**]
dead lift (kg)	137.0 ± 22.1	84.7 ± 14.2[**]
IDL (kg)[1]	79.4 ± 13.4	39.8 ± 6.6[**]
38cm uprt pull (kg)	144.6 ± 32.0	87.6 ± 18.2[**]

Significantly different from men [*] ($p<.05$), [**] ($p<.01$).
[1] incremental dynamic lift using a vertical weight stack machine

(McDaniel et al., 1983), and a 38cm upright isometric pull. Both the individual strength values and the sum of the individual values were used to develop the prediction equation.

Two-person lifting capacity was measured using a 68.6 kg lifting device (Figure 1). Following stretching, a warm up was provided by having each team lift the unloaded bar to knuckle height three times. Weight was then added to the device following each lift, in increments of 15-40 kg. The increments were reduced as the team approached their maximum load. When a team was unable to complete a lift, the load was reduced in increments of 3-10 kg until a lift was completed. Each team arrived at their maximum load in five or six lifts. A team reached its' maximum load when the members

judged the weight as too heavy, could not physically complete the lift, or could not maintain a safe lifting technique. Two technicians served as spotters and helped lower the load.

A two way analysis of variance was used to compare all male, all female and mixed gender two-person team lifts. Tukey Honestly Significant Difference Tests were performed to determine significant differences between means. Forward stepwise multiple regression analyses were used to generate two prediction equations for two-person teams.

Regression equations by Karwowski (1988) and Fox (1982) were compared with our equation using data collected in this study. The most significant variable in Karwowski's equation was the maximum acceptable lifting capacity of the heavier team member (MLH, Table 3). When this equation was used with our data, the maximal deadlift of the heavier team member was used as the MLH. The most significant variable in Fox's equation was the lightest individual maximum acceptable weight of lift (IMAWL, Table 3). When Fox's equation was used with our data, the lightest maximum deadlift was used in place of the lightest IMAWL. Both Karwowski's and Fox's equations were examined with the gender groupings on which their predictions were based, and with all gender combinations in the present study. A paired t-test was used to determine statistical differences between the actual load determined in this study and

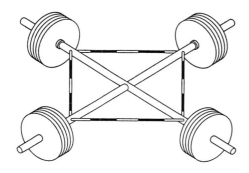

Figure 1. Two-person team lifting device.

the predicted load obtained using Karwowski's (1988) and Fox's (1982) equations.

RESULTS

Table 2 contains the maximum two-person lifting capacity by gender. All team configuration lifts were significantly different from each other (p < 0.01).

Table 2. Maximum team-lifting capacity in two-person teams.

Team Configuration	Lift Capacity (kg) mean ± SD and (n)
Men	252.9 ± 32.8 (26)
Women	155.8 ± 15.7 (24)
Mixed-Gender	183.5 ± 24.1 (25)

All team configuration lifts were significantly different from each other (p < 0.01).

Table 3 presents the two prediction equations developed from this study and those of Karwowski (1988) and Fox (1982). Equation one (EQ 1) resulted from a forward stepwise multiple regression analysis. Although the R^2 is high, the drawbacks of this equation are that both maximum deadlift and 38cm upright pull must be determined for team members, and the equation does not account for gender.

A second equation was developed forcing the following independent variables: gender, the smallest 1RM dead lift, and the sum of both team members 1RM deadlifts (Table 3, EQ 2). This equation yielded an acceptable R^2 and can be determined solely from team members deadlift strengths.

Six men and six women were used to develop Karwowski's (1988) equation to predict the maximum permissible lifting capacity (MPLC) of single gender teams of two men or two women. Table 4 shows the results when Karwowski's equation was applied to our data. The R^2 is high when predicting for single gender teams, however the predicted value is only 46% of the actual value (actual = 198.7 kg). When using Karwowski's equation to predict all team configurations, the R^2 was quite low and the predictd lifting capacity again is 46% of the actual value (actual = 206.3 kg).

Table 4 also presents the results of using Fox's (1982) equation with our data. When predicting for male teams only, the R^2 was low, and the equation overpredicted team lifting strength by 15%. When used with all team configurations, the R^2 was high and overpredicted by 17%.

DISCUSSION

EQ 2 is the first team lifting equation developed based on the inclusion of both single and mixed gender teams. The equation is easy to use and only deadlift strength is needed to predict two-person team lifting capacity to knuckle height.

Karwowski (1988) found the best predictor of MPLC was the maximum lifting capacity of the heavier team member (body weight). This would seem to indicate that the stronger team member assumes some of the weight for the lower strength team member. Our results indicate that the teams' strength is determined by the weaker team member. When used with both single gender teams only and with all team configurations, Karwowski's model underpredicted team lifting capacity by 100 kg. The disparity between the results of Karwowski's equation with his data and with our data may be due to the difference in task requirements, as well as the team configuration. Our task required a one time maximum effort, while Karwowski's (1988) required participants to select a maximum acceptable load and lift at the infrequent rate of one lift per 30 minutes. It has been shown that subjects select lighter loads for more frequent lifts (Snook and Ciriello, 1991). In addition, we had subjects lift to the knuckle height of the shortest team member, while Karwowski had participants lift from the floor to a height of 89 cm. This is 5 cm below wrist height of the 50% male soldier and 10 cm above wrist height of the 50% female soldier (Gordon, et al., 1989). It is likely that more upper body involvement may have been required of women participating in Karwowski's study, resulting in a lower load being lifted.

Fox (1982) examined male two-person repetitive lifting from the floor to a height of 80 cm. Participants lifted at a rate of four lifts per minute. The only variable that was significant in the prediction model was the smallest individual maximum acceptable load and the model had little predictive value. Although Fox's model was developed for male, repetitive, two-person lifting during a maximum acceptable lift protocol, it was more accurate with our

Table 3. Regression equations for the prediction of maximum two-person lifting capacity (kg).

	Regression Equation	R^2	SEE
(1) Present Study[1]	MTL = 13.9 + 0.20(T38cm) + 1.4(SM)	0.94	16.2
(2) Present Study[2]	MTL = 0.73 + 7.0(A) + 1.2(B) + 1.54(SM) + .2(TDL)	0.90	16.0
(3) Karwowski, 1988[3]	MPLC = 21.04 + 16.79 (sex) + 21.63 ln (MLH)	0.90	2.4
(4) Fox, 1982[4]	BMAWL = (1.85 x lighter IMAWL) + 51.43	0.35	*

[1] MTL = maximum team lift (kg), T38cm = the sum of 38cm upright pull, SM = smallest individual dead lift.
[2] TDL = total of individual dead lifts; male team (A = 0, B = 0), female team (A = 0, B = 1), mixed team (A = 1, B = 0).
[3] MPLC = maximum permissible lift capacity (kg), sex = 0 for women, 1 for men; MLH = maximum lifting capacity (kg) of the heavier team member.
[4] BMAWL = bimanual maximum acceptable weight of lift (kg), IMAWL = individual maximum acceptable weight of lift.
* SEE not reported.

Table 4. Use of existing equations to predict two-person team lifts with data from present study.

Equation	n	predicted load	difference pred - act	T-test p value	R^2	SEE
Present Study	75					
Karwowski, 1988 single gender only[1]	50	91.0 ± 14.5 kg	-115.3 ± 42.5 kg	< 0.001	0.84	22.4
Karwowski, 1988 all team combinations[2]	75	95.1 ± 13.4 kg	-103.6 ± 40.8 kg	< 0.001	0.43	36.6
Fox, 1982 male teams only[3]	26	289.6 ± 22.9 kg	36.8 ± 21.8 kg	< 0.001	0.56	22.2
Fox, 1982 all team combinations[2]	75	233.7 ± 46.2 kg	35.0 ± 17.3 kg	< 0.001	0.87	17.4

[1] single gender teams: 198.7 ± 49.3 kg
[2] all three team combinations: 206.3 ± 55.4 kg.
[3] men only: 252.9 ± 32.8

data when used with all team configurations. The accuracy may be partially due to the lower lift requirement of 80 cm, which was closer to the knuckle heights of soldiers. In addition, in mixed gender teams, the weaker team member may limit the teams lifting capacity to a greater degree than during single gender lifts. The Fox model was more accurate than the Karwowski model when used with our data. As seen in Figure 2, the Fox model overpredicted, while the Karwowski model underpredicted team lifting strength.

In summary, this research presents the first team lifting prediction equation for both single and mixed

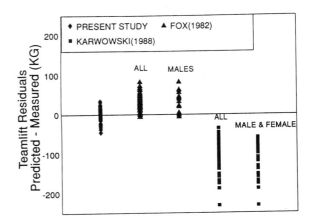

Figure 2. Comparison of actual MAL with predicted MAL when applying Karwowski (1988) and Fox (1982) to present data.

gender teams. The similar findings between the two equations based on lifts to knuckle height or below seem to indicate that the lifting capacity of the weaker individual determines the teams' maximal lifting capacity. Table 5 contains percentile norms for occasional single lifts of maximum loads (compact boxes, with handles). The majority of the population when lifting in two-person teams should be able to lift 193.5, 131.0, and 154.6 kg for two men, two women, and mixed gender teams, respectively.

Table 5. Percentile values for 1RM two-person team lifts (kg).

Percentile of population	5%	50%	95%
Men	**193.5**	245.2	308.6
Women	**131.1**	156.7	188.3
Mixed-gender	**154.6**	178.6	221.0

ACKNOWLEDGEMENTS

We acknowledge Bob Mello, M.S. and Tania Williamson, M.S. for their contributions during data collection.

REFERENCES

Fox, R.R., 1982. A psychophysical study of bimanual lifting. Masters in Industrial Engineering Thesis. Texas Tech. Lubuck, TX, 1-84.

Goldman, R.F., and Buskirk, E.R., 1961. Body volume measurment by underwater weighing: Description of a method. In: Techniques for measuring body composition. J. Brozek and A. Henschel (eds.). National Academy of Sciences-National Research Council, Washington, D.C., pg. 78-89.

Gordon, C.C., Churchill, T., Clauser, C.E., Bradtmiller, B., McConville, J.T., Tebbetts, I., and Walker, R.A., 1989. 1988 anthropometric survey of U.S. Army personnel: Summary statistics interim report. Natick, MA: U.S. Army Natick Research, Development, and Engineering Center, Technical Report 89/027.

Karwowski, W., 1988. Maximum load lifting capacity of males and females in teamwork. Proceedings of the Human Factors Society, 680-682.

McDaniel, J. W., R. J. Skandis, and S. W. Madole, 1983. Weight lift capabilities of Air Force basic trainees. US Air Force Aerospace Medical Research Laboratory Technical Report 83-001, Wright-Patterson Air Force Base, OH.

Naval Publications and Forms Center, 1989. Military standard human engineering design criteria for military systems, equipment and facilities (MIL-STD-1472D). Philadelphia, PA.

Snook, S. H., and V. M. Ciriello, 1991. The design of manual handling tasks: Revised tables of maximum acceptable weights and forces. Ergonomics, 34(9):1197-1213.

MAXIMUM ACCEPTABLE LOAD FOR LIFTING AND CARRYING IN TWO-PERSON TEAMS

Marilyn A. Sharp, Valerie J. Rice, Bradley C. Nindl, and
Robert P. Mello
Occupational Physiology Division
U.S. Army Research Institute of Environmental Medicine
Natick, MA 01760-5007

The purpose of this study was to determine and verify the maximum acceptable load for lifting and carrying (MAL-L&C) in single- and mixed-gender two-person teams. Participants lifted and carried a box 7.2 meters and placed it on a 132 cm high shelf, at rates of 1 x/min and 4 x/min. All male teams lifted and carried significantly (p<.05) more weight than all female teams or mixed-gender teams, and mixed-gender teams carried more than all female teams (p>.05). Our findings demonstrate that 1) individuals working alone or in teams can accurately estimate their ability to lift and carry loads for an hour, and that 2) when working in pairs, team MAL-L&C are approximately equal to the sum of their individual MAL-L&C. Percentile norms for MAL-L&Cs are provided for male, female, and mixed gender teams.

INTRODUCTION

It is often necessary for individuals to work in teams to accomplish heavy manual materials handling tasks. In addition, working in teams of two or more may enable lower strength individuals to perform tasks they could not execute alone.

Fox (1982) examined the maximum acceptable load (MAL) for two men lifting from floor to 91 cm at the rate of 4 x/min. The subjects determined their MAL jointly. Fox (1982) hypothesized that the MAL would be equal to two times the lowest individual MAL for the subjects making up the team. The team MAL was 41% greater than expected, and was significantly greater than two times the heavier individual MAL of the team members. In the only study that examined the MAL for lifting and carrying in two-person teams, Johnson and Lewis (1989) found men lifted and carried

33% more in pairs than they did individually at a rate of 2 x/min.

Fox (1982) and Johnson and Lewis (1989) did not include women or mixed-gender teams. The purpose of this study was to determine and verify the MAL for a lift and carry task performed by teams of two men, two women, and one man with one woman and to determine the relationship between the sum of individual MALs and team MAL.

METHODS

Subjects

Twelve male and nine female soldiers were recruited, medically screened and gave their written informed consent.

Maximal Lift and Carry Strength

A test of one-repetition maximum box lift and carry capacity (1RM-L&C) was made using a metal box with handles (31 cm w x 46.5 cm l x 23 cm h). The box was lifted from the floor to knuckle height, carried 7.2 m and lifted onto a 132 cm high platform. Subjects carried the box in front of them, and were instructed to use safe lifting and carrying techniques. Following a warm-up, weight was added with each attempt until the subject was unable to complete the task. After a failed attempt, weight was removed to yield an intermediate load to assess 1RM-L&C to the nearest 1.0 kg. A minimum three-min rest was provided between near maximum attempts.

1RM-L&C was also measured for two-person teams. A 62 cm w x 46.5 cm l x 23 cm h box was used. Following warm-up, the load was gradually increased until subjects determined their maximum load, or until the experimenter judged the safety of the subjects to be at risk. Subjects were allowed to select the carry method that was most comfortable for them. All subjects chose to reach across their body with one hand, hold two handles on the same side of the box and walk facing forward with one subject on either side of the box.

Maximal Acceptable Load for Lifting and Carrying (MAL-L&C)

Subjects performed a lift and carry task individually and in pairs. Pairs consisted of two men, two women, and one man with one woman. Each subject participated in approximately four same-gender teams and two mixed-gender teams. Pairings were made on a random basis.

The task for the MAL-L&C was similar to the 1RM-L&C except that it was performed repeatedly at rates of 1 and 4 x/min. After the box was placed on the platform, technicians moved the box onto a conveyor with rollers, and the box rolled to the starting position. Subjects practiced individual and team lifting and carrying for six sessions during the two weeks prior

to data collection. When carrying in pairs, subjects switched sides at their own discretion.

The MAL-L&C was determined during two-20 min sessions. Subjects were given a box that was expected to be either too light or too heavy. They were asked to adjust the load until it was the maximum load they could handle for one hour and still be able to perform all other duties of their job for the remainder of an eight hour day. They were instructed to discuss all load changes with their partner when working in pairs and encouraged to make many adjustments. After a 20 min rest period, another box was presented (heavy if the first box was light and light if the first box was heavy). The order of presentation (light and heavy) was randomized. Unmarked, non-standardized 1-5 kg bags containing steel shot were available to adjust the load. The mean of the two sessions was recorded as the MAL-L&C for that rate. If the two loads selected were not within 15% of one another, the selection process was repeated on another day. To verify the MAL-L&C (VMAL-L&C), subjects lifted and carried the load selected for one hour on a different day. Subjects were allowed to adjust the load at any time. The load at the end of the hour was the VMAL-L&C.

Statistical Analysis

T-test (age, height, weight and 1RM-L&C) and one-way analysis of variance (team 1RM-L&C) were used to examine gender differences in descriptive measures. To determine if there were differences between the load selected (MAL-L&C) and the verification load carried for one hour (VMAL-L&C), repeated measures analyses of variance (RMANOVA) were conducted for individual and team data separately. The dependent variables included two levels of carry rate (1 x/min and 4 x/min), two levels of task (MAL-L&C and VMAL-L&C), gender as a grouping factor (man and woman for individual L&C; two men, two women, and one man with one women for team L&C) and the load carried (kg) as the independent measure.

Subsequent analyses were conducted without the level of task to examine gender and carry rate effects on the VMAL-L&C. The sum of individual VMAL-L&C was compared to the team VMAL-L&C using a two-way RMANOVA grouped for gender.

RESULTS

Men were taller and heavier (179.6 ± 5.8 cm, 81.0 ± 9.7 kg) than women (162.5 ± 6.3 cm, 59.3 ± 5.0 kg, $p<0.01$). There was no difference in age between men (20.7 ± 2.8 yr) and women (20.7 ± 3.3 yr). The individual 1RM-L&C was greater for men (67.9 ± 11.5 kg) than women (35.6 ± 6.4 kg, $p<0.01$). The team 1RM-L&C was greatest for two men (125.2 ± 19.1 kg). Mixed-gender teams lifted 69% as much as two men (86.6 ± 13.5 kg, $p<0.01$). Two women lifted 51% as much as two men (64.1 ± 5.9 kg, $p<0.01$) and 74% as much as mixed-gender teams ($p<0.05$).

Table 1. Individual and Team Loads Selected (MAL-L&C) and Verified (VMAL-L&C) (kg, $\bar{X}$ ± SD).

	Rate	MAL-L&C	VMAL-L&C
	INDIVIDUAL:		
Men	1	35.7 ± 4.7	35.7 ± 4.6
	4	25.5 ± 2.9	26.0 ± 2.8
Women	1	23.7 ± 3.8	23.7 ± 3.8
	4	18.4 ± 3.0	17.9 ± 2.8
	TEAM:		
Men	1	72.5 ± 15.0	71.9 ± 14.6
	4	53.6 ± 7.9	53.2 ± 9.6
Women	1	46.8 ± 6.4	46.4 ± 6.2
	4	34.3 ± 6.3	32.0 ± 5.9
Mixed	1	55.9 ± 9.4	56.8 ± 8.5
	4	43.4 ± 6.2	42.6 ± 5.1

The loads selected during the individual and team MAL-L&C and VMAL-L&C are listed in Table 1. There were no differences between the MAL-L&C and the VMAL-L&C within genders for individuals or teams at either work rate. All subsequent analyses are based on the load handled for the full hour (VMAL-L&C).

Individuals (1P) and teams (2P) selected heavier loads ($p<0.01$) when working at a rate of 1 x/min (1P=30.6 kg, 2P=59.7 kg) vs 4 x/min (1P=22.5 kg, 2P=43.8 kg). The individual VMAL-L&C of men (30.9 kg) was heavier than that of women (20.8 kg, $p<0.01$). All men teams selected heavier loads (62.5 kg, $p<0.01$) than either mixed-gender teams (49.7 kg) or all-women teams (39.1 kg), and VMAL-L&C for mixed-gender teams was heavier than for all-women teams ($p<0.05$).

Figure 1 presents the team VMAL-L&C and the sum of the individual VMAL-L&Cs for each gender at 1 x/min and 4 x/min. There were no significant differences between the sum of individual VMAL-L&C and the team VMAL-L&C at 1 x/min or at 4 x/min for any of the three gender groups.

DISCUSSION

Men selected heavier loads than women, both individually and in teams, and mixed-gender teams selected heavier loads than all-women teams. These findings concur with team lifting research which demonstrated that the maximum load lifted increased with the number of men on the team (Sharp, et al., 1993).

Subjects were able to accurately estimate the maximum acceptable load they would lift and carry for one hour during two-20 min periods, as no differences were found between the MAL-L&C and VMAL-L&C. This is in agreement with previous data of Sharp and Legg (1988) who found male soldiers capable of performing one hour of lifting with the load selected

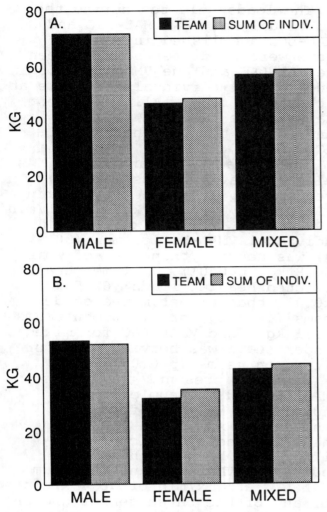

Figure 1. Team VMAL-L&C and sum of individual MAL-L&C by gender for: A. 1 x/min and B. 4 x/min.

during two-20 min periods. This supports use of the psychophysical methodology for determining maximal two-person loads for high intensity tasks lasting up to one hour.

The team VMAL-L&C was not greater than the sum of the individual VMAL-L&Cs as shown by Fox (1982) and Johnson and Lewis (1989). Despite the differences in the tasks, equipment and instructions to the subjects, two men selected between 50-53.2 kg for tasks with a rate of 2 x/min or more for all three studies (current; Fox, 1982; Johnson and Lewis, 1989). Although they carried the box further and lifted it higher, the individual VMAL-L&C was higher for men in the current

study (26.0 kg @ 4 x/min), than for Fox (1982, 17.8 kg @ 4 x/min) or Johnson and Lewis (1989, 20.4 kg @ 2 x/min). Our instructions were to set the maximum load to lift and carry for one-hour, and still be able to perform all other duties for an 8 hour day. Fox (1982) and Johnson and Lewis (1989) both asked subjects to set the load for an eight-hour day, but to avoid becoming unusually tired, weakened, overheated, or out of breath. While this difference in instructions may explain why individual MALs are higher for this study, it does not explain why the relationship between the sum of individual MALs and team MAL differed among studies.

The findings in this study also contrast with prior research findings that 1RM team lifting strength was less than the sum of individual 1RM lifting strength (Karwowski, 1988; Karwowski & Mital, 1986; Karwowski & Pongpatanasuegsa, 1988; and Sharp, et al., 1993). In the team lifting studies, there was little difference in technique between individual and team lifting. In contrast, there were observable technique differences between individual and team lifting and carrying. Individuals carried the load in front of the body, with elbows flexed to prevent the box from interfering with leg movement. Teams walked forward, but carried the box to one side, with straight arms, thus reducing the upper body muscle strength demands. During the lift to 132 cm, individuals lifted the box to chest height and bi-manually pushed it onto the platform. Teams tended to maintain stride and swing the load onto the platform, pushing with the rear hand once the edge of the box was on the platform. A biomechanical analysis of differences in lifting and carrying technique is warranted to further examine differences between individual and team performance.

The instructions used in the

current study resulted in a load greater than the 8 h MAL and less than the one repetition maximum load. The relationship between the sum of individual loads and the team load seems to fall between the two as well.

The majority of the population should be able to handle the fifth-percentile norms for lifting and carrying as listed in Table 2.

Table 2. Percentile values for VMAL-L&C (kg) for two-person teams.

	Rate 1 x/min		
Percentile	5	50	95
Male	49.0	71.2	95.8
Female	36.1*	46.4	50.6
Mixed	40.9	58.2	71.8

	Rate 4 x/min		
Percentile	5	50	95
Male	30.6	51.8	67.5
Female	24.5	31.6	40.8
Mixed	28.2	42.8	51.0

*Due to small sample size, this value is the 10th percentile.

REFERENCES

Fox, R.R. (1982). A psychophysical study of bi-manual lifting. Unpublished Master's Thesis, Texas Technical University, Lubbock, TX.

Johnson, S.L. and Lewis, D.M. (1989). A psychophysical study of two-person manual materials handling tasks. Proceedings of the Human Factors Society 33rd Annual Meeting, (pp. 651-653). Santa Monica, CA: Human Factors Society.

Karwowski, W. (1988). Maximum load lifting capacity of males and females in teamwork. In Proceedings of the Human Factors Society 32nd Annual Meeting, (pp. 680-682). Santa Monica, CA: Human Factors Society.

Karwowski, W. and Mital, A. (1986). Isometric and isokinetic testing of lifting strength of males in teamwork. Ergonomics, 29, 869-878.

Karwowski, W. and Pongpatanasuegsa, N. (1988). Testing of isometric and isokinetic lifting strengths of untrained females in teamwork. Ergonomics, 31, 291-301.

Sharp, M.A. and Legg, S.J. (1988). Effects of psychophysical training on maximal repetitive lifting capacity. American Industrial Hygiene Association Journal, 49, 639-644.

Sharp, M.A., Rice, V.J., Nindl, B.C. and Williamson, T.L. (1993). Maximum team-lifting capacity as a function of team size. Technical Report T94-2. Natick, MA: U.S. Army Research Institute of Environmental Medicine.

THE EFFECTS OF BOX DIFFERENCES AND EMPLOYEE JOB EXPERIENCE ON TRUNK KINEMATICS & LOW BACK INJURY RISK DURING DEPALLETIZING OPERATIONS

W. Gary Allread, William S. Marras, Kevin P. Granata, Kermit G. Davis, Michael J. Jorgensen
The Ohio State University
Columbus, Ohio

Workers from a local food distribution center were studied depalletizing boxes from a pallet. The objectives of this study were to determine the change in trunk kinematics associated with selecting boxes having different characteristics and to observe if there was a relationship between trunk kinematics and employee job experience. The boxes varied in terms of: size; presence/absence of handles; weight; and location on a pallet. Worker job experience also was recorded. Kinematic trunk motions and subsequent risk of low back disorder (LBD), assessed using a risk model, were studied as dependent measures. Results indicated that the weight and layer conditions influenced most of the kinematic variables. The size and handle conditions influenced fewer dependent measures. All main effects but the handle condition had an influence on LBD risk. Most of the significant interaction effects were related to layer, illustrating the tremendous influence that box location on a pallet had on trunk kinematics and LBD risk. In fact, at the bottom pallet layers, LBD risk was the same regardless of the weight lifted or the size of the box. In studying job experience, inexperienced workers were found to have LBD risk values that were, on average, 5% higher than the experienced group. This study has been successful at pinpointing which box parameters are worthy of consideration to include in a food distribution environment for the purposes of reducing the risk of work-related LBD.

INTRODUCTION

It is widely known that the order selection task in food distribution centers places the worker at risk of occupationally related low back disorders (LBDs). This job is associated with one of the greatest incidence rates of LBD in the United States. The National Association of Wholesale Grocers of America (NAWGA) and the International Foodservice Distribution Association (IFDA) disclosed that 30% of the injuries reported by food distribution warehouse workers were attributable to back sprains/ strains (Waters, 1993). In addition, over a five-year period, it was found that back injuries accounted for nearly 60% of lost work days (NIOSH Interim Report, HETA 91-405, March 1992). Hence, grocery item selectors have an incidence of low back pain that is at least as severe as other manual materials handling (MMH) jobs.

One approach to controlling this risk consists of manipulating the characteristics of the object or box to be handled in the food distribution center. A committee organized by the Food Marketing Institute is currently considering the various options available to them in order to help mediate the risk of work-related LBD in these food distribution centers. Among the options considered: 1) reducing the weight of the boxes; 2) reducing their size; or 3) incorporating handles into the boxes. However, it is currently unknown what effect these changes to the box characteristics would have on trunk kinematics and the subsequent LBD risk.

As with many MMH jobs, grocery warehouse work requires a significant amount of knowledge and expertise. It has been the experience of the authors that there are different ways in which warehouse workers perform the same job. These observations led to the hypothesis that personal characteristics, namely, job experience, may be related to motions patterns as well.

Hence, the objectives of this study were: 1) To determine the change in both trunk kinematics and LBD risk associated with selecting boxes that varied as a function of *size*, existence of *handles*, box *weight*, and the *layer* at which the box was located on the pallet; and 2) To observe if there was a relationship between trunk kinematics and employee job experience.

METHODS

Subjects

Ten experienced, male order selectors, ranging from 19 to 49 years of age, were recruited from a local food distribution center. The average (SD) weight and stature of the workers were 80.0 (8.4) kg and 180.3 (7.1) cm, respectively. Their work experience in warehouse settings ranged from 0.25 to 23 years.

Experimental Task

During the different experimental trials, the box size, coupling (handles), and box weight conditions were varied. Subjects were instructed to depalletize the entire compliment of boxes so that they could be observed picking from all locations on the pallet. They were instructed to completely pick from one layer of the pallet before unloading a new layer. While the workers were lifting boxes they were being continuously monitored so that trunk loading could be assessed.

Boxes were stacked on a standard wooden pallet as they generally would be found in a warehouse. Small boxes were placed on a double-stacked pallet to allow their handles to correspond to the heights of the large boxes, which were stacked on a single pallet.

The depalletizing task started when the subject grasped the box and ended when he crossed an imaginary line that coincided with the point where the subject was upright and facing the "palletizing" pallet. Data were collected for only this interval of time, although subjects completed the task. The lifting rate for all subjects was set at 166 boxes handled hourly. This frequency was selected for the study, as it was the minimum acceptable work level at which subjects have to perform on their jobs to "make rate."

Experimental Design

The experimental design consisted of a four-way, within-subject design. The *independent variables* included box size, handle usage, box weight, and pallet layer. Subjects served as a random effect. Two sizes of boxes were evaluated, representing "small" boxes and "large" boxes found in typical warehouse environments. Box dimensions were 20.3×40.6×30.5 cm (H×W×D) for small boxes and 27.9×49.5×30.5 cm (H×W×D) for large boxes. The position and size of the handles were similar to those commonly found on boxes in warehouse environments. Box weights studied were 18.1, 22.5, and 27.2 kg. These were at the upper percentiles of typical box weights common in warehouse settings, and they were chosen to evaluate the effects of heavier loads on trunk motions.

Each of the pallets was divided into three layers. Figure 1 shows a schematic view of these layers on a standard pallet. The bottom of each box remained at a constant level from the floor, corresponding to a height of 123.0 cm for the top layer, 89.0 cm for the middle layer, and 41.0 cm for the bottom layer. The handle cut-outs were 5.1 cm below the top of each box.

A between-subjects analysis also was performed to assess the effects of subject job experience in combination with the aforementioned box characteristics. Subjects who had one year or less experience working in the distribution center were grouped as "inexperienced" while all others were categorized as "experienced." However, one subject reported working in the warehouse for 13 months, so he was placed in the inexperienced group. This resulted in five subjects in the inexperienced group (ranging from 0.25 to 1.08 years of experience) and five subjects in the experienced group (2.58 to 23.00 years of experience).

The *dependent variables* consisted of trunk kinematics (positions, velocities, and accelerations) of the low back in

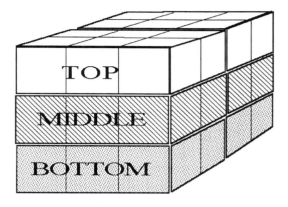

Figure 1. The three layers of a standard pallet.

the three cardinal planes of the body (lateral, sagittal, and transverse). The maximum external trunk moment during lifting was included as a dependent measure and determined using a tape measure. An assessment of LBD risk (probability of high risk group membership), as determined by Marras et al (1993), also was determined. Three kinematic parameters (maximum sagittal flexion, maximum lateral velocity, and average twisting velocity), along with lifting frequency and the maximum moment generated, comprised this risk assessment.

Apparatus

A Lumbar Motion Monitor (LMM) was used to collect kinematic information about the trunk. The LMM is essentially an exoskeleton of the spine in the form of a triaxial electrogoniometer that assessed instantaneous position, velocity, and acceleration of the trunk in three-dimensional space. The light-weight design of the LMM allowed the data be collected with minimal obstruction to the subject's movements. For more information on the design, accuracy, and application of the LMM, refer to Marras et al (1992).

RESULTS

The results of the analysis of variance (ANOVA), which evaluated the influence of the box characteristics and location upon trunk kinematics, are shown in Table 1. It lists the main effects and interactions along the first column; subsequent columns list the kinematic parameters studied. Also included are effects on the measured moment generated during box depalletizing and the LBD risk.

In general, Table 1 shows that, of the four main effects (size, handle condition, weight, and pallet layer), the weight and layer conditions influenced nearly every dependent measure. Box size mostly affected transverse plane motions but none of the lateral plane variables and only average sagittal velocity. Handle presence/absence influenced most most of the sagittal plane variables but none of the transverse plane kinematics and only average lateral velocity.

Table 1.   Summary of the ANOVA results. Main effects and interactions are listed in the first column; dependent measures are listed in subsequent columns. An asterisk (*) indicates that the effects were significant at $\alpha=0.05$.

	Lateral Plane				Sagittal Plane					Transverse Plane				Moment	LBD Risk
	ROM	Avg Vel	Max Vel	Max Acc	Flex Pos	ROM	Avg Vel	Max Vel	Max Acc	ROM	Avg Vel	Max Vel	Max Acc		
Size (S)							*			*	*	*		*	*
Handle (H)		*			*	*		*	*					*	
Weight (W)	*	*	*	*	*	*		*	*	*	*	*	*	*	*
Layer (L)	*	*	*	*	*	*	*	*	*	*	*	*	*	*	*
S×H					*										
S×W															
S×L	*	*	*		*	*	*	*		*	*	*	*	*	*
H×W	*														
H×L	*	*	*		*	*	*			*	*	*	*	*	*
W×L	*					*	*			*	*	*		*	*
S×H×W	*									*	*	*	*		*

Note: ROM = Range of Motion

Table 2.   Summary of the ANOVA results, including worker job experience. An asterisk (*) indicates which effects were significant at $\alpha=0.05$.

	LBD Risk
Size (S)	*
Handle (H)	
Weight (W)	*
Layer (L)	*
Experience (E)	*
S×H	
S×W	
S×L	*
S×E	
H×W	
H×L	*
H×E	
W×L	*
W×E	
L×E	*

Both size and handle conditions significantly affected moment, but only the size condition affected LBD risk; the handle presence/absence main effect had no influence on LBD risk. Table 1 shows that several interaction effects were present, and most were related to interactions involving pallet layer. This finding illustrates the tremendous influence that box location on a pallet had on trunk kinematics and the subsequent risk of LBD. For example, though the handle main effect was not significant, the handle×layer interaction was significant for most kinematic variables and LBD risk. Post-hoc comparisons (not shown here) found LBD risk to be significantly higher (5%) when lifting without handles from the middle pallet layer. Handle presence/absence had no effect at the top or bottom layers.

LBD risk results from Table 1 are shown graphically in Figure 2. Several features of this chart are of note. First, LBD risk increased as depalletizing occurred at lower layers. On average, this risk was 10% higher for depalletizing from the middle layer as compared with the top layer, and another 10% higher when lifting from the bottom layer as opposed to the middle layer. Second, LBD risk increased as more weight was lifted. This amounted to about a 4% increase in risk with each 10 lb added to the load. Third, lifting small boxes produced significantly higher LBD risk values than large boxes, but only at the top layer. This accounted for the significant size×layer interactions found across several of the dependent variables. Finally, at the top and middle layers, increasing the weight handled also increased the LBD risk of the task; but at the bottom pallet layer, risk was essentially the same (71%-75%). This suggests that load weight is less of a factor than location in determining risk when the loads were placed at more extreme (i.e., lower) positions.

Table 2 shows the added effects of worker job experience on LBD risk. This table indicates that the two experience groups produced a significantly different risk value ($\alpha=0.05$). In addition, the only significant two-factor interactions involved the layer condition. No higher-order interactions were found to be significant.

The effects of job work experience on LBD risk are presented in Figure 3. Results are shown for the main

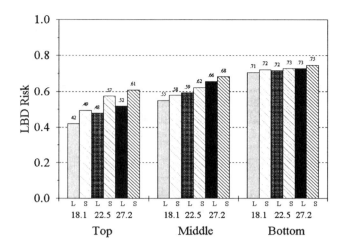

Figure 2.   Effects of box size, weight, and location on a pallet on LBD Risk.

effects. Figures 2(a) and (b) show that inexperienced subjects produced significantly higher risk (about 5%) across both box size and handle conditions. In Figure 2(c), only during lifts of 22.5 kg loads did inexperienced subjects produce higher risk. For the pallet layer condition, Figure 2(d) shows that only at the bottom layer was risk higher for the inexperienced group. This difference was over 10%. These results show that methods used by experienced workers produced less trunk motion and lower LBD risk. Also, as shown in Figure 2(d), the effects of work experience was most evident at pallet locations (bottom layer) traditionally presenting the highest risk of injury.

DISCUSSION

This work has contributed to our understating of the specific situations in a depalletizing operation that contribute to differences in trunk kinematics and the resulting risk

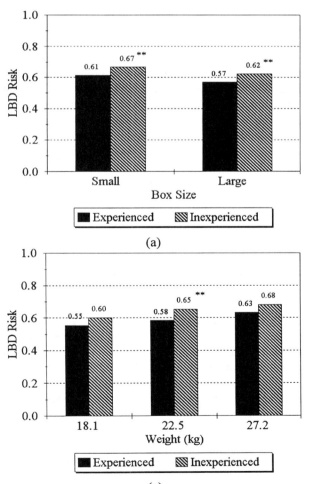

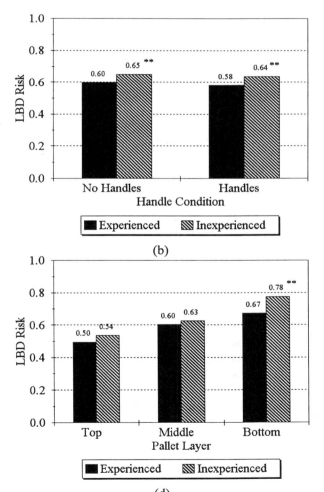

Figure 3.   Effects of job experience on LBD risk, in combination with box size (a), handle condition (b), weight (c), and pallet layer (d). The asterisks (**) indicate where inexperienced subjects produced a significantly higher LBD risk value than experienced subjects (post-hoc tests with a family-wise error rate of $\alpha=0.05$).

of experiencing a LBD. In general, conditions where a worker reached to near floor increased LBD risk. This finding is not new; however, with this study we have been able to document and quantify more specific conditions that contributed to LBD risk. The study also has advanced our knowledge of how work experience may affect trunk motions and the subsequent risk of low back injury.

In terms of specific lifting conditions, this study found that the weights of the loads lifted and their pallet locations most affected trunk kinematics. Box size and the handle type main effects were less significant factors, though handle presence did result in a lower LBD risk value at the middle pallet layer. The interaction of box size and pallet layer on LBD risk indicates the difficulty of handling small boxes at the top pallet layers. This result questions the use of the layer-by-layer approach to depalletizing, which is standard policy in many warehouse environments. Further study is needed to determine if alternative approaches (i.e., a pyramiding scheme) may affect injury risk when handling boxes. The large and consistent LBD risk values found at the bottom pallet layers (above 70%) indicate that neither box weight nor size had as much influence as box location in determining risk. This finding more specifically suggests where redesign efforts should first be focused to minimize LBD risk, namely, at raising load levels.

This effort also has facilitated our understanding of the effects of job experience on trunk motions and injury risk. Across most conditions studied here, experienced workers produced lower LBD risk than those having about a year's experience or less. This difference was most dramatic at the bottom pallet layer, in which inexperienced subjects produced an LBD risk value of over 10% more than those experienced doing the work. This suggests that more experienced workers have learned ways of doing the job that reduce their risk of injury. This result also could be used to design better and more specific training for new workers on the conditions of their jobs that pose the greatest risk.

CONCLUSIONS

This study has been successful at pinpointing which box parameter variables are worthy of consideration for inclusion in the food distribution environment for the purposes of reducing the risk of work-related LBD. In general, the following conclusions were drawn.
- LBD risk increased linearly as box weight increased.
- The greatest risk occurred during lifts from the bottom layers of the pallet; however, risk was 10% lower as loads were depalletized from the middle layers of the pallet, and another 10% lower when lifted from the top layers. LBD risk values were consistent at the bottom pallet layers regardless of the box's size or weight.
- The presence of box handles produced a lower LBD risk value, but only at the middle pallet layer. Considering that raised box heights have been found to reduce

trunk motions and injury risk, and that these findings may result in MMH being done at higher levels, use of handles may indeed play a role in reducing LBD risk.
- Experienced subjects produced lower LBD risk values (about 5%) for most of the study conditions. The difference between experience levels was most dramatic at the bottom pallet layers (inexperienced subjects had a risk value over 10% greater than the experienced group), which is commonly a high risk factor regardless of work experience.

In conjunction with the companion paper to this article, these findings provide some practical solutions to the design of the distribution center. Estimates of risk were assessed both from a biomechanical and an epidemiological perspective, and similar results were achieved. It is clear that, given the success of this study, further studies have the potential to further reduce risk through the investigation of other box features.

LIMITATIONS

This study used a relatively small sample size (ten subjects). A larger sample of workers may have produced other results. Also, even though the laboratory was designed to simulate a warehouse environment, kinematics of the trunk could be different if subjects were tested at their place of work. However, experimental constraints made this impractical.

ACKNOWLEDGEMENTS

The authors would like to thank the Food Marketing Institute for funding this study.

REFERENCES

Marras, W.S., Fathallah, F.A., Miller, R.J., Davis, S.W., and Mirka, G.A., (1992). "Accuracy of a Three-Dimensional Lumbar Motion Monitor for Recording Dynamic Trunk Motion Characteristics," Int J Ind Ergo, **9**, 75-87.

Marras, W. S., Lavender, S. A., Leurgans, S., Rajulu, S., Allread, W. G., Fathallah F. and Ferguson, S .A., (1993). The Role of Dynamic Three Dimensional Trunk Motion in Occupationally-Related Low Back Disorders: The Effects of Workplace Factors, Trunk Position and Trunk Motion Characteristics on Injury, Spine, **18**(5), 617-628.

Waters, T.R., Putz-Anderson, V., Garg, A., and L.J. Fine (1993). Revised NIOSH equation for the design and evaluation of manual lifting tasks. Ergonomics, **36**(7): 749-77.

THE EFFECT OF LIFTING VS. LOWERING
ON SPINAL LOADING

Kermit G. Davis
The Ohio State University
Columbus, Ohio

In industry, workers perform tasks requiring both lifting and lowering. During concentric lifting, the muscles are shortening as the force is being generated. Conversely, the muscle lengthens while generating force during eccentric lowering. While research on various lifting tasks is extensive, there has been limited research performed to evaluate the lowering tasks. Most of the research that does exist on lowering has investigated muscle activity and trunk strength. None of these studies have investigated spinal loading. The current study estimated the effects of lifting and lowering on spinal loads and predicted moments imposed on the spine. Ten subjects performed both eccentric and concentric lifts under sagittally symmetric conditions. The tasks were performed under isokinetic trunk velocities of 5, 10, 20, 40, and 80 deg/s while holding a box with weights of 9.1, 18.2, and 27.3 kg.. Spinal loads and predicted moments in three dimensional space were estimated by an EMG-assisted model which has been adjusted to incorporate the artifacts of eccentric lifting. Eccentric strength was found to be 56 percent greater than during concentric lifting. The lowering tasks produced significantly higher compression forces but lower anterior-posterior shear forces than the concentric lifting tasks. The differences in the spinal loads between the two lifting tasks were attributed to the internal muscle forces and unequal moments resulting from differences in the lifting path of the box. Thus, the differences between the lifting tasks resulted from different lifting styles associated with eccentric and concentric movements

INTRODUCTION

Manual material handling (MMH) tasks have been associated with lower back injuries (Snook et al., 1978; Bigos et al., 1986). A typical MMH task requires both lifting and lowering of items that have a variety of shapes, sizes and weights. Lowering involves a different mechanism of muscle operation, in which the muscles elongate as the force is generated while concentric lifting entails muscle shortening. Drury et al, (1982) found that 52% of the MMH tasks in industry were eccentric lowering while concentric lifting consisted of only 32%. While research on various concentric tasks is extensive, there has been limited research performed to evaluate the eccentric task.

Many researchers have found that the muscle activity of the trunk extensor muscles were lower for eccentric than concentric tasks despite the generation of greater external torque (Kumar and Davis, 1983; Marras and Mirka, 1989; Cresswell and Thorstensson, 1994; De Looze et al., 1993). Henriksson et al. (1972) found that the perceived exertion for lowering tasks was less than for lifting tasks. The strength of the various trunk muscles have been evaluated as a function of eccentric and concentric lifting by several authors. In general, individuals have greater strength during eccentric than concentric lifting (Marras and Mirka, 1989; Reid and Costigan, 1987; Smidt et al., 1980).

The objective of the current study was to compare both eccentric and concentric lifting. The evaluation of the two tasks consisted of estimating the moments and loads acting on the spine through the use of an EMG-assisted model that has been adjusted for the eccentric behavior of the muscle.

METHODS

Subjects

The ten subjects who participated in the study were male students with no prior history of low back pain. The ages of the subjects ranged from 22 to 34 years. The averages (sd) for height and weight were 181.0 cm (6.6) and 79.3 kg (12.6), respectively.

Task

Subjects performed the concentric and eccentric exertions while positioned in a pelvic support structure (PSS). The subjects started the eccentric lifts in an upright position (0 degrees of flexion) and lowered the box to a posture of 40 degrees of flexion. On the other hand, the concentric lifts had the subjects start at 40 degrees of flexion and lift until they reached a fully upright position.

Experimental Design

This study was a three-way, within-subject design. The independent variables included the following: box weight (9.1, 18.2, and 27.3 kg.), trunk isokinetic velocity (5, 10, 20, 40, and 80 deg/s), and lifting task. These weights and velocities were chosen to reflect the values commonly found in industry (Marras et al. 1993). In order to account for variability between the subjects, subjects were used as a random effect. The lifting tasks consisted of both eccentric and concentric lifting of the box.

The dependent variables of this study were maximum moments imposed on the spine and spinal loading. All these measurements were computed using the dynamic EMG-assisted model developed at the Ohio State University over the past decade (Marras and Reilly, 1988; Marras and Sommerich, 1991a, b; Granata and Marras, 1993; and Granata and Marras, 1995a, b; Marras and Granata, 1995). The model used the kinematic and electromyographic information about the trunk to estimate the spinal loads as well as the moments imposed on the spine in three dimensional space. The spinal loads estimated were compression, anterior-posterior shear, and lateral shear on the L_5/S_1 joint. The predicted moments were sagittal bending moments about the lumbosacral joint since the lifting tasks were performed sagittally symmetric.

The relationships for length-strength and force-velocity were empirically determined by calculating the error between the predicted to measured torque at the instantaneous lengths and velocities of the muscle during the lifts. A best fit curve was used to estimate both relationships for eccentric and concentric lifting separately.

Apparatus:

The Lumbar Motion Monitor (LMM) was used to collect the trunk motion variables. The LMM is essentially an exoskeleton of the spine in the form of a triaxial electro-goniometer that measured the instantaneous three-dimensional position, velocity, and acceleration of the trunk. For more information on the design, accuracy, and application of the LMM, refer to Marras et al. (1992).

Integrated electromyographic (EMG) activity was monitored through the use of bi-polar electrodes spaced approximately 3 cm apart at the ten major trunk muscle sites. The ten muscles of interest were: right and left erector spinae; right and left latissimus dorsi; right and left internal obliques; right and left external obliques; and right and left rectus abdominis. For the standard locations of the electrode placement for these muscles, refer to Mirka and Marras (1993).

A force plate (Bertec 4060A) was used to measure the kinetic variables of the lifts. The subject was positioned into a pelvic support structure (PSS) that was attached to the force plate. The PSS restrained the subject's pelvis and hips in a fixed position. Also, the relative position of L_5/S_1 to the center of the force plate remained constant for the entire experiment. By knowing the position of L_5/S_1, the forces and moments measured at the center of the force plate were translated and rotated to L_5/S_1.

All signals from the above equipment were collected simultaneously through a customized Windows™-based software developed in the Biodynamics Laboratory. The signals were collected at 100 Hz and recorded on a 486 portable computer via an analog-to-digital board. The data were saved by the computer for subsequent analysis.

An additional computer was used to display the instantaneous sagittal angular velocity recorded by LMM in real time. A target region was provided by displaying two lines at a given slope that corresponded to the velocity of interest. The lines allowed the subject to have a tolerance of 3 percent deviation for the target velocity. The signal was transferred from the LMM to the computer through an analog-to-digital board and converted into velocity by customized software. The computer monitor was positioned directly in front of the PSS in direct view of the subject.

Procedure:

Upon arriving at the Biodynamics Laboratory, subjects completed a consent form and anthropometric measurements were taken. After proper application of the electrodes, a set of maximum exertions was performed for normalization procedures. Subjects were then positioned into the PSS and the LMM was attached to the back. Velocities were controlled by the subject by following a trace through a given region displayed on a computer screen while lifting the box. All subjects were allowed to practice the different velocities until they were able to remain within the tolerances. If the subject's trace fell outside the tolerance levels, the lift was repeated.

RESULTS

The length-strength relationship for both eccentric and concentric exertions was found to be the same. Under the concentric conditions, the force-velocity modulation was found to be an exponential function, as seen by other researchers (Hill, 1938; Wilkie, 1950; Granata, 1993). Conversely, the eccentric force-velocity modulation factor was determined to be constant. The value of the constant was set to be equal to the ratio between the eccentric and concentric gains, thus, resulting in a constant inter-subject gain. This ratio was found to be 1.56.

The EMG-assisted model was then used to predict the external moments and spinal loads with the adjustment for eccentric lifting. As a result, there was no difference between the gains for the two types of lifting. Additionally, the model performance was slightly better for eccentric lifting. The average r^2 between the predicted and measured external moments for eccentric and concentric exertions were 0.95 and 0.88, respectively.

A summary of the analysis of variance (ANOVA) for the maximum predicted moments and spinal loads are in Table 1. Notice, tasks refers to the type of lifting being performed. Under eccentric lifting conditions, the maximum sagittal predicted moment (hereafter referred to as moment) was larger than when lifting concentrically, as seen in Figure 1. The values of moment for eccentric and concentric lifting were 140.5 Nm and 113.1 Nm, respectively (a difference of 27.4 Nm). As the weight of the box increased, the moment also increased. The maximum sagittal moments for the 9.1, 18.2 and 27.3 kg. weights were 105.1, 126.5, and 148.8 Nm respectively. Therefore, the maximum moment increased about 25 Nm for every 9.1 kg. of increased box weight. For sagittal predicted moment, the velocity of 5 degrees per second (deg/s) had a significantly smaller moment than the other velocities (10, 20, 40, and 80 deg/s). The Velocity*Task interaction indicated that for the 5 deg/s velocity, the task type had no effect on the moment, however, for the other velocities, eccentric lifting produced larger moments than the concentric lifts.

Table 1: Summary of Significant Effects for the EMG-assisted Model Outputs.

Effect	Maximum Sagittal Predicted Moment	Maximum Lateral Shear Force	Maximum A/P Shear Force	Maximum Compression Force
Task (T)	*		*	*
Weight(W)	*	*	*	*
Velocity(V)	*		*	*
W*T		*	*	
V*T	*		*	*
W*V				
W*V*T				

* indicates significant at $p < 0.05$.

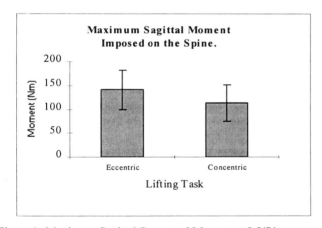

Figure 1: Maximum Sagittal Supported Moment at L5/S1 as a Function of Lifting Task.

The results for the maximum lateral shear forces indicated that the Weight main effect was the only significant effect where an increase in weight corresponded to an increase in lateral shear forces. A 9.1 kg. increase in box weight resulted in approximately 22 N of additional lateral shear force on the spine.

The maximum anterior-posterior (A-P) shear forces placed on the spine during concentric lifting were larger than when lifting eccentrically (Figure 2). The difference between the two types of lifts was approximately 135 N. An increase box weight of 9.1 kg. also produced an increase in A-P shear of about 130 N. The 5 deg/s velocity resulted in lower A-P shear forces than the 20 and 80 deg/s velocities, but was not different than the 10 and 40 deg/s lifts. Similarly, the A/P Shear forces for the 80 deg/s lifts were significantly larger than the 10 deg/s lifts but not the 20 and 40 deg/s velocities.

The A-P shear forces for the 5, 10, 20, 40, and 80 deg/s velocities were 707, 748, 765, 752, and 768 N, respectively. Additionally, the Weight*Task and Velocity*Task interaction indicated that the effects of type of lifting on A-P shear depended upon the weight lifted or the velocity of the lift. There was a larger increase in A-P shear forces with increases in weight for the concentric lifts. The concentric lifts had higher A-P shear forces at all velocities, but a larger difference was seen for the 80 deg/s lifts than the other lifts.

The maximum compression forces were greater during eccentric lifting than when lifting concentrically (Figure 2). The difference in loading between eccentric and concentric lifting was close to 600 N. As with the lateral and A-P shear forces, an increase in loading was experienced when the box weight increased. The type of task produced a larger difference in the maximum compression forces than when the box weight was increased by 9.1 kg. The 5 deg/s lifts produced less compression force than the other velocities. There was no difference between the 10, 20, and 40 deg/s lifts, but a difference was found between the 80 deg/s lifts and 10 deg/s lifts. For velocities at 10, 20, 40 and 80 deg/s, the maximum compression forces were lower for concentric lifting compared to eccentric lifting while no difference was found for the 5 deg/s velocity.

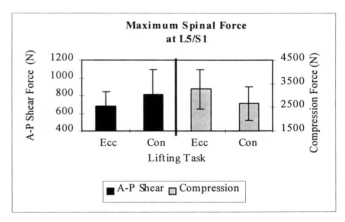

Figure 2: Maximum Spinal Loading at L5/S1 as a Function of Lifting Task.

Many of the differences between the two types of tasks were confounded by the differences in the sagittal predicted moment. Thus, the results for A-P shear force, and compression force were normalized to predicted sagittal moment. This would eliminate any variance attributed to the differences in moment. The results of the ANOVA indicate that there was still a difference between the two types of lifting (Table 2).

The spinal loading for the two types of lifting are shown in Figure 3. The A-P shear force per unit of moment during eccentric lifting was significantly lower than during concentric lifting. This would indicate that the difference between A-P shear loading for the two types of tasks resulted from not only increased moment but also from an increase in muscle loading of the spine (via coactivity). The difference between the two types of lifting was actually reversed for compression when normalized to moment. Initially, eccentric lifting had larger compression forces than concentric lifting, but when normalized to the moment, eccentric lifting was lower. This would indicate that the sagittal moment was the main reason for the increase in compression forces placed on the spine during eccentric lifting.

Table 2: Summary of Significant Effects for the Estimated Spinal Loads Per Unit Moment.

Effect	Maximum A-P Shear Force Per Unit Moment	Maximum Compression Force Per Unit Moment
Task (ecc vs. con)	*	*
Weight		*
Velocity		*
W*T		*
V*T	*	
W*V		
W*V*T		

* indicates significant at $p < 0.05$.

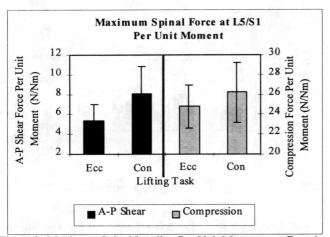

Figure 3: Maximum Spinal Loading Per Unit Moment as a Function of Lifting Task.

Additionally, the interaction between the type of lifting and velocity remained significant for A-P shear forces when normalized to external moment. The amount of box weight significantly influenced the compression forces even after normalization for moment. The internal forces resulting from the muscles contributed more at the lower weight (18.2 kg.) than the heavy weight (27.3 kg.). The compression per unit moment was only significant for the 80 deg/sec condition. Thus, the difference in compression forces between the velocities was attributable to the external moment, except for the 80 deg/sec velocity. For the 80 deg/sec velocity, the internal forces from the muscles contributed additional compression force. The interaction between task and weight was significant for compression per unit moment. The

concentric lifts had a larger increase across the different weights than eccentric lifting. Again, it appears that the difference between the two tasks at the various levels of weight resulted from both external moment and internal force differences.

DISCUSSION

The results were not affected by the fidelity of the EMG-assisted model. The gains for each of the subjects were constant between the two types of lifting. The values of r^2 between the predicted and measured moments (validity) did not differ significantly between the eccentric and concentric lifts. The ratio eccentric and concentric gains was 1.56. Reid and Costigan (1987) found that the ratio of eccentric to concentric strength was 1.2. The difference between the two ratios was that the present study was based on internal estimation of the external torque which reflects both agonistic and antagonistic muscle inputs. Additionally, their ratio was determined by using a KIN/COM to control the exertion, possibly resulting in different motions from free dynamic lifting. The importance of the difference is that the relationship between concentric and eccentric strength is not necessarily universal.

The present results show that the maximum sagittal moment was substantially higher for the eccentric lifts. De Looze et al., (1993) and Gagnon and Smyth (1991) found that concentric lifting produced larger peak moments than eccentric tasks. The difference in the studies can be explained by the different techniques used to control the subject's posture. For the present study, the subjects performed the exertions in a structure that fixed the pelvis and hips in one position. By allowing the hips and legs to move, as in the other studies, the subjects could have adopted a different lifting style for eccentric and concentric exertions. The individuals might have moved into the load during the eccentric lifts.

In the present study, the maximum sagittal moments were equivalent for all velocities except the 5 deg/s velocity. The subjects appeared to move the box to a maximum distance away from the body independently of the velocity. For the 5 deg/s, subjects seemed to keep the box closer to the body throughout the exertion. Granata and Marras (1993) found that the average sagittal moment reduced with increased velocity. These results indicate that the subjects might move the box to the same maximum horizontal distance but bring it closer to the body during the rest of the exertion for the faster velocities.

In the present study, eccentric lifting produced maximum A-P shear forces that were 135 N lower than the concentric tasks. On the other hand, the eccentric tasks resulted in higher compression values (almost 600 N) than concentric lifts. There appears to be a trade-off in the nature of the loading between eccentric and concentric exertions. However, it was hypothesized that the resulting loads occurred from the differences in sagittal moment. When the A-P shear and compression forces were normalized with respect to the sagittal moment, there was still a difference between the two lifting tasks. When normalized to moment, the difference in A-P shear force between the two types of lifting remained after normalization indicating that there was a difference in the mechanism that produced the forces to counter-balance the external forces. The A-P shear per unit moment was attributed to the difference in coactivity between the different lifting motions resulting in higher loads during concentric lifting. For compression forces per unit moment, the eccentric lifts actually produced lower compression forces per unit moment than the concentric lifts indicating that the difference resulted from the subjects using different lifting styles along with a change in the internal force mechanism.

The peak A-P shear and compression forces were the smallest for the 5 deg/sec velocity. There was no difference between A-P shear forces for the faster velocities. The differences between the velocities for A-P shear was explained by the external moment lifted under the various conditions. Thus, the A-P shear forces resulted from changes in the path of movement for the various velocities. Also, there was only a difference between the 10 deg/s and 80 deg/s lifts for compression. Granata and Marras (1993) found that average spinal forces changed as a function of velocity. These researchers found that velocity effects both the path of movement and amount of internal loading. The present study indicated that compression forces per unit moment increased only for the 80 deg/sec conditions. Once again, it appears that the subjects adopted a lifting pattern that resulted in the same peak forces for the lower velocities but the internal loading increased during the fast velocity. Hence, the dynamics of the exertion also influences the lifting style and ultimately the loading on the spine.

CONCLUSION

The present study has shown that eccentric strength was 56 percent greater than during concentric lifting. The type of lifting performed by an individual seemed to effect the nature of the lift and loading on the spine. It was also determined that the dynamics of the lift played an import role in the lifting style adopted under both eccentric and concentric lifts.

REFERENCES

Bigos, S.J., D.M Spengler, N.A. Martin, J. Zeh, L. Fisher, A. Nachemson, and M.H. Wang, (1986), Back Injuries in Industry: A Retrospective Study. II. Injury Factors. Spine, vol. 11(3), 246-251.

Cresswell, A.G. and A. Thorstensson, (1994), Changes in Intra-Abdominal Pressure, Trunk Muscle Activation and Force during Isokinetic Lifting and Lowering. European Journal of Applied Physiology, vol. 68, 315-321.

De Looze, M. P., H.M. Toussaint, J.H. Van Dieen and H.C.G. Kemper, (1993), Joint Moments and Muscle Activity in the Lower Extremities and Lower Back in Lifting and Lowering Tasks., Journal of Biomechanics, vol. 26(9), 1067-1076.

Drury, C.G., C. Law, and C.S. Pawenski, (1982), A Survey of Industrial Box Handling. Human Factors, vol. 24(5), 553-565.

Gagnon, D. and M. Gagnon, (1992), The Influence of Dynamic Factors on Triaxial Net Muscular Moments at the L5/S1 Joint During Asymmetric Lifting and Lowering. Journal of Biomechanics, vol. 25(4), 891-901.

Gagnon, M. and G. Smyth, (1991), Muscular Mechanical Energy Expenditure as a Process for Detecting Potential Risks in Manual Material Handling. Journal of Biomechanics, vol. 24(3/4), 191-203.

Granata, K.P., (1993), An EMG-Assisted Model of Trunk Loading During Free-Dynamic Lifting. Unpublished Doctoral Dissertation, the Ohio State University, Columbus, Ohio.

Granata, K.P. and W.S. Marras, (1993), An EMG-Assisted Model of Loads on the Lumbar Spine During Asymmetric Trunk Extensions. Journal of Biomechanics, vol. 26(12), 1429-1438.

Granata, K.P. and W.S. Marras, (1995a), An EMG-Assisted Model of Trunk Loading During Free-Dynamic Lifting. Journal of Biomechanics, vol. 28(11), 1309-1317.

Granata, G. P., and W. S. Marras, (1995b), The Influence of Trunk Muscle Coactivity Upon Dynamic Spinal Loads, Spine, 20(8), 913-919.

Henriksson, J., H.G. Knuttgen, and F. Bonde-Peterson, (1972), Perceived Exertion During Exercise with Concentric and Eccentric Muscle Contractions. Ergonomics, vol. 15(5), 537-544.

Hill, A.V., (1938), The Heat of Shortening and the Dynamic Constants of Muscle, Proceedings of the Royal Society of Biology, vol. 126, 136-195.

Kumar, S. and P.R. Davis, (1983), Spinal Loading in Static and Dynamic Postures: EMG and Intra-abdominal Pressure Study. Ergonomics, vol. 26(9), 913-922.

Marras, W.S., F. Fathallah, R.J. Miller, S.W. Davis, and G.A. Mirka, (1992), Accuracy of a Three Dimensional Lumbar Motion Monitor for Recording Dynamic Trunk Motion Characteristics. International Journal of Industrial Ergonomics, vol. 9, 75-87.

Marras, W. S. and K. P. Granata, (1995), A Biomechanical Assessment and Model of Axial Twisting in the Thoraco-Lumbar Spine, Spine, 20(13), 1440-1451.

Marras, W.S., S.A. Lavender, S.E. Leurgans, S.L. Rajulu, W.G. Allread, F.A. Fathallah, and S.A. Ferguson, (1993), The Role of Dynamic Three-Dimensional Motion in Occupationally-Related Low Back Disorders. Spine, vol. 18(5), 617-628.

Marras, W.S, and G.A. Mirka, (1989), Trunk Strength During Asymmetric Trunk Motion. Human Factors, vol. 31(6), 667-677.

Marras, W. S., and C. H. Reilly, (1988), Networks of Internal Trunk Loading Activities Under Controlled Trunk Motion Conditions, Spine, 13(6), 661-667.

Marras, W.S. and C.M. Sommerich, (1991a), A Three Dimensional Motion Model of Loads on the Lumbar Spine: I. Model Structure. Human Factors, vol. 33(2), 129-137.

Marras, W.S. and C.M. Sommerich, (1991b.), A Three Dimensional Motion Model of Loads on the Lumbar Spine: I. Model Validation. Human Factors, vol. 33(2), 139-149

Reid, J.G. and P.A. Costigan, (1987), Trunk Muscle Balance and Muscular Force. Spine, vol. 12(8), 783-786.

Schmidt, G.L., L.R. Amundsen, and W.F. Dostal, (1980), Muscle Strength at the Trunk. The Journal of Orthopaedic and Sports Physical Therapy, vol. 1, 165-170.

Snook, S.H., R.A. Campanelli, and J.W. Hart, (1978), A study of Three Preventive Approaches to Low Back Injury. Journal of Orthopeadic Medicine, vol.20(7), 478-481.

Wilkie, D.R., (1950), The Relation Between Force and Velocity in Human Muscle. Journal of Physiology, vol. 110, 249-280.

THE ROLE OF POWER IN PREDICTING LIFTING CAPACITY

Patrick G. Dempsey
Liberty Mutual Research Center for Safety and Health
71 Frankland Road
Hopkinton, Massachusetts 01748 USA

M.M. Ayoub
Department of Industrial Engineering
Texas Tech University
Lubbock, Texas 79409-3061 USA

Preplacement strength testing is one of the most viable supplements to ergonomic job design. Previous studies have examined the roles of isometric, isokinetic, and some isoinertial measures in predicting lifting capacity. The goal of the current study was to examine the role of maximal lifting power in predicting maximum acceptable weight of lift (MAWL) relative to previously used isometric, isokinetic, and isoinertial tests. Twenty five male subjects participated in an experiment involving two isometric tests, peak isokinetic strength at velocities between 0.1 and 0.8 $m \cdot sec.^{-1}$, and isoinertial lifting capacity and peak power measured on an incremental lifting machine. Peak isoinertial power was the measure most strongly correlated with MAWL, followed by isokinetic strength measured at $0.1 \ m \cdot sec.^{-1}$. Overall, the results support previous studies which have shown that dynamic strength measures are superior to static strength measures for the purposes of predicting maximum acceptable weights. Results of regression analyses and measures of prediction accuracy reported further support the use of dynamic measures.

INTRODUCTION

Despite the tremendous amount of research that has been conducted in the area of preventing injuries associated with manual materials handling (MMH), these injuries continue to represent a significant burden to employees, employers, and society. While job design is the most powerful and effective control tool available, it has not alleviated the problem of MMH-related injuries. Given the ineffectiveness of preplacement radiography (e.g., Gibson, 1987) and the confusion surrounding the benefits of training and education (Kroemer, 1992), preplacement strength tests appear to be the most attractive supplement to job design techniques. Additionally, strength tests can be used to prioritize engineering controls, i.e., if a mismatch is found between job demands and worker capacity then the workplace can be altered to reduce job demands to a level consistent with the capacity of the workforce. This latter technique is more desirable than job selection, but is not always an option.

A conclusion of a panel of experts convened by the National Institute for Occupational Safety and Health (NIOSH) was that there is a need for research to "concentrate on improving those worker evaluation methods that are directly related to specific job requirements, as opposed to generic physical performance tests of strength, flexibility and endurance" (NIOSH, 1992). Thus, in order to develop a test for predicting safe lifting capacity, the test should be as close to the lifting task as possible with respect to postural constraints, kinematic constraints, kinetic properties and task geometry. Past research, as well as the results reported here, supports such congruency between the criterion task and predictor test.

Previous studies have indicated that dynamic strength is superior to static strength for the purposes of predicting maximum acceptable weight of lift (MAWL) (e.g., Aghazadeh and Ayoub, 1985; Ayoub et al., 1982, 1987; Duggan and Legg, 1993; Mital, 1985; Mital and Karwowski, 1985). There has been considerable comparison of the predictive abilities of dynamic tests to those of static tests, but there have been few comparisons amongst dynamic measures. Given that fact that lifting is an isoinertial activity, it follows that the closer a predictor test resembles the kinematic and kinetic nature of the lifting task, the more accurate the predictions will be.

Having a preplacement test that closely resembles the task the test was designed for is also important from the perspective of The Americans with Disabilities Act [Public Law 101-336]. Preplacement tests must be job-related, which means they must be a legitimate measure for a specific job. The test must also possess either criterion-related, content, or construct validity properties [See 29 Code of Federal Regulations Part 1607 - Uniform Guidelines on Employee Selection Procedures].

During the past several decades, power capabilities have been studied with considerable rigor by exercise scientists seeking to determine the sources of performance variation for a variety of athletic events (e.g., Coyle et al., 1979; Dowling and Vamos, 1993; Gregor et al., 1979; Margaria et al., 1966; McCartney et al., 1983; Santa Maria et al., 1985; Sewall and Lander, 1992). The rationale for the interest has primarily been driven by the fact that many athletic contests require short, but very powerful, bursts of work to be performed. Although industrial lifting is not a competitive event, there is a great deal of inter-individual variability in MMH-capabilities and many benefits can be achieved by gaining a better understanding of which attributes are responsible for this variation. Power has a great deal of intuitive appeal as a source of furthering this information. Power tests can be developed that are free from the limiting "iso" constraints that have been used previously. MMH activities, particularly lifting, require force exertions under conditions which are by no means isokinetic or isometric. Thus, the goals of this study were to investigate the role of maximal power in predicting lifting capacity and to compare the predictive ability of power to previously used isometric, isokinetic, and isoinertial measures.

METHODS

Subjects

Twenty five healthy male subjects volunteered to participate in the experiment, and were paid for their participation. Table 1 summarizes the characteristics of the subjects. Most subjects were university students, as indicated

Table 1. Anthropometric characteristics of subject sample.

Variable	Mean	SD	Minimum	Maximum
Age	21.64	2.46	18	26
Height (cm.)	179.63	5.81	170.2	195.6
Weight (kg.)	78.45	13.52	61.2	112.5

SD = standard deviation

by the fairly narrow range of ages.

Variable Selection

While the primary goal of the study was to investigate the role of power in predicting lifting capacity, the secondary goal was to compare power to previously used measures. For this reason, a fairly wide array of predictor tests was used.

Two isometric measures were chosen: maximum lifting strength at 15 cm. and 75 cm. Fifteen cm. represents strength at the origin of the lift for the psychophysical task involved, and this strength measure has been used to define capacity in previous research (e.g., Chaffin and Park, 1973). The reason for selecting 75 cm. is that the NIOSH equations (NIOSH, 1981; Waters et al., 1993) use a height of 75 cm. as the vertical dimension for the "standard lifting location."

The velocities chosen for the isokinetic strength tests (0.1, 0.2, 0.4, 0.6, and 0.8 m. • sec.$^{-1}$) represent a fairly wide range of velocities. A velocity of 1.0 m. • sec.$^{-1}$ was going to be used, but it was found that this velocity was too fast for the limited vertical range (0 - 95 cm.) of the isokinetic apparatus. The X-factor incremental lifting test was used due to its popularity in previous research (e.g., Ayoub et al., 1982, 1987; Kroemer 1983, 1985). Finally, a test of maximal power was devised. The X-factor machine was also used for this test because of its safety and ease of attaching transducers. The authors felt that the only way to measure maximum power was by having subjects lift a load as quickly as possible. The 25 kg. load used for the power test was the lightest load that could be used. Heavier loads were not selected for safety reasons.

The dependent measure chosen was MAWL at a frequency of one lift per eight hours. The low frequency was chosen because the predictor measures investigated are appropriate for low frequency tasks. Low frequency tasks are more taxing to the musculoskeletal system than are high frequency tasks, which are taxing to the cardiovascular

system. Thus, all measures were associated with strength and not cardiovascular capacity.

Apparatus

Isometric strength and isokinetic strength were measured using a Cybex® II isokinetic dynamometer, cables, and handles guided by linear bearings. The midpoints of the handles were 35.5 cm. apart, which is equal to the separation of the handles used for the psychophysical procedure.

An aluminum wheel was attached to the isokinetic dynamometer. By attaching a cable to the wheel, guiding the cable with pulleys, and attaching the cable to the handles, the isokinetic dynamometer was converted from rotary to linear (vertical) motion. A load cell was attached between the cable and the handles to measure forces.

A velocity transducer was affixed to the apparatus. The transducer was used so that a digital readout could be used to set the various velocities, rather than using the less accurate analog readout on the control box of the Cybex® II. The transducer was affixed to the apparatus and the signal was fed to the analog-to-digital (A/D) board, as were the signals from the load cell.

This apparatus was also used for collecting the isometric strength data. This was accomplished by placing the handles at the proper vertical height (15 or 75 cm.) and setting the velocity of the dynamometer to zero.

The isoinertial apparatus consisted of the X-factor incremental lifting machine fitted with a load cell to measure vertical forces applied to the handles and a velocity transducer. This apparatus was used for the incremental lifting test and the isoinertial power measurements. For the power measurements, velocity and force were sampled from the respective transducers at 100 Hz. The LabView® software program, which was used to control all data acquisition, converted all digital outputs to appropriate units (N [force], m. • sec.$^{-1}$ [velocity], or m. [displacement]). Linear regression parameter estimates from calibration procedures were used in the conversions. LabView® also was used to calculate power values (dot product of force and velocity). Since the X-factor machine only permits vertical motion, the power measurements represent vertical power.

The apparatus used for the psychophysical assessments consisted of a 30.5 cm. × 30.5 cm. × 30.5 cm. wooden box with handles, lead weights of various shapes and sizes, and a stationary 76 cm. high shelf.

Procedure

Subjects were given a basic questionnaire about various health issues and filled out a consent form. All subjects were examined by a physician prior to participation. The purpose of the examination was to ensure that subjects were free from any musculoskeletal disorders that would predispose them to injury.

The experimentation was divided into 4 sessions with at least 24 hours between sessions. Session 1 involved determination of MAWL for a frequency of one lift per eight hours. The starting load of the box was alternately light (2-18 kg.) or heavy (32-45 kg.) (Ciriello et al., 1990). The subjects were read instructions specific to a frequency of 1 lift per 8

Ergonomics and Musculoskeletal Disorders

hours (Ciriello and Snook, 1983), and performed two repetitions of the psychophysical procedure. If the two values were not within ±15% of each other, then the procedure was repeated on another day as was done by Snook and Ciriello (1991).

The remainder of the first session involved practice with the X-factor isoinertial lifting machine. Subjects lifted the 25 kg. carriage to a height of at least 183 cm. once per minute for 15 minutes. Subjects then practiced lifting the carriage as quickly as possible once per two minutes for 20 minutes. After a 10-minute rest period, subjects practiced the incremental lifting test. Subjects lifted the 25 kg. carriage to a height of 183 cm., then a 4.5 kg. weight was added and the subject lifted the weight to 183 cm. The experimenter continued to add 4.5 kg. increments until subjects could not lift the load to 183 cm.

During the second session, the isoinertial power measurements (ISOPOW) were taken. Subjects were instructed to lift the 25 kg. carriage as quickly as they safely could, and the score on the test was the peak power value during the exertion. Subjects performed three repetitions of this test, separated by five-minute rest periods. In some cases, a fourth trial was run. Some subjects appeared to need a "warm-up" trial, indicated by a relatively low value on the first trial. For these subjects, the fourth trial replaced the first. If one of the three trials appeared to be significantly lower than the other trials, another trial was run. Following a 10 minute rest period, subjects performed the X-factor incremental lifting test (XFAC). The score on the test was the largest load the subject could lift to at least 183 cm.

The third session involved practicing the isometric and isokinetic strength measurements. Subjects practiced the isometric measurements at 15 cm. (IMET15) and 75 cm. (IMET75) for three trials each separated by three-minute rest periods. Subjects also practiced the isokinetic measurements at speeds of 0.1 m. • sec.$^{-1}$ (IKIN0.1), 0.2 m. • sec.$^{-1}$ (IKIN0.2), 0.4 m. • sec.$^{-1}$ (IKIN0.4), 0.6 m. • sec.$^{-1}$ (IKIN0.6) and 0.8 m. • sec.$^{-1}$ (IKIN0.8) for three trials each separated by three-minute rest periods.

The fourth session involved collecting the isometric and isokinetic strength data. The isometric data were collected first. The order of collecting IMET15 and IMET75 was alternated. The experimenter had subjects build up to their maximum voluntary contraction (MVC) over a period of two to three seconds. Once subjects achieved their MVC, based on the analog torque meter on the Cybex®, the experimenter initiated the collection of 75 samples over three seconds. The strength datum used was the mean value of the 75 samples, as suggested by Caldwell et al. (1974). If any sample was outside ±10% of the mean value, the test was rerun (Caldwell et al., 1974). Two consecutive exertions within ±15% of each other were considered acceptable, and the mean of these two values was used as the datum for a given subject. Subjects were given five minutes rest between all exertions.

Following the isometric testing, subjects performed the isokinetic testing. The order of presentation of the five velocities was random. Due to a lack of instructions and protocols for isokinetic testing in the literature, subjects completed three repetitions at each velocity. Subjects were instructed to exert maximum force on the handles until the handles reached a height of 95 cm., at which point the apparatus restricted the handles from moving any further. The

peak force value during the trial was used as the datum for that trial. If one datum appeared to be considerably lower than the other two, the trial was repeated. There was a five-minute rest between all exertions.

Statistical Analyses

For the correlation and regression analyses, a single value for each predictor variable and the dependent measure was needed. Two or three trials were run for each type of test, so the multiple samples for each subject were averaged and used as the datum.

Pearson product-moment correlations were calculated between MAWL and each of the predictor variables. In order to assess the predictive ability of each independent measure, simple linear regression models were estimated for each predictor variable. The root mean square error (RMSE) and mean absolute residual (MAR) statistics were calculated for each simple regression model. These statistics are indicative of the accuracy associated with predicting MAWL with the various predictor variables.

RESULTS

The means and standard deviations of the various strength measures are presented in Table 2. The results show a clear effect of velocity on force capabilities, but also postural effects. The mean IMET15 value is lower than IKIN0.1 and IKIN0.2, whereas IMET75 is higher than all isokinetic measures. The 75 cm. height provides a distinct strength advantage over 15 cm., as expected from the literature.

Table 2. Summary statistics for dependent and predictor measures.

Variable	Mean	SD
IMET15 (N.)	802.87	190.45
IMET75 (N.)	990.29	198.42
IKIN0.1 (N.)	885.92	194.41
IKIN0.2 (N.)	843.66	201.30
IKIN0.4 (N.)	776.06	183.61
IKIN0.6 (N.)	723.24	183.68
IKIN0.8 (N.)	686.93	199.49
XFAC (kg.)	55.79	12.14
ISOPOW (W.)	1210.89	256.18
MAWL (kg.)	56.06	13.71

The correlations between MAWL and each of the predictor values are presented in Table 3. All of the correlations were significant. Table 4 provides performance measures for each of the models estimated, where MAWL was the dependent measure.

DISCUSSION

The goals of this study were to investigate if power is related to the ability to lift from the floor to knuckle height and to compare the predictive ability of power to previously investigated strength measures. The results indicate that power

Table 3. Correlations between MAWL and predictor variables. Significance levels are below the correlation coefficients.

Variable	MAWL
IMET15	0.62622
	0.0008
IMET75	0.61574
	0.0011
IKIN0.1	0.79042
	0.0001
IKIN0.2	0.62677
	0.0008
IKIN0.4	0.65531
	0.0004
IKIN0.6	0.67281
	0.0002
IKIN0.8	0.73549
	0.0001
XFAC	0.49835
	0.0112
ISOPOW	0.83849
	0.0001

Table 4. Performance measures for simple OLS regression models. MAWL is the response variable for each model.

Predictor Variable	RMSE (kg.)	MAR (kg.)
IMET15	10.92	8.24
IMET75	11.03	8.38
IKIN0.1	8.58	6.61
IKIN0.2	10.91	8.04
IKIN0.4	10.58	7.97
IKIN0.6	10.36	7.25
IKIN0.8	9.49	7.28
XFAC	12.14	8.58
ISOPOW	7.63	5.46

is significantly related to lifting capacity and that power, as investigated here, results in more accurate predictions of MAWL than the isometric, isokinetic, and isoinertial predictor variables investigated.

The performance measures in Table 4 provide an indication of the accuracy of low-frequency lifting capacity prediction using single strength tests. With the exception of XFAC, the trend is clearly that dynamic measures of strength outperform static measures. The reason for the poor performance of the XFAC test may be because the precision of the test is limited to 4.5 kg. A second reason may be that the vertical range of the test is 0 - 183 cm. versus 0 - 76 cm. for the MAWL determination. The isokinetic tests had a range of 0 - 95 cm and the isometric tests were within the 0 - 76 cm. range. The disparity in vertical range between XFAC and MAWL, coupled with the fact that failure on the XFAC test usually occurs at waist height or higher, indicate that the 183 cm. XFAC test may be more suited for a floor-to-shoulder prediction.

IKIN0.1 performed almost as well as ISOPOW in terms of the correlations, but predictions of MAWL using ISOPOW were approximately one kg. more accurate in terms of the

performance measures in Table 4. Many researchers have only reported r^2 values for similar studies. While this measure is important, prediction accuracy is also important. Prediction accuracy and the proportion of variation a model accounts for (r^2) are correlated, but the relationship is not perfect. Thus, both types of measures should be used to assess models of this type.

Earlier, issues of validity were mentioned. The results reported here would be useful for predicting an individual's capacity for low-frequency floor-to-knuckle lifting. Since the predictor tests are significantly correlated with the lifting task, these tests have criterion-related validity.

The research reported here further supports literature indicating that dynamic tests are better predictors of lifting capacity than are static measures. Combined, the results of the present study and previous research clearly indicate that isometric testing for predicting lifting capacity should be replaced with dynamic testing when practical and feasible. The XFAC test was the exception to the trend, and the reason for the poor correlation between XFAC and MAWL were discussed above. Also, the low correlation may simply be a sampling artifact, i.e. another sample of 25 subjects might produce a higher, or even lower, correlation.

One limitation of the results is that only one type of lifting task was investigated. Thus, the validity of the results is limited to these conditions. The results should not be extrapolated to lifting tasks with different ranges or frequencies. Future research will be needed to investigate the relationships for a wider array of task conditions. Additionally, the subjects were college-aged males and future research will be needed to investigate the relationships studied here among a more heterogeneous subject pool. Preferably, the study should be replicated using an industrial subject pool composed of males and females from a wider age range.

Many researchers have developed tests to predict the handling capacity of workers. Unfortunately, few of the tests have been validated epidemiologically. Examples of field validations of such techniques include the investigations by Chaffin and Park (1973) and Liles et al. (1984). Unfortunately, these techniques are based upon capacity definitions that use isometric strength. Certainly there is a need for more accurate techniques based upon dynamic strength measures.

Determining the value of preplacement strength testing with respect to reducing MMH-related injuries can be very beneficial. Given the ineffectiveness of education and training, and the enormous costs that continue to be associated with MMH, it is apparent that job design techniques are not sufficient.

Based on studies such as Liles et al. (1984), it is fair to assume that preplacement strength testing techniques can have a net positive economic benefit. However, determining the benefits associated with these techniques will require several steps. First, further research will be needed to define appropriate predictor variables and equations for numerous handling activities. These equations should utilize dynamic predictors. Second, the predictions will have to be tested with epidemiological studies. Typically, such studies have been performed by looking at the ratio of task demands to operator capacity. Statistically modeling the relationships between such ratios and the incidence and severity of associated injuries can be very beneficial for loss prevention purposes. These models would permit the prediction of the expected cost and incidence rate for a particular task demands to operator capacity ratio.

Although the strength tests described here have been primarily discussed within the context of preplacement screening, it should be noted that another effective use of such techniques can be prioritizing design changes and allocating resources, such as ergonomics personnel, etc. This aspect of strength testing is often overlooked. The advantage of such an approach is that no selection of personnel is required, which completely alleviates the potential for litigation associated with some personnel selection procedures. Additionally, the engineering changes will be permanent and will benefit all workers performing the job.

ACKNOWLEDGMENTS

This project was supported by grant R03 OH03335 from the National Institute for Occupational Safety and Health of the Centers for Disease Control and Prevention and was completed while the first author was at Texas Tech University.

REFERENCES

Aghazadeh, F., and Ayoub, M.M. (1985). A comparison of dynamic- and static-strength models for prediction of lifting capacity. *Ergonomics*, 28(10), 1409-1417.

Ayoub, M.M., Denardo, J.D., Smith, J.L., Bethea, N.J., Lambert, B.K., Alley, L.R., and Duran, B.S. (1982). *Establishing Physical Criteria for Assigning Personnel to Air Force Jobs*. Final Report, Contract No. F49620-79-006, Air Force Office of Scientific Research.

Ayoub, M.M., Jiang, B.C., Smith, J.L., Selan, J.L., and McDaniel, J.W. (1987). Establishing a Physical Criteria for Assigning Personnel to U.S. Air Force Jobs. *Am Ind Hyg Assoc J*, 48(5), 464-470.

Caldwell, L.S., Chaffin, D.B., Dukes-Dobos, F.N., Kroemer, K.H.E., Laubach, L.L., Snook, S.H., and Wasserman, D.E. (1974). A Proposed Standard Procedure for Static Muscle Strength Testing. *American Industrial Hygiene Association Journal*, 34(4), 201-206.

Chaffin, D.B., and Park, K.S. (1973). A longitudinal study of low-back pain as associated with occupational weight lifting factors. *American Industrial Hygiene Association Journal*, 34(12), 513-525.

Ciriello, V.M., and Snook, S.H. (1983). A Study of Size, Distance, Height, and Frequency Effects on Manual Handling Tasks. *Human Factors*, 25(5), 473-483.

Ciriello, V.M., Snook, S.H., Blick, A.C., and Wilkinson, P.L. (1990). The effects of task duration on psychophysically-determined maximum acceptable weights and forces. *Ergonomics*, 33(2), 187-200.

Coyle, E.F., Costill, D.L., and Lesmes, G.R. (1979). Leg extension power and muscle fiber composition. *Medicine and Science in Sports and Exercise*, 11, 12-15.

Dowling, J.J., and Vamos, L. (1993). Identification of Kinetic and Temporal Factors Related to Vertical Jump Performance. *Journal of Applied Biomechanics*, 9, 95-110.

Duggan, A., and Legg, S.J. (1993). Prediction of maximal isoinertial lift capacity in army recruits. In R. Nielsen and K. Jorgensen (eds.) *Advances in Industrial Ergonomics and Safety V*. London: Taylor and Francis.

Gibson, E.S. (1987). The Value of Preplacement Screening Radiography of the Low Back. *SPINE: State of the Art Reviews*, 2(1), 91-107.

Gregor, R.J., Edgerton, V.R., Perrine, J.J. and DeBus, C. (1979). Torque-velocity relationships and muscle fiber composition in elite female athletes. *Journal of Applied Physiology: Respiratory, Environmental and Exercise Physiology*, 47(2), 388-392.

Kroemer, K.H.E. (1983). An Isoinertial Technique to Assess Individual Lifting Capability. *Human Factors*, 25(5), 493-506.

Kroemer, K.H.E. (1985). Testing Individual Capability to Lift Material: Repeatability of a Dynamic Test Compared with Static Testing. *Journal of Safety Research*, 16, 1-7.

Kroemer, K.H.E. (1992). Personnel training for safer material handling. *Ergonomics*, 35(9), 1119-1134.

Liles, D.H., Deivanayagam, S., Ayoub, M.M., and Mahajan, P. (1984). A Job Severity Index for the Evaluation and Control of Lifting Injury. *Human Factors*, 26(6), 683-693.

Margaria, R., Aghemo, P., and Rovelli, E. (1966). Measurement of muscular (anaerobic) power in man. *Journal of Applied Physiology*, 21, 1661-1669.

McCartney, N., Heigenhauser, J.F., and Jones, N.L. (1983). Power output and fatigue of human muscle in maximal cycling exercise. *Journal of Applied Physiology: Respiratory, Environmental and Exercise Physiology*, 55(1), 218-224.

Mital, A. (1985). Use of anthropometry and dynamic strength in developing screening and placement procedures for workers. In H.J. Bullinger and H.J. Warnecke (eds.) *Toward the Factory of the Future*. Berlin: Springer-Verlag.

Mital, A., and Karwowski, W. (1985). Use of Simulated Job Dynamic Strengths (SJDS) in Screening Workers for Manual Lifting Tasks. In *Proceedings of the Human Factors Society 29th Annual Meeting*. Santa Monica, CA: Human Factors Society.

NIOSH (1981). *Work Practices Guide for Manual Lifting*. HEW(NIOSH) Report No. 81-122. Cincinnati, OH: Author.

NIOSH (1992). *A National Strategy for Occupational Musculoskeletal Injuries: Implementation Issues and Research Needs* DHHS(NIOSH) Pub. No. 93-101. Cincinnati, OH: Author.

Santa Maria, D.L., Grzybinski, P., Hatfield, B. (1985). Power as a function of load for a supine bench press exercise. *National Strength and Conditioning Association Journal*. 6(6), 58.

Sewall, L.P., and Lander, J.E. (1992). Biomechanical components of the vertical jump and an analogous task involving the upper body. *Journal of Human Movement Studies*, 23, 77-93.

Snook, S.H., and Ciriello, V.M. (1991). The design of manual handling tasks: revised tables of maximum acceptable weights and forces. *Ergonomics*, 34(9), 1197-1213.

Waters, T.R., Putz-Anderson, V., Garg, A., and Fine, L.J. (1993). Revised NIOSH equation for the design and evaluation of manual lifting tasks. *Ergonomics*, 36(7), 749-776.

THREE-DIMENSIONAL SPINAL LOADING DURING COMPLEX LIFTING TASKS

Fadi A. Fathallah[*], William S. Marras, and Mohamad Parnianpour

Biodynamics Laboratory
The Ohio State Univerity
Columbus, Ohio

[*]Currently at the Liberty Mutual Research Center for Safety & Health
Hopkinton, Massachusetts

Knowledge of the complex *three-dimensional* loads imposed on the spine during typical manual materials handling (MMH) tasks could provide more insights about the mechanical etiology of low back injuries in occupational settings. Comprehensive treatment of such information has been lacking. Most previous studies quantified spinal loading in terms of compressive forces alone. However, there is enough empirical and epidemiological evidence to indicate that the shear forces imposed on the spine may be more important than mere compression. Hence, the purpose of this study was to assess, *in-vivo*, the three-dimensional complex spinal loading associated with lifting tasks. Subjects performed simulated lifting tasks with varying workplace characteristic. An EMG-assisted model provided the continuous three-dimensional spinal loads. Asymmetric (complex) lifting tasks showed distinctive loading patterns from those observed under symmetric conditions. Simultaneous occurrences of spinal loads in all three directions (compression and shear forces) were patterns unique under the "risky" asymmetric lifting conditions. These situations could be identified and abated through proper workplace design. In conclusion, this approach allow the determination of the magnitudes and temporal occurrence(s) of complex spinal loading, and assess the sensitivity of these loading patterns to workplace characteristics.

INTRODUCTION

Pope et al. (1991) best described the state of low back pain and its effect on industry as: "Low back pain (LBP) has been called the nemesis of medicine and the albatross of industry." Back pain has been established to be one of the most common and significant musculoskeletal problem in the US, leading to substantial levels of disability, morbidity, and economic loss (Webster and Snook, 1994; Praemer et al., 1992; Andersson, 1991; Hollbrook et al., 1984). Finding means to abate or at least reduce the risk of low back disorders in occupational settings would benefit workers and employers alike.

There is significant epidemiological and biomechanical evidence that implicates complex motions and loading as important risk factors for LBD. Several epidemiological studies have shown that risk increases under these *combined* risk factors conditions. Magora (1973) has found that twisting and lateral bending were significant risk factors only when occurring simultaneously with sudden movements (dynamic activities). Kelsey et al. (1984) have also indicated that occupational LBD risk increases in jobs involving lifting activities when the lift is combined with twisting action. Other epidemiological studies have also indicated combined (asymmetrical) motions as potential risk factors for developing low back disorders (e.g., Marras et al., 1993; Drury et al., 1982; Damkot et al., 1980; Brown, 1975). The effect of

complex motions on the spinal structure *in-vivo* is not well understood. These complex spinal motions generate complex loading patterns on the spinal elements (e.g., combined lateral shear and compressive loading). Pope et al. (1991) have expressed the importance of quantifying such complex combined loads; however, the authors indicated that such a task is difficult to investigate under both field and laboratory situations. Biomechanical changes in spine tolerance are also expected to occur when these risk factors occur simultaneously. Shirazi-Adl (1994, 1991, 1989) has demonstrated how strains in spinal disc annulus fibers are dramatically increased under combined lateral and twisting conditions, reaching levels that may exceed the tissue tolerance limits. These latter studies as well as other *in-vitro* studies have implicated combined loading of the spinal structure as a mechanism for back injury. Therefore, for prevention of LBD, it is essential to quantify the types and magnitudes of the three-dimensional mechanical loading experienced by the spinal structure when subjected to dynamic combined or complex motions. Knowledge of this type of information allows us to identify, in detail, the situations that compromise the integrity of the structure, and thereby help reduce injuries resulting from these conditions. Hence, the main objective of this study was to investigate the magnitudes and patterns of the three-dimensional spinal loading under complex lifting conditions and assess their potential risk.

METHODS

Subjects

Eleven healthy male subjects volunteered to participate in this experiment. Mean (standard deviation) age was 28.4 (4.4), mean stature and weight were 180.7 cm (3.7) and 78.6 kg (10.8), respectively. A questionnaire was administered to each subject to ensure that there was no recent history of back disorders.

Apparatus

An EMG system collected signals from ten pairs of bipolar silver-silver/chloride surface electrodes affixed over the specific locations of the ten muscles of interest. Five pairs of muscles were studied: 1) Right/left latissimus dorsi, 2) right/left erector spinae, 3) right/left rectus abdominus, 4) right/left external obliques, and 5) right/left internal obliques.

The EMG signals were first amplified 1000x by preamplifiers placed at short distances from the muscle sites (less than 25 cm). The signals were further amplified in the main amplifier between 30x to 55x, depending on the muscle and the subject under consideration. Also, to eliminate undesired signal artifacts, the signals were low pass filtered at 1000 Hz. The filtered signals were rectified and processed via a 20 ms moving average window (integration constant). The window was moved at 2 ms increments. An asymmetric reference frame (Marras and Mirka, 1992) was used to solicit static maximum voluntary contractions (MVCs) in six directions (flexion, extension, left and right twist, left and right lateral bending).

Three-dimensional continuous position, velocity, and acceleration of the trunk were determined using the Lumbar Motion Monitor (LMM) developed at the Ohio State University Biodynamics Laboratory. Three-dimensional external forces and moments about L5/S1 were monitored by the combination of a Bertec 4060A force plate (Bertec, Worthington, Ohio) and two eloctrogoniometers used to determine the continuous location and orientation of the L5/S1 joint in three-dimensional (3-D) space (Fathallah, 1995). This system provided a mechanism to monitor torque or moment about L5/S1 during a lift. The EMG-assisted model provided estimates of the internal moments required to achieve the balanced equilibrium condition(s) (e.g., Granata and Marras, 1995).

During the experiment the subject lifted a wooden box that was filled with the proper weight of the prescribed condition. For each condition, the subject was provided with an auditory signal (loud tone) indicating the lift pace by identifying the start and end of each lift phase. This tone was necessary to control the lift speed (duration). All the analog signals gathered from the devices described above were collected at 100 Hz via a 12-bit 32-channel Analog-to-Digital (A/D) converter connected to a 386-based microcomputer. Figure 1 shows a subject performing a symmetric lift.

Figure 1. *A subject shown performing a typical symmetric lift.*

Design

The experiment consisted of a three-way within-subject design. The independent variables included: 1) Lift type, 2) speed of lift (duration of movement), and 3) weight handled. These variables were chosen to solicit complex motion conditions under varying workplace parameters similar to those observed in industrial settings (Marras et al., 1993). The lift types investigated were symmetric lifts and asymmetric (complex) lifts. Lift speed was set at three levels: low (2 sec per lift), medium (1.5 sec per lift) and high (1 s per lift). Three weight levels were considered: low (22 N), medium (67 N), and high (156 N).

The dependent variables consisted of either "average maximum" (peak) or "continuous" three-dimensional spinal loading magnitudes at the L5/S1 level in terms of the compressive, anterior/posterior shear, and medial/lateral shear forces.

Procedure

The subject was provided with written instructions detailing various parts of the experiment. The subject first performed the six *static* MVCs in a randomized order. There was a two-minute rest period given between two exertions. After all six MVCs were collected, the subject was ready to perform the dynamic lifting tasks. The subject was first fitted with the LMM along with the proper attachments of other equipment. The subject was given time to familiarize himself with the tasks, and to ask any pertinent questions. Each task consisted of two lifts and one lowering with tones indicating the start and end of each part.

In the symmetric condition, the box (weight) was placed on a platform in front of the subject just above knee height, and at a horizontal distance equal to his arm length. At the onset of the tone, the subject was asked to lift the box from its location to a position as close as possible to the body. For the complex (asymmetric) conditions, the box was placed in front of the subject in the same manner as the symmetric condition; however, in this case the subject was asked to set the box down at another platform placed to his right at an angle perpendicular to the mid-sagittal plane. The platform height was set at the level of the subject's iliac crest and was placed at about arm length distance horizontally. Within a given type of lift (symmetric or asymmetric), the three weights and speeds were presented in random order to control for carryover effects.

Data Analysis

In order to assess the three-dimensional loading on the spine, the current study utilized an extension of the EMG-assisted modeling approach developed in Biodynamics Laboratory at the Ohio State University (e.g., Marras and Reilly, 1988; Granata and Marras, 1995). In order to investigate the effects of weight, speed and lift type on the internal loading of the spine, multivariate analysis of variance (MANOVA) was conducted. The internal spinal loading was characterized by average maximum compression, anterior shear, and lateral shear. The *simultaneous* occurrence of combined loading was quantified by bivariate distributions of lateral and anterior shear forces. Note that, in this paper, only the lifting portions of the tasks were considered in all analyses. The bivariate distributions were statistically compared among conditions using the two-dimensional Kolmogorov-Smirnov test. (2-D K-S) (Fasano and Farnceschini, 1987).

RESULTS

The results of the MANOVA for the *magnitudes* of maximum three-dimensional loading revealed a significant interaction effect between weight and lift type ($p < 0.0001$) (Table 1). This interaction was significant for both compression and lateral shear ($p < 0.0001$), but not for anterior shear. Figure 2 depicts such interactive pattern. With the exception of lateral shear under symmetric lifts, maximum spinal loading significantly increased with increases of the box weight ($p < 0.001$). For the low weight condition, only lateral shear showed a significant difference between the two lifting types. With increase in the lift weight, the anterior shear was significantly different between the lifting types; and compression was significantly different only for the medium weight condition (Newman-Keuls; $p < 0.05$).

Figures 3a and 3b depict the continuous three-dimensional loading profiles of a typical "high weight" symmetric and asymmetric lifts, respectively. Both types of lifts had increased compressive and anterior shear at the beginning of the lift with low levels of lateral shear. This corresponded to the phase where the subject lifted the load off the first platform. However, as the subject progressed in executing the task, loading magnitudes in the symmetric condition continued to systematically decrease until reaching considerably low loading levels (see Figure 3a). On the other hand, the asymmetric lift exhibited rather complex loading patterns

TABLE 1

Multivariate analysis of variance (MANOVA) and analysis of variance (ANOVA) for maximum spinal loading : Compression (COMP), lateral shear (LATSHR), and anterior shear (ANTSHR). Type I error probabilities are shown only for significant effects.

Factor	MANOVA	ANOVA COMP	ANOVA LATSHR	ANOVA ANTSHR
L	0.00001	0.00001	0.00001	0.00002
W	0.00001	0.00031	0.00001	----
S	----	----	----	----
L X W	0.00001	0.00001	0.00001	----
L X S	----	----	----	----
W X S	----	----	----	----
L X W X S	----	----	----	----

L = Lift Type; W = Weight; S = Speed

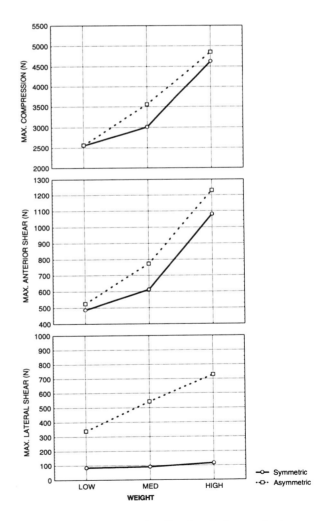

Figure 2. *Average maximum loading in all three spinal loading directions under each weight*

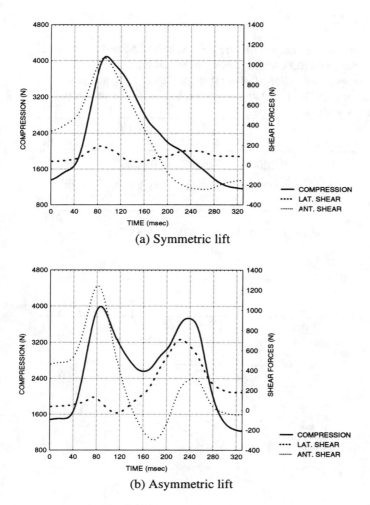

(a) Symmetric lift

(b) Asymmetric lift

Figure 3. *Example of three-dimensional continuous profile for a "high weight" symmetric lift (a), and asymmetric lift (b).*

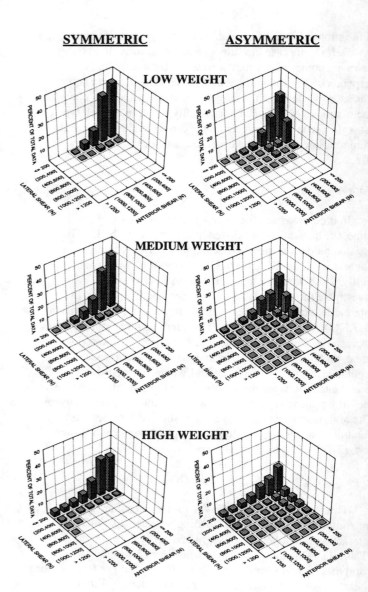

Figure 4. *Bivariate lateral and anterior shear distributions for each weight and lift type (symmetric/asymmetric). Note that under each weight condition, the distributions were statistically different between the two lifting types (2-D K-S, p < 0.001)*

especially in the last phase of lift (most asymmetric region). In that phase, there was a notable period of **simultaneous** occurrence of compressive and shear forces (see Figure 3b). To fully explore the combined (complex) loading conditions, the simultaneous continuous combined anterior and lateral shear forces were investigated. Figure 4 shows the bivariate distributions of anterior and lateral shear under each of the three weight levels for all subjects. This figure provided a means to investigate the nature of complex loading on the spine. As shown earlier, the magnitude of spinal loading significantly increased with increase in the weight; however, for each weight level, the patterns of combined loading differ between the symmetric and asymmetric lifts. Under symmetric lifts, lateral shear forces were maintained at rather low levels. For all three weights, over 90% of the time the lateral shear force was maintained under 200 N during symmetric lifts. In contrast, during asymmetric lifts, lateral shear was observed at increased levels for a considerable proportion of the time. For instance, for both the medium and high weight asymmetric conditions, about 40% of the total data was observed at lateral

shear levels exceeding 200 N with some instances reaching higher than 1200 N **combined** with substantial magnitudes of anterior shear loading.

DISCUSSION AND CONCLUSIONS

This study provided us with estimates of the combined three-dimensional loads experienced at the L5/S1 spinal unit during various lifting conditions. Although the compressive forces reported in this study were within the range reported by several studies that shared similar experimental conditions (e.g., Adams and Dolan, 1995; Cholewicki et al., 1995), very few dynamic lifting studies have quantified the <u>three-</u>

dimensional loading experienced on the spinal structures. Most studies report only the compressive loads without quantifying the shear forces. Also, the study is unique due to its capability to account not only for the effects of dynamic motion variables and inertial forces on the net external moment, but also the effects on the internal loading and muscular recruitment parameters.

The complex asymmetric lifting conditions in general and those with increased weight in particular exhibited elevated levels of complex spinal loading patterns. More specifically, elevated magnitudes of compressive loading **combined** with lateral and anterior shear forces seemed to be a key element in distinguishing such "risky" situations. This approach allows the identification of loading patterns that expose the spinal structures to loading levels which may exceed their tolerance limits and hence, could lead to injury. These situations could be identified and abated through proper workplace design. In sum, the level of detail provided by the present approach allows us to determine the **magnitudes and temporal occurrence(s)** of complex or combined loading, and assess the sensitivity of these loading patterns to workplace characteristics.

ACNOWLEDGEMENTS

Partial funding for this study has been provided by the Ohio Bureau of Workers' Compensation, Division of Safety and Hygiene. The authors would like to thank Dr. Kevin Granata for his invaluable assistance in the experimental phase of this study.

REFERENCES

Adams, M.A., and Dolan, P. (1995). Recent advances in lumbar spinal mechanics and their clinical significance. *Clinical Biomechanics,* 10,3-19.

Andersson, G.B. (1991). The epidemiology of spinal disorders. In J.W. Frymoyer (Ed.), *The adult Spine* (pp. 107-146). New York: Raven Press.

Brown, J.R. (1975). Factors contributing to the development of low back pain in industrial workers. *Amer Ind Hyg Assoc J,* 36, 26-31.

Cholewicki, J., McGill, S.M., and Norman, R.W. (1995). Comparison of muscle forces and joint load from an optimization and EMG assisted lumbar spine model: Towards development of a hybrid approach. *J. Biomechanics,* 28,321-331.

Damkot, D.K., Pope, M.H., Lord, J., and Frymoyer, J.W. (1984). The relationship between work history, work environment and low back pain in men. *Spine,* 9(4), 395-399.

Drury, C.G., Law, C.H., and Pawenski, C.S. (1982). A survey of industrial box handling. *Human Factors,* 24, 553-65.

Fasano, G., and Franceschini, A. (1987). A multidimensional version of the Kolmogorov-Smirnov test. *Mon Not R Astr Soc,* 255, 155-170.

Fathallah, F.A. (1995). *Coupled spine motions, spine loading and risk of occupationally-related low back disorders.*

Unpublished doctoral dissertation, The Ohio State University, Columbus, OH.

Granata, K.P., and Marras, W.S. (1995). An EMG-assisted model of trunk loading during free-dynamic lifting. *Journal of Biomechanics* 28, 1309-1317.

Hollbrook, T.L., Grazier, K., Kelsey, J.L., and Stauffer, R.N. (1984). The frequency of occurrence, impact and cost of selected musculoskeletal conditions in the United States. *American Academy of Orthopeadic Surgeons.* Chicago, IL, pp. 24-25.

Kelsey, J.L., Githens, P.B., White, A.A., et al. (1984). An epidemiological study of lifting and twisting on the job and risk for acute prolapsed lumbar intervertebral disc. *J. Orth. Res.,* 2, 61-66.

Magora, A. (1973). Investigation of the relation between low back pain and occupation: 4. Physical requirements: Bending, rotation, reaching and sudden maximal effort. *Scand. J. Rehabil. Med,* 5, 186-190.

Marras, W.S., and Reilly, C.H. (1988). Networks of internal trunk-loading activities under controlled trunk-motion conditions. *Spine* 13,661-667.

Marras, W.S., and Mirka, G.A. (1992). A comprehensive evaluation of trunk response to asymmetric trunk motion. *Spine,* 17,318-326.

Marras, W.S., Lavender, S., Leurgans, S., Rajulu, S.L., Allread, W.G., Fathallah, F.A., and Ferguson, S.A. (1993). The role of dynamic three-dimensional trunk motion in occupationally-related low back disorders: The effects of workplace factors, trunk position and trunk motion characteristics on risk of injury. *Spine,* 18(5), 617-628.

Pope, M.H., Andersson, G.B.J, Frymoyer, J.W., and Chaffin, D.B. (1991). *Occupational Low back pain: Assessment, treatment, and prevention.* Chicago: Mosby Year Book.

Praemer, A., Furner, S., and Rice, D.P. (1992). Musculoskeletal conditions in the United States. *American Academy of Orthopeadic Surgeons.* Park Ridge, IL, pp. 23-33.

Shirazi-Adl, A. (1989). Strain in fibers of a lumbar disc: analysis of the role of lifting in producing disc prolapse. *Spine,* 14, 96-103.

Shirazi-Adl, A. (1991). Finite-element evaluation of contact loads on facets of an L2-L3 lumbar segment in complex loads. *Spine* 16,533-513.

Shirazi-Adl, A. (1994). Biomechanics of the lumbar spine in sagittal/lateral moments. *Spine,* 19(21), 914-927.

Webster, B.S., and Snook, S.H. (1994). The cost of 1989 workers' compensation low back pain claims. *Spine* 19,1111-1116.

Three-Dimensional Functional Capacity of Normals and Low Back Pain Patients

Sue A. Ferguson*, William S. Marras* and Robert R. Crowell**

The Ohio State University *   The Ohio Spine Center **
Biodynamics Laboratory   223 East Town Street
Columbus, Ohio   Columbus, Ohio

Most current functional capacity evaluations focus on sagittal range of motion, strength testing as well as simulation of daily activities and job tasks as measures of wellness. The goal of this study was to evaluate the dynamic functional motion capacity of normals and low back pain patients in the three cardinal planes of the body. The hypothesis was that injury would not only affect sagittal motion but also lateral and twisting motions that would load the spine in a different manner. Trunk motion parameters of range of motion (ROM), velocity and acceleration were measured in all three planes of the body as subjects performed three separate experimental tasks, eliciting motion in each plane of the body. The MANOVA results showed a significant difference between the patients and normals. The final discriminant function model predicting membership for the two groups contained ROM, velocity, and acceleration parameters with two measures from each experimental task. The cross-validation error rate for the model was 4%. These results provide insight into new methods for functional capacity evaluation of low back pain patients which may influence return to work decisions.

INTRODUCTION

Previous history of back pain is one of the most frequently found risk factors for the occurrence of low back pain (Bigos et. al. 1986; Riihimaki 1994). The prominence of this risk factor indicates our measures of recovery may not be sensitive enough. According to Haddod (1987) nearly 90% of low back pain patients recover within 2 to 8 weeks. However, Nordin and Frankel (1989) discussed animal studies where deconditioned ligaments did not reach previous strength and stiffness for 12 months. Thus one may question whether or not we have a realistic understanding of the recovery process or the knowledge to evaluate when one may successfully return to work.

Functional capacity testing is often used to indicate readiness to return to work after a low back injury. The direct measurement aspects of functional capacity evaluations consist predominantly of range of motion (ROM) and strength testing (Nelson et. al. 1988; Tramposh 1988; Mayer et. al. 1988). Klein and associates (1991) found that isometric strength and mobility could not be used to identify individuals with low back pain. Mandell et al. (1993) found no difference between patients and postal workers in isometric strength, isokinetic strength or ROM. Marras and Wongsam (1986) found that sagittal velocity was more effective at distinguishing between normals and low back pain patients than sagittal ROM. Marras and colleagues (1995) have found that during the rehabilitation process sagittal plane ROM recovered first, followed by recovery of velocity and acceleration. The velocity and acceleration capabilities of patients which have been neglected in current functional capacity evaluations may provide valuable information in determining true recovery and successful return to work.

Since activities of daily living as well as workplace activities may require trunk motion in all three planes of the body, it is hypothesized that functional capacity testing should incorporate evaluation of all three planes of the body. The goals of this study were: 1. to develop and implement a test that elicits maximum functional motion capacity in the three cardinal planes of the body, 2. to quantify differences in normals and low back pain patients, and 3. to develop a model that distinguishes between the two groups incorporating all motion parameters (ROM, velocity and acceleration) in all three planes of the body.

METHODS

Approach

It has been shown that velocity and acceleration distinguish between normals and low back pain patients in the sagittal plane, thus the concept of this study was to determine whether or not similar results occur in tasks that require primarily frontal and transverse plane motion. Since trunk motion is three-dimensional, it is theorized that differences in motion parameters due to low back pain would occur in all three planes. The goal of this study was to elicit a maximum voluntary motion in each plane of the body and measure ROM, velocity and acceleration in all three planes. The experimental tasks were designed to elicit motion in a single plane. However motion rarely occurs in a single plane, there is generally some coupled movement (Twomey and Taylor 1994). Thus, the goal wsa to quantify both primay motion as well as coupled motion. White and Panjabi (1990) have defined motion in two categories; 1) main motion and 2) coupled motion as a function of the axis about which force is applied. Applying White and Panjabi's definition of motion to the current study, main motion is in the same plane as the maximum voluntary motion and coupled motion is defined as motion that occurs in the other two planes of the body. Specifically, during lateral bending coupling is defined as motion in the sagittal and transverse planes of the body and the main motion is in the frontal plane. During sagittal bending, the main motion is in the sagittal plane with coupled motion defined as frontal and transverse plane motion. Finally during a twisting task, the main motion is in the transverse plane and coupled motion is defined as motion in the frontal and sagittal planes.

Subjects

Twenty-six normals with no history of low back pain volunteered for the study. The subjects ranged in age from 20 to 60. Twenty-six low back pain patients from a secondary referral practice were gender matched and age matched within the decade to the normals for a total of 52 subjects in the study.

Experimental Design

There were two independent variables in the study. The first variable was group defined as either normal or low back pain patient. The second variable was plane defined as either sagittal, frontal or transverse.

Nine trunk motion characteristics (dependent variables) were measured for each of the three tasks. These motion characteristics include: (1) range of motion (ROM) in the frontal planes, (2) peak velocity in the frontal plane, (3) peak acceleration in the frontal plane, (4) ROM in the sagittal plane, (5) peak velocity in the sagittal plane, (6) peak acceleration in the sagittal plane, (7) ROM in the transverse plane, (8) peak velocity in the transverse plane, and (9) peak acceleration in the transverse plane.

Apparatus

The lumbar motion monitor (LMM) was used to measure trunk motion in all three planes of the body. The LMM measures instantaneous changes in the position of the thoracolumbar region of the spine. A detailed description of the design and calibration may be found in Marras et al. (1992). The signal from the LMM was sent via hard wire to the analogue-to-digital converter board, resident on a 386 microcomputer. The data collection rate was 60 Hz. The digitized signal was stored in the microcomputer for further processing.

Procedure

Upon arrival to the testing location, the subject's anthropometry was collected. The lumbar motion monitor was placed on the subject. The subject was instructed to warm up for the testing. The subjects were instructed to: (1) cross their arms in front of their chest, (2) twist from left to right as fast as you can comfortably in a functional ROM, (3) bend side to side (lateral) as fast as you can comfortably in a functional ROM and, (4) flex and extend your trunk (sagittal) as fast as you can comfortably in a functional ROM without hyperextension of the trunk. The start instruction is "ready set go" and you will be instructed to "relax" at the end. Instructions 2, 3 and 4 were completely randomized. Thus the order of the trials were completely randomized.

Data Analysis

Custom software developed in the Biodynamics Laboratory converted the electrical voltages from the LMM into position, velocity, and acceleration. The software program displays trunk position in each plane of the body and allows one to choose which plane as well as the specific part of the data to analyze. The main plane of motion was analyzed from minimum to maximum position for each cycle of motion the accessory planes of motion were also analyzed in the same section. Four cycles of motion were measured and averaged. The analysis programs determined the average ROM, and average peak velocity, average peak acceleration in the main plane of motion and in the coupled planes of motion for the four cycles.

The focus of the analysis was to distinguish between the normal and patient group. A MANOVA was performed to determine whether or not the two groups were significantly different overall. Discriminant function analyses were performed for each dependent measure separately to show how well each individual motion parameter distinguishes the two groups. Finally, a single discriminant function model was developed from all the possible twenty seven motion variables to distinguish between the two groups.

RESULTS

The MANOVA results showed a significant difference between the normal and low back pain patients. The asterisks (*) in table I indicate which parameters were significantly different for Bonferroni post hoc analysis. Descriptive statistics indicated that normals had higher functional capacity in the sagittal as well as frontal and transverse planes when compared to patients. Discriminant function analyses were performed for each dependent measure as a function of the task. The cross-validation error rates for distinguishing between normals and low back pain patients are listed in table I as a function of task and motion parameter. The error rate indicates the percentage of patients that misclassify into the normal group and the number of normals that misclassify as patients. The cross-validation processing method classifies each observation in the data set using a

discriminant function computed from the other observations in the data, excluding the observation classified. This method provides a more realistic error rate for future data compared to the resubstitution classification method. The results in table I show that velocity and acceleration motion parameters distinguish normals and low back pain patients more effectively than ROM parameters in all three planes of the body. Theoretically, it may be possible to have impairment in one plane and not in another depending on which components of the spine are injured. Therefore, it is important to measure dynamic functional motion capacity in the three cardinal planes of the body. In addition, coupled motion also distinguished between normals and low back pain patients.

A discriminant function model was developed to distinguish between normals and patients. First, a stepwise model was performed choosing from all nine dependent measures in the three tasks for a total of twenty seven candidate motion variables. Several discriminant function models were developed from the stepwise results. Additional models were developed including: a main plane motion model, frontal plane models, sagittal plane models, transverse plane models, ROM models, velocity models, and an acceleration model as well as some additional mixed models. Table II lists the motion parameters that were incorporated in the final models for the three experimental tasks. The final model was chosen

Table I. Cross-validation error rates for discriminant function and indicator (*) of significant difference (Bonferroni) between patients and normals.

Motion parameter	Frontal Task Error Rate	Sagittal Task Error Rate	Transverse Task Error Rate
Frontal plane			
ROM	42%	34% *	56%
Velocity	21% *	25% *	40% *
Acceleration	11% *	17% *	29% *
Sagittal Plane			
ROM	36%	44% *	34% *
Velocity	23% *	11% *	13% *
Acceleration	19% *	13% *	13% *
Transverse Plane			
ROM	46%	48%	27% *
Velocity	31% *	36%	19% *
Acceleration	25% *	34%	19% *

Table II. Motion parameters in the final model distinguishing between normals and patients
1. Lateral acceleration during lateral task
2. Lateral acceleration during sagittal task *
3. Sagittal velocity during sagittal task
4. Lateral range during lateral task
5. Sagittal range during twisting task *
6. Lateral acceleration during twisting task *

* coupled motion parameters

based on the cross-validation error rates. The model contains three primary planes motion parameters as indicated in the table as well as three coupled motion parameters. The model contains two parameters from each experimental task. It should also be noted that the model contains ROM, velocity, and acceleration motion parameters. The discriminant function cross-validation results for this model are shown in table III. The overall error rate for this model was 4%. It is interesting to note that the two misclassifications were normals classifying as patients and that no patients misclassified as normals. The exhaustive search for a model suggests that combinations of motion parameters from all three planes of the body, as well as main and coupled motions distinguish between normals and patients most effectively.

Table III. Discriminant function cross-validation results predicting normals and patients.

	Normals	Patients	Total
Normals	24 92%	2 8%	26 100%
Patients	0 0%	26 100%	26 100%

DISCUSSION

The results show that motions of higher order derivatives including velocity and acceleration provide valuable information for distinguishing between normals and low back pain patients. The results of this study are in agreement with previous research (Marras et. al. 1986; Marras et. al. 1995) indicating that velocity distinguishes between normals and patients more effectively than ROM. Furthermore, this investigation shows that this discriminating capability also holds for main motions in the frontal and transverse planes. The final model distinguishing between normals and low back pain patients included a combination of main motion, and coupled motion with ROM, velocity and acceleration parameters from all three tasks. This shows the importance of evaluating all three planes of the body dynamically. In addition, not only is it important to evaluate the main motion in each of the three cardinal planes of the body, but coupled motion in the accessory planes should be evaluated as well. It also shows that the combination of ROM, velocity and acceleration are all necessary in order to best distinguish between patients and normals.

The results of this study can be compared to previous research in the Biodynamic Laboratory. Marras et. al. 1995 classified 350 normals and 171 low back pain patients with a cross-validation error rate ranging from 6% to 12% depending on statistical methods. The previous protocol elicited maximum effort in the sagittal plane while controlling twisting. The main motion in the frontal or transverse planes of the body were not evaluated in the previous protocol. The protocol from the current study does elicit a maximum effort in the frontal and transverse plane and

incorporated measures from these tasks in the final model, which had a cross-validation error rate of only 4% with far fewer subjects. It is theorized that evaluating a maximum performance in all three planes of the body provides a more thorough evaluation resulting in the improved error rate.

The amount of coupling decreased in patients compared to normals. This result initially appears to be in contradiction with the results of Oxland and colleagues (1992) who found that disc injury increased coupled lateral rotation under axial torque loading. However, Oxland and colleagues performed their study with cadavers thus neuromuscular control was not considered. It is hypothesized that the reduction in coupling found in low back pain patients may be caused by an increase in coactivity of the muscles in low back pain patients or inappropriate activation patterns.

By neglecting to evaluate the dynamic components of functional capacity in all three planes one may be missing important information in evaluating recovery. It was found that models using only ROM or one plane of motion misclassified more patients as normal than models incorporating velocity and acceleration parameters in all three planes. It is theorized that these types of misclassifications may result in early release from treatment. Therefore in evaluating functional capacity of low back pain patients it is suggested that frontal and transverse plane performance should be evaluated as well as velocity and acceleration measures.

Future research is needed to link the dynamic functional capacity evaluation with the dynamic evaluation of the job. The current functional capacity testing methods that incorporate simulation of the job usually evaluate the start height of the lift and horizontal distance both from the NIOSH lifting guide as well as carrying ability (Isernhagen 1991). Previous research has shown that the motion parameter of velocity in all three planes of body indicates risk of low back disorder due to the job more effectively than start height of the lift (Marras et. al. 1995). If the simulation aspect of functional capacity evaluation is attempting to evaluate whether or not the person can perform the risky aspects of the job without symptoms then it is hypothesized that direct measure of velocity would be an important aspect to incorporated in the evaluation. This project provides a foundation indicating that there is a difference between normals and patients in all three planes of body and therefore all three planes should be assessed dynamically.

CONCLUSIONS

1.   Significant differences occur between normals and low back pain patients in all three planes of the body. Therefore measuring all three planes of the body provides a more thorough understanding of dynamic functional motion capacity of patients.

2.   Main motion components of velocity and acceleration distinguish between patients and normals more effectively than main ROM in all three planes of the body.

3.   Coupled motion parameters as well as velocity and acceleration contributed to the discriminating power between patients and normals. Therefore, these measures should be incorporated into functional motion capacity evaluations.

REFERENCES

Bigos S., Spenlger D., Martin N., Zeh J., Fisher L., and Nachemson A. (1986) Back injuries in industry: a retrospective study II. injury factors. Spine 11: 246-251.

Haddod G. (1987) Analysis of 2932 worker's comp back injury cases. The impact of the cost to the system. Spine 12:765-769.

Isernhagen S. (1991) Functional capacity evaluation and work hardening perspective. In Mayer T, Mooney V, Gatchel R. Contemporary conservative care for painful spinal disorders. Philadelphia, Lea and Febiger, 328-345.

Klein A., Snyder-Mackler L., Roy S., and Deluca C. (1991) Comparison of spinal mobility and isometric trunk extensor forces with electromyographic spectral analysis in identifying low back pain. Physical Therapy 71: 445-454.

Mandell P., Weitz E., Berstein J., et. al. (1993) Isokinetic trunk strength and lifting measure differences and similarities between low-back-injured and noninjured workers. Spine 18: 2491-2501.

Marras W. and Wongsam P. (1986) Flexibility and velocity of the normal and impaired lumbar spine. Arch Phys Med Rehab 67 :213-217.

Marras W., Fathallah F., Miller R. Davis S., and Mirka G. (1992) Accuracy of a three-dimensional lumbar

motion monitor for recording dynamic trunk motion characteristics. Int J of Ind Ergon 75-87.

Marras W., Lavender S., Leurgans S., et. al. (1993) The role of dynamic three-dimensional trunk motion in occupationally-related low back disorders: the effects of workplace factors, trunk position, trunk motion characteristics on injury. Spine 18:617-628.

Marras W., Parnianpour M., Ferguson S., et. al. (1995) The classification of anatomic and symptom based low back disorders using motion measure models. Spine 20; 2531-2546.

Nelson R. and Nester D. (1988) Standardized assessment of industrial low-back injuries: development of the NIOSH low-back atlas. Topics in Acute Care and Trauma Rehabilitation. 2: 16-30.

Nordin M. and Frankel V. (1989) Basic biomechanics of the musculoskeletal system, 2nd ed. Philadelphia: Lea and Febiger.

Riihimaki H., Viikari-Juntura E., Moneta G. Kuha J., Videman T., and Tola S. (1994) Incidence of sciatic pain among men in machine operating, dynamic physical work, and sedentary work. Spine. 19: 138-142.

Mayer T. and Gatchel R. (1988) Functional restoration for spinal disorders: the sports medicine approach. Philadelphia: Lea and Febiger 1988.

Oxland T., Crisco J., Panjabi M., and Yamamoto I. (1992) The effect of injury on rotational coupling at the lumbosacral joint a biomechanical investigation. Spine 17: 75-80.

Tramposh A. (1988) The functional capacity evaluation: measuring maximal work abilities. Occupational Medicine: State of the Art Review. 7: 113-124.

Twomey L. and Taylor J.(1994) Lumbar posture, movement, and mechanics. In:Twomey and Taylor Ed. Physical therapy of the low back 2nd ed. New York: Churchill Livingstone 57-91.

White A. and Panjabi M.   (1990) Clinical biomechanics of the spine, 2nd ed. Philadelphia: L.B. Lipincott Co.

RELATION BETWEEN BIOMECHANICAL SPINAL LOAD FACTORS AND RISK OF OCCUPATIONAL LOW-BACK DISORDERS

Kevin P. Granata, William S. Marras, Susan A. Ferguson
Biodynamics Laboratory
The Ohio State University
Columbus, Ohio

ABSTRACT

The objective of this study was to identify individual and combinations of biomechanical parameters which are associated with the probability of risk for occupationally related low-back disorders. Ten subjects performed lifting tasks simulating warehouse order selection. During the lifting exertions dynamic trunk motion, EMG, and workplace data were collected. Risk of low-back disorder was assessed from an epidemiologic model incorporating workplace factors and trunk motion data. Comparison with biomechanical results indicated that static estimates of spinal compression poorly predicts the probability of risk. Spinal compression computed from dynamic models improve the correlation, and regression models including multi-dimensional dynamic spinal loads and load rates best predict the probability of risk. The results agree with biomechanical research demonstrating vertebral failure modes are influenced by complex biomechanical interactions. This study highlights the limitations of ergonomic assessment via static compression, demonstrates the influence of biomechanical interactions upon risk of LBD, and advocate greater efforts toward understanding complex dynamic interactions in human factors research.

INTRODUCTION

Spinal compression is traditionally presumed to be the principle biomechanical mechanism associated with occupationally related low-back disorders (LBD). The NIOSH lifting guide discriminates between safe and hazardous tasks based primarily upon the static compressive loads on the spine (NIOSH 1981). Research examining the causative nature of ergonomic risk of low-back pain most often focuses on axial compressive loads associated with occupational tasks (Schultz and Andersson 1981; Freivalds et al 1984). However, epidemiologic studies indicate that repetitive twisting or lateral bending and lifting, even with relatively light loads, are significant risk factor for LBD (Kelsey et al 1984; Punnet et al 1991). These findings suggest that spinal shear and torsional loads associated with asymmetric lifting postures may be under-appreciated. Similarly, based upon the high correlation between task dynamics and the risk of LBD (Marras et al 1993), spinal and biomechanical load dynamics may be associated with the mechanism of injury.

Biomechanical literature indicates spinal injury solely due to spinal compression is unlikely. Brinkman (1986) demonstrated that compressive loads applied to in vitro lumbar vertebra failed to produce clinically relevant injuries unless pre-existing endplate damage was present. Adams et al (1987) stated that compression in the lumbar spine, in the absence of forward bending moments, cannot "injure the soft tissue without first causing gross damage to the vertebrae." Thus, tissue failure and associated LBD is more likely generated by combinations of multi-dimensional spinal loads. In a theoretical analysis, Shirazi-Adl (1989) found that bending moments and shear forces combined with axial compression significantly increased the risk of injury to the lumbar disc. Yingling et al (1995) demonstrated load rate influences the ultimate strength, stiffness, displacement, and failure mechanisms. Clearly, occupationally related low-back pain associated with vertebral tissue injury is unlikely caused by static compression alone.

The objective of this study was to identify the multi-dimensional, dynamic biomechanical factors associated with LBD risk. Specifically, answers to two questions were sought. First, does axial compression in the lumbar spine correlate well with the risk of LBD associated with a prescribed task? Second, does the inclusion of three-dimensional, dynamic spinal load and load rate data more accurately describe the probability of risk than static compression alone? Identifying possible biomechanical parameters capable of discriminating between safe and hazardous tasks may contribute to the development of more accurate and robust ergonomic analyses and reduced incidence of LBD in the workplace.

METHODS

Ten experienced warehouse order selectors 19 to 49 years of age were recruited from a local food distribution center. The weight and stature of the subjects was 80.1 ±8.4 kg. (176 lb.) and 180.3 ±7.1 cm. (71 in.) respectively. The subjects' experience as warehouse selectors ranged from 0.25 to 23 years.

In order to simulate realistic warehouse working conditions, subjects were required to lift boxes ranging from 18.2 kg. (40 lb.) to 27.3 kg. (60 lb.) from one pallet to another until the entire pallet load, an average of 35 boxes, was transferred. Twelve pallets of boxes were moved at a frequency of 166 lifts per hour, simulating a "slow" five hour work day. Dynamic, three-dimensional trunk motion data were collected from the Lumbar Motion Monitor (LMM) (Marras et al 1993) and integrated myoelectric (EMG) activity of ten trunk muscles were collected from bi-polar surface electrodes (as per Mirka and Marras 1993) during the depalletizing tasks. Prior to beginning each pallet, a set of "test" exertions were performed while standing on a force plate (Bertec 4060A) and with added electrogoniometers to measure the location and orientation of the lumbo-sacral spine relative to the center of a force plate. The test exertions were designed to permit data quality assurance and supply calibration data for biomechanical analyses.

Each lifting task was assigned a probability of being at high risk for occupationally related LBD. The assessment was achieved from a multiple logistic regression model of dynamic trunk motion parameters and workplace factors (Marras et al 1993). This epidemiologic model was developed from a database developed from on site measurements over 400 industrial workers, and incorporated factors including the lifting moment, lift rate, multi-dimensional trunk range of motion and velocities. As subjects' lifted each box, a measure of risk was assigned to that task by the epidemiologic model, and saved for comparison with a variety of biomechanical parameters including spinal loads and load rates.

An EMG-assisted biomechanical model was employed to determine the dynamic spinal loads associated with the lifting exertions (Granata and Marras, 1995). The analysis incorporates normalized EMG data, anthropometrically scaled muscle cross-sectional areas and vector directions as well as force-length and force-velocity relations to determine the force supplied by ten dynamically co-contracting muscles. Three-dimensional spinal loads were determined from the vector sum of the muscle forces, and trunk moments from the sum of vector products of muscle forces and moment arms. Direct comparison of dynamic trunk moments determined from the force plate data collected during the test exertions, with predicted trunk moments determined from the biomechanical model during those exertions provided subject dependent calibration values and model validation parameters. Model output included peak spinal loads as well as the peak value of the rate of change of spinal load, i.e. load rate.

For comparison with the dynamic biomechanical data, static compressive forces were computed for each task using the method outlined in Chaffin and Andersson (1981). External moments were determined from the product of box weight and moment arm distance from the trunk measured during each lifting exertion. Upper body mass and center of mass were determined from subject anthropometry and multiplicative coefficients cited in Chaffin and Andersson (1981). The restorative force generated by a single extensor muscle and spinal compression was determined from the muscle moment arm and total trunk moment. Thus, static estimates of spinal compression were determined for comparison with probability of LBD risk and dynamic biomechanical values.

Linear regression analyses were used to examine the association between the probability of a high risk classification and spine biomechanics factors. Correlation values were achieved for all possible models of combined biomechanical factors, including regression models of individual factors. The association between biomechanical parameters and occupationally related LBD were evaluated based upon the correlation coefficient (R^2) between the regression model of biomechanical values

RESULTS

The correlation between the probability of risk represented in the simulated warehouse tasks and independent biomechanical factors are presented in table 1. Results indicate that dynamic spinal compression represented the strongest individual correlation with probability of risk at $R^2 = 0.443$. Static compression represented the poorest individual correlation with risk, $R^2 = 0.137$. Lateral and AP shear forces were also poorly correlated with the probability of risk as individual factors. R^2 values for load rate in the compressive direction were slightly lower than spinal compression at $R^2 = 0.429$. However, shear load rate in the lateral and AP directions were better correlated with predicted risk than the shear forces.

Table 1. Correlation (R^2) between individual biomechanical factors and probability of high risk for LBD. Results indicate dynamic compression and spinal load rates are better correlated to LBD risk than static estimates of compression or dynamic estimates of shear load.

	X	Y	Z
Static Load	--	--	0.137
Dynamic Load	0.193	0.197	0.443
Load Rate	0.344	0.354	0.429

X = Lateral Shear Force, Lateral Shear Load Rate
Y = AP Shear Force,   AP Shear Load Rate
Z = Compressive Force,   Compressive Load Rate

Incorporating multi-dimensional factors and dynamic parameters dramatically improves the predictive power of the regression model. Table 2 identifies the three best regression models containing two, three and four biomechanical factors each. Models with greater than four biomechanical factors demonstrated reduced predictive power. The strongest two parameter model includes dynamic compression force and AP shear force, which improved the correlation with probability of risk to 0.473. Combined load rate data also demonstrates improved correlation with probability of risk. Predicted probability of risk is best represented by a combination of dynamic spinal compression, AP shear force, lateral shear load rate and AP shear load rate, $R^2 = 0.517$. Combinations of biomechanical factors with static compression failed to generate the levels of correlation demonstrated by equivalent models with dynamic compression.

DISCUSSION

Analyses of LBD risk has traditionally focused on static compressive loads in the spine. However, epidemiologic and biomechanical data suggest the risk associated with shear and torsional loads on the spine significantly enhance the risk of occupationally related LBD. Similarly, dynamic loading parameters may be related to LBD risk. Thus, examination of dynamic, and multi-dimensional spine biomechanics factors may improve the ability to identify hazardous occupational tasks.

Axial compressive loads on the spine determined from a fundamental static biomechanical model were significantly ($p<.001$) lower and poorer predictors of risk than dynamic spinal compression. The average static compression, 2700 N., was 20% lower than the average dynamic value of 3400 N. Freivalds et al (1984) and McGill and Norman (1985) similarly indicated static analyses underpredict dynamic spinal compression by 20 to 40%. As a single variable, dynamic compression correlated with the results from the epidemiologic risk model at $R^2 = 0.443$, much better than the statically determined compression, $R^2 = 0.137$. This demonstrates the static and dynamic determination of spinal load are largely unrelated ($R^2 = 0.31$), as illustrated by the scatter of points when plotting the dynamic compression versus the static values (Figure 1). Since the static determination of compression fails to account for the variability associated with lifting dynamics, it becomes difficult for this variable, or combinations including it, to discriminate between safe and hazardous tasks in a dynamic work environment.

Table 2. Correlation (R^2) between combinations of biomechanical factors and task probability of high risk for LBD. Only the three best regression models in each category are provided.

Regression Models	R^2
Two Factor Models	
Fy + Fz	0.473
Fx + LRz	0.466
LRy + LRz	0.433
Three Factor Models	
Fy + Fz + LRx	0.501
Fx + Fy + LRz	0.475
Fy + LRx + LRz	0.474
Four Factor Models	
Fy + Fz + LRx + LRy	0.517
Fx + Fy + LRx + LRz	0.494
Fy + LRx + LRy + LRz	0.490

F_X = Lateral Shear Force LR_X = Lateral Load Rate
F_Y = AP Shear Force LR_Y = AP Load Rate
F_Z = Compressive Force LR_Z = Compr. Load Rate

Results indicated that a regression model including multi-dimensional spinal loads improves the predictive power for LBD risk when compared to a model including only axial compression. This agrees with the theoretical assessment of Shirazi-Adl (1989) who concluded that shear forces combined with compression increases the risk of vertebral disc failure. The influence of bending and torsional moments supported by the spine may also enhance the risk of injury. In this study, the relation between passive spinal moment and risk of LBD could not be quantified. However, considering the trend demonstrating greater fidelity when shear loads are included, and the biomechanical evidence relating spinal bending and torsional moment with vertebral failure, one might be tempted to hypothesize that inclusion of those parameters would further enhance the predictive power of the biomechanical regression model.

Lifting dynamics may directly influence the load and tolerance of the spine to injury. Analyses of dynamic exertions have demonstrated spinal load increases with velocity and acceleration (Freivalds 1984; Granata and Marras, 1995). Furthermore, those increased loads must be supported by a spinal column wherein the tolerance may be influenced by the load rate (Yingling et al 1995). There are few studies examining the influence of load rate upon injury mechanisms in the lumbar spine, but our results indicate that load rate may adversely affect the safety of a lifting exertion in terms of the probability of LBD risk. Passive spine bending moment rate was not examined in this study. Considering the known response of biologic tissue to viscous loading, future research might demonstrate moment rate offers significant insight into the injury mechanism associated with occupational tasks.

The interpretation of these results is limited by the use of an epidemiologic model as a baseline. This study examined the correlation between dynamic biomechanical parameters and predicted LBD risk, which in turn represented the actual risk values measured in industry (Marras et al 1993). This method to estimate LBD risk undoubtedly introduced variability into the analyses. However, in order to achieve the appropriate biomechanical data it was necessary to simulate the warehouse selection environment, requiring an epidemiologic model to estimate probability of risk. The predictive ability of the epidemiologic model has been determined to be quite high, more than three times better than the NIOSH lifting model. Thus, the data represents an association between biomechanical factors and probability of risk, but cannot be interpreted as causative.

The research methods allow several potential interpretations of the results. The most direct conclusion is that LBD risk is better predicted from dynamic, multi-dimensional biomechanics than static assessment of compression alone. Other possibilities include potential limitations or interactions in the biomechanical and epidemiologic models. However, both these models have been exhaustively validated in separate research efforts. Thus, the enhanced ability to predict LBD risk from multi-dimensional dynamic spinal load and load rate factors is believed to be valid. Furthermore, the implication agrees with the biomechanical literature wherein spinal tolerance and material failure has been associated with multi-dimensional loads on the functional units of the spine.

The improved predictive ability associated with multi-dimensional, dynamic analyses of coactive load and load rate is generated at the high cost of biomechanical complexity. This technical cost-benefit may appear prohibitive for simple ergonomic assessments of the workplace. However, identifying the one-dimensional static parameters may ignore the injury mechanism associated with the greatest number of workplace injuries. Thus, these results advocate greater efforts toward understanding complex dynamic interactions in human factors research.

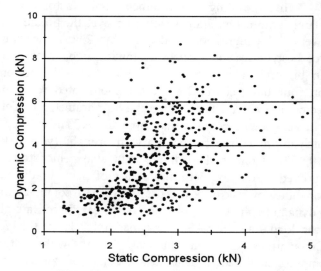

Figure 1. Plot of dynamic estimates of spinal compression versus static estimates of compression. Results indicate the static and dynamic compressive values are poorly correlated ($R^2 = 0.36$). Consequently, LBD risk factors associated with workplace dynamics may be overlooked when using static biomechanical analyses.

REFERENCES

Adams MA, Dolan P, Hutton WC. Diurnal variations in the stresses on the lumbar spine. *Spine* 1987; 12 (2), 130-137

Brinkman P. Injury of the anulus fibrosis and disc protrusions. An in vitro investigation on human lumbar discs. *Spine* 1986; 11 (2), 149-153

Chaffin D.B. and G.B.J. Anderson (1984) Occupational Biomechanics, John Wiley and Sons, N.Y.

Freivalds A, Chaffin DB, Garg A, Lee KS, A dynamic biomechanical evaluation of lifting maximum acceptable loads. *J Biomechanics* 1984; 17 (4): 251-262

Granata KP, Marras WS. An EMG assisted model of biomechanical trunk loading during free-dynamic lifting. *J. Biomechanics* 1995; 28 (11): 1309-1317,

Kelsey KL, Githens PB, White AA III, Holford TR, Walter SD, O'Conner T, Ostfeld AM, Weil U, Southwick WO, Calogero JA. (1984) An epidemiologic study of lifting and twisting on the job and risk for acute prolapsed lumbar intervertebral disc. *J Ortho Res*, 2(1): 61-66

Marras WS, Lavender SA, Leurgans SE, Rajulu SL, Allread WG, Fathallah FA, Ferguson SA. (1993) The role of dynamic three-dimensional trunk motion in occupationally-related low back disorders: The effects of workplace factors, trunk position and trunk motion characteristics on risk of injury. *Spine*, 18 (5), 617-628

McGill S.M., Norman R.W. Dynamically and statically determined low back moments during lifting. *J. Biomechanics* 1985; 8 (12): 877-885.

Mirka GA, Marras WS. A stochastic model of trunk muscle coactivation during trunk bending. *Spine* 1993; 18 (11): 1396-1409

NIOSH (1981) National Institute for Occupational Safety and Health: A Work Practices Guide for Manual Lifting Tech. Report No.81-122, U.S. Dept. of Health and Human Services (NIOSH), Cincinnati, Oh.

Punnet L., Fine L.J., Keyserling W.M., Herrin G.D. and D.B. Chaffin (1991) A case-referent study of back disorders in automobile assembly workers : The health effects of non-neural trunk postures. *Am. J. Epidemiology*,

Schultz A, Anderson G. Analysis of loads on the lumbar spine. *Spine* 1981; 6 (1): 76-82

Shirazi-Adl A. Stress in fibers of a lumbar disc, analysis of the role of lifting in producing disc prolapse. *Spine*, 1989; 14 (1): 96-103.

Yingling VR, Callaghan JP, McGill SM. The effect of load rate on the mechanical properties of porcine spinal motion segments. *Proceeding of the 19th Annual Meeting of the American Society of Biomechanics* 1995; 119-120

MAXIMUM SAFE WEIGHT OF LIFT: A NEW PARADIGM FOR SETTING DESIGN LIMITS IN MANUAL LIFTING TASKS BASED ON THE PSYCHOPHYSICAL APPROACH

Waldemar Karwowski
Center for Industrial Ergonomics
Department of Industrial Engineering
University of Louisville
Louisville, Kentucky 40292, USA
email: w0karw03@ulkyvm.louisville.edu

The main objective of this study was to introduce a new concept for setting design limits in manual lifting tasks, i.e., the notion of the maximum safe weight of lift (MSWL), and to examine the effects of different sets of subject instructions for load selection on the amount of weight determined by male subjects in the psychophysical experiment. The results revealed that the MSWL values were significantly lower than the maximum acceptable weight of lift (MAWL) values. It was proposed that the classical approach to setting design limits in manual lifting tasks, based on the psychophysical method introduced by S. Snook, be modified to emphasize the subjects' safety and to reduce the risk of musculoskeletal injury.

BACKGROUND

The psychophysical approach to determining the maximum acceptable weight of lift (MAWL) has been extensively used for almost thirty years (Snook and Irvine, 1967; Ayoub et al, 1978; Snook 1985). This approach aims to quantify human lifting capacity based on subjective perception of exertion, under the assumption that *workers are able to determine with some accuracy the highest acceptable workload* (Gamberale, 1985). The experimental procedures first proposed by Snook and Irvine (1967) and again reported by Snook (1978), require the subjects *to imagine working on the incentive basis, as hard as they can (lifting as much as they can), without straining themselves or becoming unusually tired, weakened, overheated, or out of breath*. Typically, the subjects are given control of the weight (or force) handled, and are asked to adjust the weight of the box (force) over a period of thirty/forty minutes, up to the maximum level they are willing to accept for an eight hour shift at a given frequency of task repetition.

According to Gamberale et al (1987), the assumptions of the psychophysical approach to setting limits in manual lifting tasks are as follows: 1) the individual is able to rate the perceived effort in a lifting task, 2) he/she is able to produce an individually acceptable level of performance on this task, and 3) this level of performance will be safe from manual handling injuries. Unfortunately, validity of these assumptions has never been fully examined with the required scientific scrutiny that

would warrant their wide acceptance in the research community.

SUBJECT INSTRUCTIONS IN DETERMINATION OF MAWL

The original procedure proposed by Snook and Irvine (1967) for determining what is assumed to be the "maximum acceptable weight of lift," focuses on work productivity, based on proportionate pay incentive scheme. Furthermore, the instructions do not explicitly refer to the worker's safety or injury avoidance. With only a few exceptions discussed below, all studies in the field of manual materials handing conducted, since the original study by Snook and Irvine (1967), followed the same assumed premise of self-selected load acceptability. It should be noted, however, that the subjects themselves are never asked to determine the "maximum acceptable weight of lift" or to evaluate the degree of acceptability of the selected weights with respect to lifting the entire working day. The subjects are only asked to adjust the weight of the box to the level they believe they would be willing to lift over an eight hour work day. The resulting weights selected by the subjects were then assumed by Snook and Irvine (1967) to represent the "maximum acceptable weights of lift."

Gamberale et al (1987) pointed out that the individual choice of workload is governed not only by sensory inputs (from tendons, muscles and cardiovascular system), but also the subject's assessment with regard to the risk of musculoskeletal injuries or fatigue leading to accidents. Therefore, it was concluded that assessment of acceptable workload performed using the psychophysical method would rely more on the subjects' cognitive judgement rather than on their immediate perception of exertion. These important issues, however, have not been explored in the subject literature.

OBJECTIVES

The classical psychophysical approach to setting limits in manual lifting tasks (Snook, 1978) implies that the workload or force established using this method represents a balance between health, satisfaction, and performance (Straker, 1994). According to Gamberale and Kilböm (1988), this approach also assumes that a worker is able to estimate with sufficient accuracy the maximum tolerable workload, and that the acceptable workload selected in a simulated task is "safe." However, the validity of such assumptions is to some extent unknown (Karwowski, 1989; Chaffin and Page, 1994; Straker, 1994). To the author's knowledge, no published study has examined validity of the assumption that the weights selected under Snook and Irvine's (1967) original instructions, and called the MAWL (Snook, 1978), would also be perceived by the same subjects as 'safe.' Except for a rather vague warning against overstraining, there are no references as to worker safety and avoidance of back pain and musculoskeletal injury. Therefore, in this study, a new set of instructions to determine the maximum safe weights of lift (MSWL) proposed by Karwowski (1995) was used.

The main objective of this study was to examine the effects of different sets of subject instructions for load selection (called here the MAWL instructions and the MSWL instructions) on the amount of weight determined by the subjects in the psychophysical experiment. The secondary objective of the study was to investigate human perception of load heaviness in terms of linguistic variables (natural expressions), for manual lifting tasks performed under the same experimental conditions and subjects' instructions for load selection.

METHODS AND PROCEDURES

Subjects

Ten male volunteer subjects were selected from the student population at the University of Louisville. The students were compensated for their participation in the study. Each subject filled out an Informed Consent Form, and only those subjects with no history of low back pain or musculoskeletal

disorders were allowed to participate. The subjects were asked to wear comfortable clothes, such as athletic wear, to stretch before the experiment, and refrain from any strenuous activities prior to the experimental sessions.

Experimental Design

A randomized complete block design, with subjects as blocks, was used in both experiments. Subjects were asked to determine the maximum acceptable weight of lift (MAWL) and the maximum safe weight of lift (MSWL) based on the instructions provided in the Appendix. The order of experimental sessions was chosen randomly for each of the subjects. In order to assure uniformity in the presentation of the instructions, the instructions were audiotaped and played to each subject before each experimental session. Only one session per subject was conducted on any given day.

Snook's (1978) proposed instructions for determining the MAWL were utilized. In summary, the subjects were asked to imagine working under an incentive pay scale, getting paid for the work completed, and working as hard as they could without straining themselves or becoming tired, weakened, overheated or out of breath.

The instruction set for the MSWL (for exact wording see Appendix) asked the subjects to imagine working safely over an eight hour shift, and to determine the weight they could safely lift without increasing the risk of low back pain or muscular overexertion, as proposed by Karwowski (1995). Both sets of instructions used the same wording with respect to encouraging subjects to make as many adjustments as necessary, and to caution them against perception of being in any kind of competition with others while performing their own lifting tasks.

Subjects Physical Characteristics

The anthropometric measurements were taken according to NASA (1978) procedures. For each subject, the body weight, height, shoulder height, hip height, knee height, and arm length, were measured. Four static strengths were determined for each subject. These measurements are: static arm strength, static back strength, static composite strength, and static shoulder strength, as recommended by Chaffin (1975). Two dynamic strengths, i.e. dynamic lift strength and dynamic back extension strength, were also measured for each subject. The procedures proposed by Pytel and Kamon (1981) were used for this purpose.

Procedures

Each subject was asked to repeatedly lift identical boxes with handles on the sides, from the floor and place it on the table (69 cm high) during a period of 30 minutes. The box dimensions were 33 x 33 x 26 cm, and each box had a false bottom to minimize visual cues. An electronic timer controlled the lifting frequency of 4 lifts per min (lpm). At the start and end of each experimental session, the box was confidentially weighed by the experimenters. Standard procedures for the weight adjustment process as proposed by Snook (1978) were adopted.

After selecting the preferred load (either MAWL or MSWL), the subject was asked to rank his perception(s) of both the weight just selected and the task performed as follows. The subject selected a rating of perceived exertion (RPE) during the session using the Borg scale (1982). The subject was also asked to choose one value which, in his/her opinion, would best describe heaviness of the selected weight, using one of the eight linguistic values, i.e. *very light, light, more-or-less-medium, medium, heavy, very heavy,* and *extremely heavy.* A 10 cm bipolar scale (marked with *uncertain*, on one end, and *positive* on the other end), was used to determine the subject's confidence in his choice of the linguistic category associated with the selected weight concept (MAWL or MSWL).

In addition, using a series of similar bipolar, 10 cm long scales, the subject was asked to provide the following estimates: 1) the degree to which the selected weight was perceived as acceptable for an 8-hr work-day, 2) the degree to which the selected task was perceived as acceptable for an 8-hr work-day (using the end markers named *absolutely unacceptable* and *fully acceptable*), 3) the degree to which the selected weight was perceived as safe for an 8-hr work-day, and 4) the degree to which the selected task was perceived as safe for an 8-hr work-day (using the end markers named *absolutely unsafe* and *fully safe*).

RESULTS AND DISCUSSION

The Statistical Analysis Systems (SAS) software was used to analyze the experimental data. A summary of results is given in Table 1.

Main Effects in Determination of the Preferred Weight

The analysis of variance (ANOVA) showed significant effects of subject instructions on the preferred weights. The subjects selected less weight when lifting according to the MSWL instruction versus the MAWL instruction (see Table 1). The average MSWL value was 38.3 pounds (17.36 kg), while the average value of MAWL was 46.02 pounds (20.87 kg). This 16.82% average difference (7.74 pounds or 3.51 kg) was found statistically significant (Tukey test) at p < 0.0001 level.

RPE analysis

The effect of instructions for selection of preferred weight of lift (MAWL or MSWL) on the rate of perceived exertion (RPE) was significant at p<0.001 level. Subjects reported higher average RPE values associated with the MAWL values (mean of 15.7), than those associated with the MSWL values (14.7), indicating that they elected to work at lower levels of perceived physical exertion when instructed to pay attention to their own safety.

It is important to note that the RPE values weakly correlated with the MAWL values (Pearson coefficient r = 0.51, p < 0.02), but did not correlate with the MSWL values (r = 0.22, p > 0.1), indicating that when selecting the maximum safe weight of lift, the subjects took into consideration more than just perception of their physical exertion, as they likely did when selecting the maximum acceptable weights of lift. Similar observations with respect to the lack of strong correlation between the RPE values and the preferred weights selected by the subjects during the psychophysical lifting experiments were made by Ljungberg et al (1982) and Gamberale et al (1988).

Table 1. Main effects for dependent variables.

Variable	Mean	S.D.	Range	N
Preferred Weight				
MAWL	46.02	12.12	23-74	20
MSWL	38.28	8.39	18.5-55.7	20
RPE				
MAWL	15.8	0.71	15-17	20
MSWL	14.7	0.71	14-16	20

Estimation of Load Lifted

The effect of instructions on the subjects' estimation of how much weight (ESW) they selected at the end of each trial was statistically significant at p < 0.02 level. On average, the subjects estimated to have selected 51.01 lbs while asked to determine the MAWL values, and only 40.1 lbs while asked to determine the MSWL values.

The Effect of Instructions on Perception of Load Heaviness

Of the forty trials performed in this study, 67% of the preferred loads were independently judged by the subjects as heavy. However, under the MAWL instruction set, 75% of the time the psychophysically determined loads were classified as heavy. Under the MSWL instructions, only 65% of the selected loads were judged to be heavy.

The Safety/Acceptability Scores

Upon completion of each experimental session, the subjects were asked to indicate their degree of belief using a 10 cm bi-polar scale (marked not safe & not acceptable, and absolutely safe & acceptable on each end, respectively), regarding how safe, and, therefore, how acceptable the selected MAWL and MSWL values were. On average, the preferred MAWL values were judged by the subjects to be much less safe and acceptable (p < 0.01) for continuous lifting over an 8-hr period (mean score of 5.6), than the MSWL values (mean score of 6.3).

CONCLUSIONS

Despite relative simplicity of the psychophysical method to determine 'acceptable' limits for manual lifting, which makes this approach popular, caution should be exercised with respect to interpretation and usability of the currently available design limits which were developed based on the psychophysical approach. It seems plausible that with the MAWL instructions, the subjects are mainly focusing on task performance and efficiency. On the other hand, the main focus of attention while lifting under the MSWL instructions is subject's safety. The results of this study revealed that the MSWL values were significantly lower than the MAWL values. Therefore, the classical approach to setting design limits in manual lifting tasks based on the psychophysical method should be modified, as proposed by Karwowski (1995), in order to emphasize the subjects' safety and to reduce the risk of musculoskeletal injury.

REFERENCES

Ayoub, M., Bethea, N. J., Deivanayagam, S., Asfour, S., Bakken, G. M., Lilies, D., Mital, A., and Serif, M., 1978, *Determination and Modeling of Lifting Capacity. NIOSH Report,* Grant No. 5-r01-OH-000545-02, Cincinnati, Ohio.

Borg, G. A. V., 1982. Psychophysical bases of perceived exertion. *Medicine and Science in Sports and Exercise,* 4(5), 377-381.

Gamberale, F., 1985. The perception of exertion, *Ergonomics,* 28, 299-308.

Gamberale, F. and Kilböm, A., 1988. An experimental evaluation of psychophysically determined maximum acceptable workload for repetitive lifting. In *Proceedings of the 10th Congress of the International Ergonomics Association,* A. S. Adams, R. R. Hall, B. J. McPhee, and M. S. Oxenburgh, Eds., Taylor and Francis, London, pp. 233-235.

Gamberale, F., Ljungberg, A. S., Annwall, A., and Kilbom, A., 1987. An experimental evaluation of psychophysical criteria for repetitive lifting work. *Applied Ergonomics,* 18.4, 311-321.

Karwowski, W., 1995. *The Maximum Safe Weight of Lift: A New Design Paradigm for Setting Limits in Manual Lifting Tasks.* Unpublished Technical Report, Center for Industrial Ergonomics, University of Louisville, Louisville, Kentucky, 40292, USA.

NASA/Webb (Eds), 1978. *Anthropometric Sourcebook* (3 volumes). (NASA Reference Publication 1024). Houston, TX, LBJ Space Center.

Pytel, J. L., and Kamon, E., 1981. Dynamic Strength Test as a Predictor for Maximal and Acceptable lifting. *Ergonomics,* 24, 663-672.

Snook, S. H., 1978. The Design of Manual Handling Tasks. *Ergonomics,* 21, 963-985.

Snook, S. H., 1985. Psychophysical acceptability as a constraint in manual working capacity. *Ergonomics,* 28, 331-335.

Snook, S. H., and Irvine, C. H., 1967. Maximum Acceptable Weight of Lift. *American Industrial Hygiene Association Journal,* 28, 322-329.

APPENDIX

Instructions for determination of the Maximum Safe Weight of lift (MSWL) as proposed by Karwowski (1995):

We want you to imagine that you are working over a normal 8 hour shift, and we ask you to determine the maximum level of load that you feel you could SAFELY lift every day without hurting yourself, i.e. without the risk of experiencing any low back pain, or possibly overexerting yourself.

YOU WILL ADJUST YOUR OWN WORK LOAD. You will work only when the timer beeps. Your job will be to adjust the load; that is, to adjust the weight of the box that you are lifting WITH RESPECT TO YOUR OWN PERCEPTION OF HOW SAFE IT IS FOR YOU.

Adjusting your own work load is not an easy task, only you know how safely you feel.

IF YOU FEEL THE LOAD IS TOO HEAVY, AND THEREFORE NOT SAFE, or that lifting of this load could result in low back pain and/or muscular overexertion, you should reduce the load by removing weight from the box.

HOWEVER, IF YOU FEEL THAT YOU CAN WORK HARDER, WITHOUT INCREASING THE RISK OF LOW BACK PAIN OR OVEREXERTION, you should put in more weight into the box.

DON'T BE AFRAID TO MAKE ADJUSTMENTS. You have to make enough adjustments so that you get a good feeling for what is SAFE and what is NOT SAFE for you. You can never make too many adjustments -- but you can make too few.

REMEMBER, THIS IS NOT A CONTEST. EVERYONE IS NOT EXPECTED TO DO THE SAME AMOUNT OF WORK.

WE WANT YOUR JUDGMENT ON HOW HARD YOU CAN WORK SAFELY, I.E., WITHOUT EXPOSING YOURSELF TO THE RISK OF LOW BACK PAIN AND/OR MUSCULAR OVEREXERTION WHEN PERFORMING THIS TASK OVER AN 8-HOUR PERIOD.

PHYSICAL TRAINING AND MANUAL-MATERIAL HANDLING: LITERATURE AND MILITARY APPLICATIONS

Joseph J. Knapik
U.S. Army Research Laboratory
Human Research and Engineering Directorate
Aberdeen Proving Ground, MD 21005

Manual material handling (MMH) is performed in 83% of all U.S. Army enlisted military occupations. Studies that have examined the influence of physical training on MMH can be separated into two types: those that use the same task for testing and training (task-specific training studies) and those that do not (general training studies). Reported relative improvements in maximal symmetric lifting and repetitive lifting are 26% to 99% in task-specific training studies and 16% to 23% in general training studies. Psychomotor learning probably accounts for a large proportion of performance gains in task-specific training while both psychomotor learning and muscle hypertrophy account for gains in general training studies. While both types of training are effective and currently practiced in the military, general training may be useful for improving a wide range of MMH tasks while task-specific training results in larger gains in targeted MMH tasks.

INTRODUCTION

Traditional ergonomic approaches to reducing worker job stress during manual material handling (MMH) have largely focused on redesigning the working environment through changes in equipment or task requirements (NIOSH, 1981). Until relatively recently, there has been little work examining how improving the physical capacity of the worker might influence MMH capability (Asfour, Ayoub, & Mital, 1984; Genaidy, Bafna, Sarmidy, & Sana, 1990a). Worker physical capacity can be improved by specific physical training designed to improve certain components of physical fitness, especially muscular strength and muscular endurance (Asfour, et al., 1984).

In the U.S. Army, 83% of all military occupational specialities (MOS) have manual material-handling (MMH) requirements. More than 175 MOS require occasional lifting of 44 kg or more and frequent lifting of 23 kg or more. Examples include 1) the field artilleryman who may be required to lift 44-kg artillery rounds as often as 275 times per day, 2) the cargo specialist who is required to lift

and carry 240 kg in four-soldier teams, and 3) the chemical operations specialist who must lift oil drums weighing 108 kg from the ground onto a truck in two-soldier teams (Army Regulation 611-201). Beside these occupational requirements other, infrequent physical demands are placed on all soldiers during operational exercises or deployments. These include such tasks as loading equipment onto trucks, setting up large canvas tents, covering tents and vehicles with cumbersome camouflage netting, and evacuation of casualties with litters or by body carriage.

The purpose of this paper is to review and analyze studies that have examined influence of physical training on MMH capability. Mechanisms for training-induced changes are examined and military applications of these investigations explored. Before proceeding with the literature review, it is useful to define and clarify what is meant by physical training.

Physical Training

Physical training can be defined as muscular activity targeted at enhancing the physical capacity of the individual by improving one or more of the components of physical fitness. The components of physical fitness include muscular strength, muscular endurance and cardiorespiratory endurance (aerobic capacity). Muscle strength is the ability of a muscle group to exert a maximal force (e.g., lifting as much weight as possible). Muscular endurance is the ability of a muscle group to perform short-term, high intensity physical activity (e.g., repetitively lift 44-kg artillery shells as fast as possible). Cardiorespiratory endurance is the ability of the circulatory and respiratory systems to supply fuel to sustain long term physical activity (e.g., road marching, long distance running, bicycling). (Caspersen, Powell, & Christenson, 1985). In order for physical training to effectively increase or maintain physical fitness, it must be performed regularly and be of sufficient frequency, intensity, and duration to induce changes in specific fitness components (Wenger & Bell, 1986).

Limitations in performance imposed by inadequate muscle strength and muscular endurance are probably the most common problem in military and industrial setting (Hogan, 1991). These fitness components can be developed simultaneously or individually using progressive resistance training (PRT). PRT involves exercising with resistances or loads that just fatigue particular muscle groups. The resistance is continually increased as the individual improves his fitness (hence the term "progressive"). PRT is based on exercising with "repetition maximums". A one-repetition maximum (1RM) is the maximum load that can be moved through a range of joint motion just one time. A 10-repetition maximum (10RM) is the maximum load that can be moved 10 times (the individual is not able to perform an eleventh repetition because of fatigue). There is a continuum of repetition maximums that have different effects on strength and endurance. A 1RM builds primarily strength, a 10RM a combination of strength and endurance, and a 20RM primarily muscular endurance (Fleck & Kraemer, 1987).

REVIEW OF THE LITERATURE

Studies that have examined physical training and MMH capability can be divided into two categories. The first category involve studies that use the same MMH task for testing and training (task-specific training studies). The second category of studies are those that use more generalized and traditional training programs that do not include the MMH tasks in the training program (general training studies).

MMH and Task-Specific Training Studies

In the earliest investigation, Asfour et al. (1984) had 10 male college students train for a total of 30 sessions (5 days/week, 6 weeks) doing a symmetrical lifting task. For strength training they performed three sets of a 6-repetition maximum (6RM), lifting a box to three different heights (nine exercise sets total). For muscular endurance training they performed 10 minutes continuously lifting 14 to 20 kg at rates of six to nine lifts/min. For cardiovascular endurance training, they performed cycle ergometer exercise for 30 minutes. At the end of the program, improvements in a 1RM box lift was 41% for the floor to 76-cm lift (78 to 110 kg, p<0.01), 99% for the 76- to 127-cm lift (44 to 88 kg, p<0.01) and 55% for the floor to 127 cm lift (51 to 79 kg, p<0.01). Cardiorespiratory endurance (VO$_2$max estimated from heart rate) also improved 23%.

Sharp and Legg (1988) used a psychophysical approach. Eight male soldiers selected the maximal mass they thought they could lift to a distance of 132 cm for 1 hour at a rate of six lifts/min. Subjects were trained with the self-selected loads (continuously subject-adjusted) in 20 sessions (5 days/week, 4 weeks), lifting in two 15-minute periods each session. At the end of training, the self-selected box mass had increased 26% (25 to 31 kg, p<0.05), 1RM box lift increased 7% (64 to 68 kg, p<0.05), and there was no change in perceived exertion on the psychophysical task.

A number of studies have been performed by Genaidy and coworkers (Genaidy, Davis, Delgado, Garcia, & Al-Herzalla, 1994; Genaidy, 1991; Genaidy, et al., 1990a; Genaidy, Gupta, & Alshedi,

1990b; Genaidy, Mital, & Bafna, 1989; Guo, Genaidy, Warm, Karwowski, & Hidalgo, 1992). All of these investigations used tasks involving a complex series of lifting, carrying, pushing and pulling tasks. Subjects trained for periods of 2.5 to 6 weeks (8 to 24 sessions) in the same task for which they were tested. In general, training resulted in a) progressive improvements in endurance time (time to volitional exhaustion) ranging from 46% to 1350%, b) little or no change in the rating of perceived exertion and c) a decrease in activity heart rate suggesting an improvement in cardiovascular endurance.

MMH and General Physical Training Studies

Sharp et al. (1993) trained 18 male soldiers for 36 sessions (3 days per week, 12 weeks), using 10 traditional weight training exercises. For each exercise, the men performed three to five sets of a 10RM. MMH tasks consisted of 1) 10 minutes of lifting a 41-kg box as many times as possible from floor to chest level and 2) a 1RM for the same distance. After the training program, there was a 17% improvement in the 10-minute task (79 to 92 lifts per 10 minutes) and a 23% improvement on the 1RM task (73 to 89 kg). This study was the first to demonstrate that a well-designed general training program fashioned to improve muscle strength and endurance could augment the performance of men on MMH tasks.

Knapik and Gerber (1996) trained 13 female soldiers for 36 sessions (12 weeks). The women trained with certified instructors, performing resistance training 3 days per week, and running with interval training 2 days per week. Resistance training involved nine exercises (each three sets of a 10RM) using exclusively free weights. Compared to values obtained before training, subjects increased their maximal ability to lift a box from floor to knuckle height by 19% (68 to 81 kg, $p<0.001$) and from floor to chest height by 16% (49 to 57 kg, $p<0.001$). They improved by 17% their ability to lift a 15-kg box from floor to chest height as many times as possible in 10 min (167 to 195 lifts, $p<0.001$). Cardiorespiratory endurance was also enhanced since maximum effort 2-mile run time improved 9% (20.3 to 18.4 minutes, $p<0.001$).

ANALYSIS OF SPECIFIC AND GENERAL TRAINING STUDIES

Specific physical training programs appear to effect larger changes in MMH than generalized training programs. Specific training programs resulted in improvements of 26% to 99% in maximal symmetric lifting and repetitive lifting (Asfour, et al., 1984; Genaidy, et al., 1990a; Sharp & Legg, 1988). This contrasts with relative improvements of 16 to 23% reported in generalized physical training studies (Knapik & Gerber, 1996; Sharp, Harman, Boutilier, Bovee, & Kraemer, 1993). Investigations employing endurance time as a dependent measure (and involving complex motor tasks) report improvements of 34% to 1350% (Genaidy, et al., 1994; Genaidy, 1991; Genaidy, et al., 1990a; Genaidy, et al., 1990b; Genaidy, et al., 1989; Guo, et al., 1992).

The improvements seen in specific training studies may have been largely attributed to enhanced psychomotor learning. In fact, several authors (Asfour, et al., 1984; Genaidy, 1991; Genaidy, et al., 1990a; Genaidy, et al., 1989) noted that at least some of the gains in endurance and lifting capacity were attributable to improved MMH "technique". This would be especially true of the studies of Genaidy and coworkers because the complexity of the tasks may have allowed for many adjustments to achieve longer endurance time (e.g., alterations in body position, pacing of the task, pushing force, etc.). Further, the specific training studies cited were conducted for no longer than 6 weeks, and most for 4 weeks or fewer. It has been demonstrated that neural adaptations account for the majority of strength gains in the first few weeks of resistance training, with hypertrophy becoming a more dominant factor later in training (Moritani & deVries, 1979). Early neural adaptations include fuller activation of muscle prime movers, reduced co-contraction of antagonistic muscles, improved coordination of muscle involved in the intended movement, and removal of inhibitory influences (Sales, 1988).

Improvements in MMH capability seen in the general training studies may involve both a combination of neural adaptations and improvements in muscle hypertrophy. The two general training studies performed were both 12 weeks long, allowing sufficient time for hypertrophy to become a dominant factor in training (Moritani & deVries, 1979). Muscle hypertrophy is important because absolute muscle strength and muscular endurance are proportional to the cross-sectional area of muscle tissue (Maughan, 1984).

MILITARY APPLICATIONS OF STUDIES IN MMH AND PHYSICAL TRAINING

Both general and specific physical training are currently practiced to varying degrees in the U.S. Army. General physical training is an integral part of the daily routine. Army Regulation 350-41 prescribes vigorous exercise three to five times per week during the normal duty day. There is a strong institutional pressure to adhere to this requirement. The importance of physical training is further emphasized to the individual soldier by the Army Physical Fitness Test. This three-event test (pushup, situps, and 2-mile run) must be completed and passed twice a year; promotion and retention in service are tied to the results. Generalized training programs can improve performance on many tasks provided that a wide variety of muscle groups are included in the training program. This could be important in the military (as well as occupations such as police and fire fighting) in which individuals are often called upon to perform non-routine tasks.

Task-specific training is also used in the U.S. Army but is not always of sufficient frequency, duration, or intensity to improve physical capability. Creative methods of introducing task-specific physical training could be extremely useful. A unique application of this type of training was conducted with field artillerymen (Sharp, Knapik, & Schopper, 1994). The soldiers participated in a 45-hour continuous operations exercise involving repeated manual lifting of 44-kg artillery rounds. During the course of the exercise, time required to complete fire missions was reduced 14% (28 to 24 minutes, $p < 0.01$) and energy cost was reduced 23%

(8.0 to 6.2 kcals/min, $p < 0.01$), despite increases in self-reported fatigue and perceived exertion. Thus, a short, intense exercise improved soldier performance and energy efficiency, probably reflecting improvements in psychomotor learning.

CONCLUSIONS

The advantages of regular physical activity include increased worker health and productivity with reduced absenteeism and medical costs (Shephard, 1992). Regular physical training may also help reduce injuries since strength training (Lehnhard, Lehnhard, Young, & Butterfield, 1996) and higher levels of strength and endurance (Barnes, Reynolds, Dettori, Westphal, & Sharp, 1995) are associated with fewer injuries. Besides these benefits, the present review demonstrates that both task-specific and general fitness training programs can improve MMH capability, a common occupational task in the military.

REFERENCES

Asfour, S.S., Ayoub, M.M., & Mital, A. (1984). Effect of an endurance and strength training programme on lifting capability of males. Ergonomics, 27, 435-442.

Barnes, J., Reynolds, K., Dettori, J., Westphal, K., & Sharp, M. (1995). Association of strength, muscular endurance and aerobic endurance with musculoskeletal injuries in US Army female trainees. Medicine and Science in Sports and Exercise, 27, S77.

Caspersen, C.J., Powell, K.E., & Christenson, G.M. (1985). Physical activity, exercise and physical fitness: definitions, and distinctions for health related research. Public Health Reports, 100, 126-131.

Fleck, S.J., & Kraemer, W.J. (1987). Designing Resistance Training Programs. Champaign IL: Human Kinetic Publishers.

Genaidy, A., Davis, N., Delgado, E., Garcia, S., & Al-Herzalla, E. (1994). Effects of a job-simulated exercise programme on employees performing manual handling operations. Ergonomics, 37, 95-106.

Genaidy, A.M. (1991). A training program to improve human physical capability for manual handling jobs. Ergonomics, 34, 1-11.

Genaidy, A.M., Bafna, K.M., Sarmidy, R., & Sana, P. (1990a). A muscular endurance program for symmetrical and asymmetrical manual lifting tasks. Journal of Occupational Medicine, 32, 226-233.

Genaidy, A.M., Gupta, T., & Alshedi, A. (1990b). Improving human capabilities for combined manual handling tasks through a short and intensive physical training program. American Industrial Hygiene Association Journal, 51, 610-614.

Genaidy, A.M., Mital, A., & Bafna, K.M. (1989). An endurance training programme for frequent manual carrying tasks. Ergonomics, 32, 149-155.

Guo, L., Genaidy, A., Warm, J., Karwowski, W., & Hidalgo, J. (1992). Effects of job-simulated flexibility and strength-flexibility training protocols on maintenance employees engaged in manual handling operations. Ergonomics, 35, 1103-1117.

Hogan, J. (1991). The structure of physical performance in occupational tasks. Journal of Applied Psychology, 76, 495-507.

Knapik, J.J., & Gerber, J. (1996). Influence of physical fitness training on the manual material handling capability and road marching performance of female soldiers (Technical Report No. No. ARL-TR-1064). Aberdeen Proving Ground, MD: Human Research and Engineering Directorate, U.S. Army Research Laboratory.

Lehnhard, R.A., Lehnhard, H.R., Young, R., & Butterfield, S.A. (1996). Monitoring injuries on a college soccer team: the effect of strength training. Journal of Strength and Conditioning Research, 10, 115-119.

Maughan, R.J. (1984). Relationship between muscle strength and cross-sectional area. Sports Medicine, 1, 263-269.

Moritani, T., & deVries, H.A. (1979). Neural factors versus hypertrophy in the time course of muscle strength gain. American Journal of Physical Medicine, 58, 115-130.

National Institute of Occupational Safety and Health (1981). Work practice guide for manual lifting (Report No. No. PB82-178948). Cincinnati, OH: U.S. Department of Health and Human Services, National Institute for Occupational Health and Safety.

Sales, D.G. (1988). Neural adaptation to resistance training. Medicine and Science in Sports and Exercise, 20, S135-S145.

Sharp, M.A., Harman, E.A., Boutilier, B.E., Bovee, M.W., & Kraemer, W.J. (1993). Progressive resistance training program for improving manual materials handling performance. Work, 3, 62-68.

Sharp, M.A., Knapik, J.J., & Schopper, A.W. (1994). Energy cost and efficiency of a demanding combined manual materials-handling task. Work, 4, 162-170.

Sharp, M.A., & Legg, S.J. (1988). Effect of psychophysical lifting training on maximal repetitive lifting capacity. American Industrial Hygiene Association Journal, 49, 639-644.

Sharp, M.A., Rice, V., Nindl, B., & Williamson, T. (1993). Effects of gender and team size on floor to knuckle height one repetition maximum lift. Medicine and Science in Sports and Exercise, 25, S137.

Shephard, R.J. (1992). A critical analysis of work-site fitness programs and their postulated economic benefit. Medicine and Science in Sports and Exercise, 24, 354-370.

Wenger, H.A., & Bell, G.J. (1986). The interaction of intensity, frequency and duration of exercise training in altering cardiorespiratory fitness. Sports Medicine, 3, 346-356.

PULLING ON SLIPPERY VERSUS NON-SLIPPERY SURFACES: CAN A LIFTING BELT HELP?

Steven A. Lavender[1], Sang-Hsiung Chen[2], Yi-Chun Li[3], Gunnar B.J. Andersson[1]
[1]Rush-Presbyterian-St. Luke's Medical Center, Chicago, Illinois
[2]Chang Gung Memorial Hospital, KaoHsiung, Taiwan
[3]University of Illinois at Chicago

Biomechanical research has shown that pull forces are largely generated by leg extension and by leaning (using one's body mass). However, this pulling technique requires sufficient friction in the foot floor interface. When pulling tasks are performed on slippery surfaces the center of gravity must be kept over the feet, thereby increasing the trunk's role in force generation. The purpose of this study was to investigate using electromyography (EMG) the muscle recruitments in the torso when pulling tasks are performed on slippery versus non-slippery surfaces. It was hypothesized that if a lifting belt acts to stiffen the torso, this would reduce the muscle activities detected with the EMG. Twelve subjects pulled at 40% of their maximal exertion values in four postures, with a lifting belt either tight or very loose, and under good footing conditions or extremely slippery footing conditions. When averaged across subjects all EMG activities increased under the slippery conditions. These increases ranged between 25 and 131 percent, although for some muscles the response to the slippery conditions was mediated by the pulling posture. The lifting belt had no effect on any of the muscle activities and created no differential muscle response due to the footing conditions or the pulling postures employed.

Epidemiologic studies frequently cite occupational tasks that require lifting, pushing and or pulling as contributing to low-back disorders (LBD's) (Andersson, 1991). Most of the biomechanics research to date has focussed on the lifting aspect of the manual material handling problem. Perhaps this is because, for most workers, pulling tasks are performed less frequently than lifting tasks. Yet pulling tasks can still create considerable bending moments and compressive forces on the spine (Chaffin et al., 1983; Dempster, 1958).

In order to develop a pulling force from a standing posture the body must be capable of transferring the reaction force from the floor all the way to the hands. This requires the trunk act as a rigid link in the kinetic chain. It is has been demonstrated that lifting belts may act to increase the stiffness of the trunk, particularly in the frontal and transverse planes (Lavender et al. 1995; McGill et al., 1994). Thus, it is possible that a lifting belt will facilitate the force transfer, thereby requiring less internal force from the trunk musculature during whole body pulling tasks.

Much of the work done when pulling is accomplished through the moments created by gravitational forces acting on the leaning body mass and by the leg muscles. Therefore, standing pulling tasks often result in postures that are extremely dependent upon the quality of the foot-floor interface. Hence, when pull forces are generated under slippery

conditions the trunk extensor musculature must play a bigger role as the body's center of gravity is now maintained over the feet. Further, the uncertainty of pulling objects under slippery footing conditions may lead to increased co-contraction of the antagonistic muscles, thereby, further accentuating spinal loading.

The objective of the current study was to test the following hypotheses with regard to isometric pulling tasks on non-slippery and slippery flooring conditions:
1. A lifting belt reduces the trunk extensor muscle forces required to generate a specified pull force.
2. Slippery conditions result in greater trunk muscle forces as determined via electromyography.
3. The lifting belt reduces the co-contraction of the antagonistic trunk muscles when pulling under slippery conditions.

METHODS

Subjects

Twelve subjects, ten males and two females, without a history of LBD volunteered for the study. Their mean age was 29.6 years (range 22 - 46 years), height was 1.77 m (range 1.56 - 1.88 m), and weight was 82.7 kg (range 55 - 132 kg).

Experimental Design

A repeated measures experiment was designed to evaluate three independent variables: the pulling posture, the lifting belt and the footing condition, during controlled submaximal exertions. The dependent measures were the electromyographic data (EMG) obtained from the following bilateral muscle groups: Erector Spinae (ERSR and ERSL), Latissimus Dorsi (LATR and LATL), External Oblique (EXOR and EXOL), and Rectus Abdominus (RABR and RABL).

Apparatus

Subjects pulled on a 30.5 cm long, 4.5 cm diameter bar that was connected to a uni-axial dynamometer via cable attached in the middle. The dynamometer was positioned 22 cm off the floor. This system allowed to handle height to float as the subjects achieved postures that allowed them to maximize their force output. The dynamometer was angled such that the axis was in-line with the anticipated cable orientations based on pilot testing and was adjusted for each defined pulling posture tested.

A reference frame apparatus was used to obtain EMG data during maximal isometric exertions in a standing posture. The apparatus included a floor-mounted stand to affix each subject's pelvis. The subjects were fitted with a chest harness that had attachment points in the center and 15 cm each side of center on the front and the rear.

The lifting belts were supplied by Chattanooga Group, Inc (Chattanooga, TN). The belts had a double layer construction with the inner mesh layer containing sewn in vertical plastic stiffeners across the lumbar section. The inner layer attached anteriorly with velcro. The outer layer was comprised of a two 10 cm wide elastic bands. The bands were offset such that the elastic region was 15 cm wide in the center of the back where they were anchored to the inner mesh layer and narrowed to 10 cm at the velcro attachment points.

Surface EMG data were collected using disposable Niko bipolar electrodes (model 4533, Distributed by Salberg Medical, Minneapolis) with a 9 mm diameter sensor. The inter-electrode distance was 3 cm. The electrodes were connected to small preamplifiers with a gain of 1000 attached with velcro to the lifting belt. These were connected to amplifiers the had a gain of 57. The amplified signals were rectified and integrated with a time constant of approximately 100 ms. The band pass frequencies of the system were 30 to 1000 hz. A Gateway P5-60 computer sampled the EMG data entering through the A/D card at 120 Hz.

Procedure

Electrodes were applied to the each of the four

bilateral muscle pairs along the line of action for a given muscle. The specific electrode placements were as follows: (a) ERSR and ERSL: centered halfway between the L3 and L4 spinous processes and approximately 3 to 4 cm lateral from the midline over the belly of the muscle; (b) LATR and LATL: over the muscle belly at the level of T7 and approximately 13-15 cm lateral from the midline; (c) EXOR and EXOL: at the level of the umbilicus and approximately halfway between the iliac crest and the anterior superior iliac spine; (d) RABR and RABL: level of the umbilicus 2 cm lateral from the midline.

Once instrumented with electrodes the chest harness was applied and the subject was strapped to the pelvic support in the reference frame. The harness was connected to dynamometers via turnbuckles in the front or rear for maximal trunk extension and flexion exertions, respectively. Two dynamometers were connected at the 15 cm lateral connection point, one front and one rear, to provide resistance and measure the peak force during maximal trunk twisting efforts. The subjects were instructed to provide maximal exertions over a five second period. Exertions were repeated as necessary after a two minute rest period to insure maximal values were obtained in each of the four directions: Extension, flexion, twist left, and twist right.

Following the maximal exertions the electrodes were surrounded with a 1 cm dense foam material taped to the skin. These pads prevented the lifting belt from applying direct pressure to the electrodes. Next the subject was fitted with a lifting belt. The subject was instructed that during the "belt" trials the belt should be tightened, first, by tensioning the under layer, and second by stretching the elastic binders as far anteriorly as possible. During the "No Belt" trials the subjects were instructed to remove the tension from the elastic binders and loosen the under layer so that their hand could easily be slid between their belly and the lifting belt.

The subjects were instructed in how to attain the four pulling postures shown in figure 1. The instructions also stressed that foot positions could not be changed during the exertions. All pulling exertions were performed barefoot so as remove potential confounding due to variation in shoe tread material and design. In each posture the subjects were encouraged to lean away from the dynamometer during the pulling exertion if they felt it would increase their force output. The postures were defined in terms of the number of hands used and placement of the feet as follows:

1. "Sagittally Symmetric, Bar over Ankles" (SSBOA), Front pull with both hands on the bar which was positioned over the ankles. Both feet were placed so that tips of the big toes were along a designated line. The cable length was adjusted so that the handle was approximately over the ankles.

2. "Sagittally Symmetric, One Leg Back" (SSOLB), Front pull with both hands on the bar which was positioned over the anterior ankle. One foot (of the subjects choosing) was placed with the big toe along a designated line. The other foot was shifted posteriorly to a position where the subject felt comfortable.

3. "Asymmetric, Two Handed" (ATH), The subject was turned such that the dynamometer was in the mid-frontal plane through the ankles. The left foot was placed such that the long axis was in a sagittal plane perpendicular to the cable connecting the handle to the dynamometer. The right foot was positioned approximately shoulder width apart from, and generally parallel to, the left foot. Subjects were permitted to vary the right foot's orientation if they believed greater pull force could be achieved, although, they were instructed to repeat this placement through all ATH trials. The subject was instructed to twist to the left so that both hands could be placed on the bar.

4. "Asymmetric, One Handed" (AOH), same foot stance as above in the ATH posture, however, the bar was only grasped with the left hand in the center. The subjects were instructed to hold their right hand across their chest.

In the first phase subjects pulled as hard as they could against the dynamometer in each of the four described postures. A two minute rest period was given between each of the pulls. Peak dynamometer readings were recorded by the investigators but these values were not shared with the subjects. Pulls were repeated only if the subject felt that they had not provided a maximal effort and could do better.

In the second phase the subjects were required to generate a pull force that was 40 percent of their lowest maximal exertion value measured in the first phase. For example, if a subject's maximal pull values ranged between 430 and 580 N, the trials during the second phase of the experiment required pull forces of 172 N. The pull forces were controlled using a video display connected to the computer sampling the pull force data. The analogue display provided the subject with feedback regarding the current level of force applied relative to a target level. The subject performed the eight exertions comprising the combinations of the belt and pull posture condition while maintaining designated 40 percent pull force value in a randomized sequence. EMG data were sampled for three seconds once the subject achieved the proper pull force.

The eight trials just described were then repeated in a new randomized sequence but with the subject standing on a slippery floor surface. Two pieces of vinyl flooring material were mounted on the

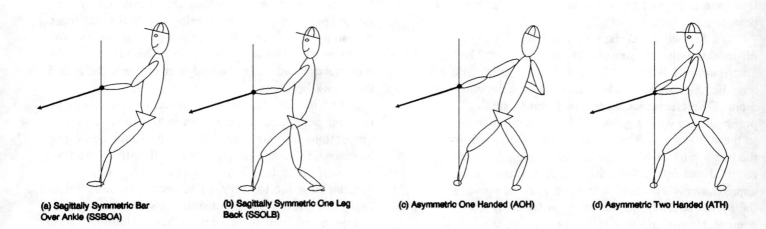

(a) Sagittally Symmetric Bar Over Ankle (SSBOA)   (b) Sagittally Symmetric One Leg Back (SSOLB)   (c) Asymmetric One Handed (AOH)   (d) Asymmetric Two Handed (ATH)

Figure 1. The four pulling postures used in this study.

floor and covered with a generous amount of baby oil. Subjects were assisted onto the platforms so they would not fall. The data were collected as described above.

Data Treatment

The integrated EMG were normalized with respect to maximal and minimal values as follows:

$$NEMG(i) = \frac{IEMG(i) - Min\ EMG(I)}{Max\ EMG(i) - Min\ EMG(i)} * 100 \quad (1)$$

where:
i =   muscle 1 through 8
IEMG=   integrated EMG signal for muscle i
Min EMG =   minimum signal observed for muscle i
Max EMG=   maximum signal observed for muscle i

The NEMG signal is expressed in terms of a percentage of the muscles maximal voluntary contractile capacity.

RESULTS

The peak pull force measured by the dynamometer varied significantly (p<.05) between the four postures used. The greatest force, 681 N, was generated in the SSBOA posture followed by the SSOLB (621 N), ATH (564 N), and the AOH (504 N) postures. For most subjects the AOH posture resulted in smallest maximal pull force. Therefore, on the average the subjects exerted 184 N (sd = 26 N) during the submaximal (40 percent) trials.

Significant differences were observed in the analyses of peak EMG data. Table 1 summarizes the five muscles which changed as a function of the pulling posture employed. All of the eight muscles showed a significant increase in their recruitment when the floor surface was oily (Table 2). On the average the oily surface lead to a 73 percent increase in the normalized EMG of the four posterior muscles and a 51 percent increase in the EMG measured from the anterior muscles. The ERSR and the LATL showed a differential response in the four postures depending on whether the oil was present (Figure 2). More specifically, the LATL increased by a factor of 2.8 for pulls made from the AOH and SSBOA postures on the oily surface versus the non-oily surface. The SSOLB posture on the oily surface only increased the LATL by a factor of 1.6. The ERSR, on the other hand, showed no increase electromyographic activity with the ATH posture when oil was present. Although, this

same muscle's activity was 1.4 times greater on the average during pulls in the other three postures when oil was present.

Table 1. Relative levels of muscle activity across the four postures tested in Experiment 1 (A>B>C). Postures with the same letter are not significantly different with regard to peak EMG levels for the tested muscle.

| | POSTURE | | | |
MUSCLE	SSBOA	SSOLB	AOH	ATH
LATL	B	B	B	A
ERSR	C	C	B	A
ERSL	B	A	D	C
EXOR	C	C	A	B
EXOL	B	B	B	A

Table 2: The mean and standard errors of the mean for the eight muscles as a function of the presence of the oil on the flooring surface.

Muscle	No Oil	No Oil SE	Oil	Oil SE	% Change
LATR	10.5	0.9	21.1	1.8	101.0
LATL	13.5	1.0	31.2	2.2	131.1
ERSR	37.6	2.0	47.0	1.9	25.0
ERSL	25.5	2.1	34.6	2.1	35.7
EXOR	9.1	1.0	11.5	1.1	26.4
EXOL	5.8	0.3	9.8	0.8	69.0
RABR	3.1	0.2	4.9	0.7	58.1
RABL	4.9	0.4	7.4	0.8	51.0

There were no positive or negative changes in the EMG data from the eight trunk muscles due to the lifting belt. Likewise, the belt had no differential effect on the muscle recruitments, positive or negative, due to the four postures employed, or the presence of oil on the floor.

DISCUSSION

This study has shown that pulling strength is greatest in sagittally symmetric postures. In our paradigm the symmetric postures allow subjects to use their leg strength and their body weight to assist in force generation. The EMG results indicate that there are significant changes in the way pulling forces are generated when working on a slippery surface. The need for stability shifts the role of the trunk muscles from just maintaining the rigid link between the upper and lower extremities to that of force generation.

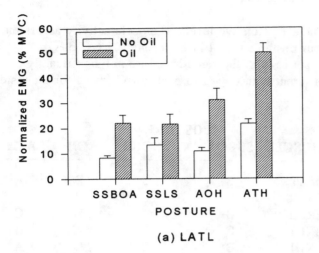
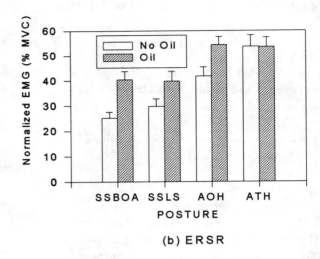

Figure 2. The mean response levels of the LATL (a) and the ERSR (b) as a function of the four postures tested and the presence of oil on the floor surface.

Associated with the pull force generation in the torso under slippery conditions is the uncertainty regarding the overall stability of the posture. When such uncertainty exist, the co-contraction of antagonistic muscles has been shown to occur (De Luca and Mambrito, 1987). It is interesting to note that the increase in the EMG sampled from the anterior muscles due to the oil was independent of the posture condition. Thus, even though the external oblique muscles were more heavily recruited in the asymmetric postures, this co-contraction did not preclude the need for additional spine stability from these muscles under slippery conditions.

The consequences of this co-contraction are most evident when the internal forces contributing to spine loading are considered. Chaffin et al. (1983) estimated that the spinal compression for their sagittally symmetric pulls with low handles were approximately 3600 N. However, this model may severely underestimate the compression due to the co-contraction. Take for example the simple SSBOA posture. Under optimal pulling conditions the back muscles just stiffen the torso to maintain the rigidity of the link (Dempster, 1958). Recognizing that the internal muscle forces are the primary contributor to spine compression, the poor footing and the associated need for postural stability increases the spine loading for three reasons. First, the posterior trunk muscles become responsible for generating the pull force as pulling techniques that rely on the body's mass and leg strength cannot be utilized to the same degree. Second, the high slip potential leads to the recruitment of the antagonistic anterior trunk muscles, again increasing the spine compression. And third, this

co-contraction then requires an even further increase in the posterior muscle recruitment as the net internal trunk moment must be sufficient to generate the desired external force. It was hypothesized that the lifting belt would assist in stiffening the trunk, possibly alleviating the need for the co-contraction response. If this had happened the net result would be less compression on the spine as the second and third reasons mentioned above would be reduced or eliminated. This, however, did not occur when our entire group of subjects was evaluated as a whole. But given the variability in human muscle recruitment strategies, further analysis is needed to identify whether our sample can be subdivided into those who utilized the belt for this purpose and those who did not.

REFERENCES

Andersson, G.B.J. (1991). The epidemiology of Spinal Disorders. In J.W. Frymoyer(ed.) The Adult Spine: Principles and Practice. (pp. 107-146). New York: Raven Press.

Chaffin, D.B., Andres, R.O., Garg, A. (1983). Volitional postures during maximal push/pull exertions in the sagittal plane. Human Factors, 25, 541-550.

De Luca, C.J. and Mambrito, B. (1987). Voluntary control of motor units in human antagonistic muscles: coactivation and reciprocal activation. J. Neurophysiology, 58, 525-542.

Dempster, W.T. (1958). Analysis of two-handed pulls using free body diagrams. J. Applied Physiology, 13, 469-480.

Lavender, S.A., Thomas, J.S., Chang, D., Andersson, G.B.J. (1995). Effect of lifting belts, foot movement, and lift asymmetry on trunk motions. Human Factors, 37, 844-853

McGill, S., Seguin, J., and Bennett, G. (1994). Passive stiffness of the lumbar torso in flexion , extension, lateral bending, and axial rotation: Effect of belt wearing and breath holding. Spine, 19, 696-704.

THE EFFECTS OF BOX WEIGHT, SIZE, AND HANDLE COUPLING ON SPINE LOADING DURING DEPALLETIZING OPERATIONS

William S. Marras, Kevin P. Granata, Kermit G. Davis, W. Gary Allread, and Mike J. Jorgensen
Biodynamics Laboratory
The Ohio State University
Columbus, Ohio

It is widely known that the order selection task in food distribution centers places the worker at risk of occupationally-related low back disorders (LBDs). One approach to controlling this risk consists of manipulating the characteristics of the object or box to be handled in the food distribution center. However, it is currently unknown what effect these changes to the box characteristics would have on the loading of the spine and the subsequent risk of low back disorder. Hence, the *objectives* of this study were to determine the change in spine loading at L5/S1 associated with selecting boxes that varied as a function of: 1) weight (40, 50, and 60 lbs), 2) size (2681 or 1584 cu. in.), and 3) the existence of handles or hand holds. In addition, these variables were explored as a function of where the box was on a pallet. Ten experienced order selectors were recruited from a local food distribution center and were evaluated as they selected boxes of different characteristics from a slot (bin) on to a pallet jack. Workers were monitored for their trunk motion characteristics as well as the EMG activity of 10 trunk muscles as they performed the task. The kinematic and EMG information was used as inputs to an EMG-assisted model that was used to predict the three-dimensional spine loadings that occurred during the task. The results indicated that conditions where a worker must reach to a low level of the pallet significantly increases spinal load. Thus, spinal loads were significantly great and only of a magnitude that would be expected to lead to low back disorders when workers lifted form the lowest layer of the pallet. Handles had the affect of reducing the spinal loading by an amount that was equivalent to a reduction in box weight of about 10 pounds. This effort has also facilitated our basic understanding as to why spine loading increases under the various conditions studied in this experiment. Nearly all differences in spinal loading can be explained by a corresponding difference in coactivation of the trunk musculature. This in turn significantly increase the synergistic forces supplied by each muscle to the spine and resulted in an increase in spinal loading.

INTRODUCTION

It is widely known that the order selection task in food distribution centers places the worker at risk of occupationally-related low back disorders (LBDs). This job is associated with one of the greatest incidence rates of LBD in the United States. The National Association of Wholesale Grocers of America (NAWGA) and the International Foodservice Distribution Association (IFDA) disclosed that 30% of the injuries reported by food distribution warehouse workers were attributable to back sprains/strains (Waters, 1993). In addition, over a five year period, it was found that back injuries could account for nearly 60% of lost work days (NIOSH Interim Report, HETA 91-405, March 1992). Hence, grocery item selectors have an incidence of low back pain that is at least as severe as other manual materials handling jobs.

One approach to controlling this risk consists of manipulating the characterisitcs of the object or box to be handled in the food distribution center. A committee organized by the Food Marketing Institute is currently considering the various options available to them in order to help mediate the risk of work related LBD in these food distribution centers. Among the options considered are: 1) reducing the weight of the boxes, 2) reducing the size of the boxes, or 3) incorporating handles into the boxes. However, it is currently unknown what effect these changes to the box characteristics would have on the loading of the spine and the subsequent risk of low back disorder.

Hence, the *objectives* of this study were to determine the change in spine loading at L5/S1 associated with selecting boxes that varied as a function of: 1) weight (40, 50, and 60 lbs), 2) size (2681 or 1584 cu. in.), and 3) the existence of handles or hand holds. In addition, these variables were explored as a function of where the box was on a pallet.

METHODS

Subjects

Ten experienced order selectors ages 19 to 49 years of age were recruited from a local food distribution center and were evaluated as they selected boxes from a slot (bin) on to a pallet jack. The average (SD) weight and stature of theworkers was 176.3 lbs (18.5 lbs) and 71 in. (2.8) respectively. Their work experience in warehouse settings ranged from 0.25 to 23 years.

Experimental Task

During the different experimental trials the box weight, box size, and box coupling (handles) conditions were varied. Workers were instructed to pick the entire compliment of boxes from the pallet so that they could be observed picking from all locations on a pallet. While the workers were lifting boxes they were being continuously monitored so that trunk loading could be assessed.

In order to simulate a "realistic" warehousing depalletizing task, subjects transferred boxes from one pallet to another. The depalletizing task started when the subject grasped the box and ended when he crossed an imaginary line that coincided with the point at which the subject was upright and facing the "palletizing" pallet. Data were collected for only this interval of time, although subjects completed the task. The lifting rate for all subjects was set at 166 boxes handled per hour and was determined from the minimum loading rate required at the local warehouse where the subjects were employed. The actual lifting cycle was one box lifted every 10 seconds (360 per hour) which was signaled by a computer tone; however, the actual lifting rate was adjusted to 166 lifts/hour by including any down time (e.g., moving pallets, filling out body part discomfort surveys by the subjects, lunch, and additional rest breaks).

Experimental Design

The experimental design consisted of a four-way, within subject design. The *independent variables* included box size, box weight, box weight and pallet region. Subjects served as a random effect. Two sizes of boxes were evaluated to represent "small" boxes and "large" boxes in typical distribution center environments. The "small" box dimensions were 8 in. by 16 in. by 12. in (H x W x D), whereas, the "large" box dimensions were 11 in. by 19.5 in. by 12 in. (H x W x D) which corresponded to volumes of 1584 in^3 and 2681 in^3, respectively. Handle conditions consisted of boxes with cut-out handles and boxes without handles. The cut-out handles were 3.5 in (8.9 cm) wide and 1 in (2.5 cm) high, positioned at the center of the sides of the boxes, 2 in (5.1 cm) below the top of the box. The position and size of the handles were similar to those commonly found on boxes in warehouse environments. The weights of

the boxes in this study were 40, 50, and 60 lbs. (18.2, 22.7 and 27.3 kg.). These weights were at the upper percentiles of typical box weights in a common warehouse setting. Therefore, these weights were chosen to evaluate the effects of heavier loads on the low back and the subsequent risk of LBD.

Each of the pallets were divided into six regions corresponding to front-top, back-top, front-middle, back-middle, front-bottom, and back-bottom areas. Figure 1 shows a schematic view of these six regions on a standard pallet. The handles of the boxes in each of the regions remained at a set level corresponding approximately to: regions A and B at a height of 52.7 in (133.8 cm) from the floor, regions C and D at a height of 37.5 in (95.3 cm) from the floor, and E and F at a height of 18.75 in (47.6 cm) from the floor. The number of boxes in each region depended on the size of the box. A pallet of large boxes had four boxes in region A, three boxes in region B, and seven boxes in regions C, D, E, and F, while the pallets of small boxes had eight boxes in all six regions. In order to simplify the analysis, the six regions were combined into three layers. The top layer contained regions A and B, the middle layer contained regions C and D, and the bottom layer contained regions E and F.

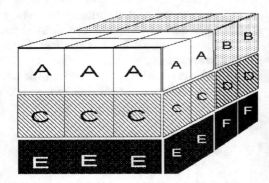

Figure 1. Six Regions of a Pallet.

Dependent variables consisted of the spine loading at the lumbosacral junction (L5/S1). Spine loading variables of interest included compression, lateral shear, and anterior-posterior (A-P) shear forces. These forces were determined via the Biodynamic EMG-assisted model developed at the Ohio State University over the past decade (Marras and Reilly, 1988; Reilly and Marras, 1989; Marras and Sommerich, 1991a; 1991b; Granata and Marras, 1993; Mirka and Marras, 1993; Granata and Marras, 1995a; 1995b; Marras and Granata, 1995). The model uses kinematic information about the trunk along with electromyographic information about the trunk musculature to estimate spinal loads as well as predict the moments imposed on the spine in three dimensional space. The spinal loads estimated in this study were the maximum values of compression force, anterior-posterior shear and lateral shear forces on the lower back at the lumbosacral joint. The trunk moments supported during the lifts were also included as dependent measures. The maximum values of sagittal bending, lateral bending, and axial twisting moments were considered in this study.

Apparatus

The Lumbar Motion Monitor (LMM) was used to collect kinematic information about the trunk. The LMM is essentially an exoskeleton of the spine in the form of a triaxial electro-goniometer that measured instantaneous position, velocity, and acceleration of the trunk in three dimensional space. The light-weight design of the LMM allowed the data be collected with minimal obstruction to the subject's movements. For more information on the design, accuracy, and application of the LMM, refer to Marras et al. (1993).

Electromyographic (EMG) activity was monitored via bi-polar surface electrodes spaced approximately 3 cm apart at the ten major trunk muscle sites. The ten muscles of interest were: right and left erector spinae; right and left latissimus dorsi; right and left internal obliques; right and left external obliques; and right and left rectus abdominis. For the standard locations of electrode placement for these muscles, refer to Mirka and Marras (1993).

A force plate (Bertec 4060A) and a set of electro-goniometers measured the external loads and moments placed on L_5/S_1 during the various calibration exertions that permitted one to "tune" the model for the individual subject. The purpose of the calibration exertions was to determine the individual gain to be used in the "open-loop" exertions. The term "open-loop" referred to exertions that use a predetermined gain to calculate internal moments and forces, rather than calculating a specific gain for each exertion. The electro-goniometers measured the relative position of L_5/S_1 with respect to the center of the force plate, along with the subject's pelvic angle. The forces and moments were translated and rotated from the center of the force plate to L_5/S_1 in this manner (Fathallah, 1995). The internal moments were adjusted to equal the external moments through the use of a gain factor. The value of the gain represented the force output of the muscles per unit area for the particular subject.

All signals from the above equipment were collected simultaneously through customized Windows™-based software developed in the Biodynamics Laboratory. The signals were collected at 100 Hz and recorded on a 486 portable computer via an analog-to-digital board.

The boxes were stacked on a standard pallet generally found in a warehouse. The pallet was constructed of wood with a width of 40 in. (101.5 cm) and a depth of 44 in. (112 cm). The small box conditions used a double-stacked pallet to allow the handles of the boxes in various regions to correspond to the heights of the large boxes which were stacked on a single pallet. The small boxes contained 5 lb boxes of salt while the large boxes contained plastic bottles of water. In order for the boxes to have the desired weight, specific amounts of material were removed from the

Table 1. Significance Summary of Biomechanical Variables.

	SPINE LOAD			MOMENT				EMG
	Shear (Lat)	Shear (A/P)	Compression	Sagittal	Lateral	Twisting	Resultant	Coactivity
Size (S)		*						
Handle (H)		*	*	*		*	*	*
Weight (W)	*	*	*	*	*	*	*	*
Region (R)	*	*	*	*	*	*	*	*
S x H								
S x W								
S x R	*	*	*	*				*
H x W								
H x R	*	*	*	*			*	*
W x R		*	*	*	*	*	*	
S x H x W								

* Significant at $\alpha \leq 0.05$

inside containers (i.e., water was drained from the center bottle). The pallets of small boxes contained six rows of ten boxes, while pallets of the large boxes comprised five rows of seven boxes.

RESULTS

The results of the analysis of variance (ANOVA) which evaluates the influence of the box characteristics and box location upon spinal loading are shown in Table 1. For the most part, compression, lateral shear, and A-P shear responded significantly to the same box characteristics. This analysis indicates that spinal loading responded primarily to changes in handle conditions, weight, and the position of the box on the pallet. It is was particularly interesting to note that all significant interaction terms involved the pallet region variable. Table 1 also shows that these same variables significantly influenced the imposed moment about the trunk. The analysis also indicated that these same three variables increased trunk muscle coactivity which, in turn, increased spinal loading.

A summary of maximum spine compression is shown as a function of the significant independent variables in Figure 2. Spine shear forces responded in a similar manner to spine compression but with a lower magnitude. Several insights can be gained from examining Figure 2. First, position on the pallet has a dramatic effect on maximum spine compression. When lifting from the bottom layers of the pallet, regions E and F in Figure 1, spinal compression is greatest. In fact, few compression forces over 3400 N (typically less than 30% of the observations regardless of box weight) were observed at the top or middle

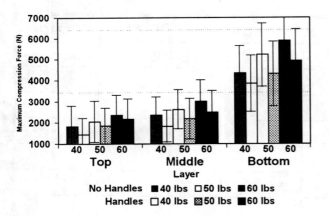

Figure 2. Maximum Spinal Compression Forces as a Function of Handle Condition, Box Weight and Pallet Layer.

layer lifts. However, depending upon the weight and handle conditions of the box between 60% and 97% of the lifts from the lower layer of the pallet resulted in spine compression values above 3400 N and would be expected to increase the risk of an occupationally related low back disorder. This figure also shows that the increases in spine compression for

every 10 lb increase in box weight was more dramatic at the bottom level of the pallet (W x R interaction).

Second, as expected as box weight increases, spine loading increases on average by about 470 N for every 10 lb increase in box weight. However, as shown in Figure 2, the increase in loading is far more dramatic when one considers the point on the pallet from where one is lifting (weight x region interaction). For example the average increase in compression for each 10 lb increase in box weight in region F (far bottom layer) was greater than 655 N or a 130 percent increase in spine compression per 10 lb increase in box weight over what was seen in region A (top layer). Thus, the position from which one lifts is far more important to spine loading than the mere weight of the box.

Third, handles had a significant load relieving effect on the spine. In general, when handles were included in the box the spine compression was equivalent to a box weight which was 10 pounds less than that actually lifted. However, as with box weight, the region on the pallet significantly influenced the effects that handles had on spine compression loading (H x R interaction). In region F, over all box weights, handles reduced spine compression by an average of 770 N, whereas, in region A the reduction was only of the order of 25 N.

Table 1 also indicated a significant influence of box size on A-P shear. Even though this was statistically significant this finding was of little practical value. Post-hoc analyses indicated that the significance of this variable was dictated by changes in A-P shear at the top level of the pallet. However, none of the shear loads at this level were large enough to exceed the spine tolerance to shear. Thus, there is little impact of this finding.

DISCUSSION

This work has facilitated our understating of which specific conditions in a depalletizing operation contribute to excessive spinal loading. In general, conditions where a worker must reach to a low level increases spinal load. This fact in itself is not new. However, we have now been able to document that within these box weight ranges the lowest level of the pallet is the only level that presents a substantial risk. In addition, we have now been able to quantitatively assess the degree or magnitude of spine loading that occurs once one bends down to the lowest level of a pallet. This work has also been able to delineate the added benefit that one could receive from the existence of hand holds or handles in a box. On average, they have the effect of reducing spine loading by an amount equivalent to about a 10 pound reduction in box weight. We have also been able to determine the collective influence of pallet location and box handles on spinal loading. Thus, this work can serve as a guide for workplace design involving pallet unloading such as is done in distribution centers and warehouses.

This effort has also facilitated our basic understanding as to why spine loading increases under the various conditions studied in this experiment. Nearly all differences in spinal loading can be explained by a corresponding difference in coactivation of the trunk musculature. When workers bend to the lower levels of the pallet they must co-contract their muscles in order to increase trunk stability. This increase in cocontraction causes the muscles within the trunk to contract simultaneously, thereby opposing each other and increasing their inefficiency. This process, in turn increases the loading of the spine. This logic adds an additional aspect to the traditional beliefs that forward bending was hazardous because of the reduction in muscle strength that occurs when flexed due to the length-strength relationship of the muscle. A combination of these two effects may explain why risk of low back disorder increases at these low levels of lifting. This same coactivation principle can explain why handles reduce the apparent loading of the spine. The effects of including hand holds and in a box is to effectively require the worker to bend to a lesser flexion angle. This, in turn, reduces the amount of coactivation as discussed previously which reduces the spine loading.

CONCLUSIONS

This study has been successful at pinpointing which box parameter variables are worthy of consideration for inclusion in the food distribution environment for the purposes of reducing the risk of work-related LBD. In general, the following conclusions can be drawn:

- Risk of LBD increases linearly as box weight increases.
- The greatest risk and loading of the spine occur during lifts from the bottom layers of the pallet. The other layers of the pallet pose acceptable lifts regardless of the box weight.
- Box size has a significant effect on risk but the difference has no practical meaning. There is no reason to control box size based upon the range of sizes explored in this study.
- Handles has a significant effect upon spine loading. The effect is particularly significant when lifting from the lowest levels of the pallet. The 40 pound box when combined with handles represents the condition with the lowest level of risk even when lifting from the lowest level of the pallet. Handles have the effect upon spine loading of lowering the box weight by 10 pounds. Thus, a 50 pound box with handles can reduce spine loading to the level of a 40 pound box without handles.

These findings provide some practical solutions to the design of the distribution center. It is clear that given the success of this study, further studies have the potential to further reduce risk through the investigation of other box features.

REFERENCES

Granata, K.P. and Marras, W.S.. (1993), An EMG-Assisted Model of Loads on the Lumbar Spine During Asymmetric Trunk Extensions. *Journal of Biomechanics*, 26(12), 1429-1438.

Granata, G. P., and Marras, W.S., (1995a), W. S., An EMG-Assisted Model of Trunk Loading During Free-Dynamic Lifting, *J. Biomechanics*, 28(11), 1309-1317.

Granata, G. P., and Marras, W. S, (1995b), The Influence of Trunk Muscle Coactivity Upon Dynamic Spinal Loads, *Spine*, 20(8), 913-919.

Marras, W. S. and Granata, K. P., (1995b), A Biomechanical Assesment and Model of Axial Twisting in the Thoraco-Lumbar Spine, *Spine*, 20(13), 1440-1451.

Marras, W.S., Lavender, S.A, Leurgans, S., Rajulu, S., Allread, W.G., Fathallah F. and Ferguson, S.A., (1993), The Role of Dynamic Three Dimensional Trunk Motion in Occupationally-Related Low Back Disorders: The Effects of Workplace Factors, Trunk Position and Trunk Motion Characteristics on Injury, *Spine*, 18(5), 617-628.

Marras, W. S., and Reilly, C. H., (1988), Networks of Internal Trunk Loading Activities Under Controlled Trunk Motion Conditions, *Spine*, 13(6), 661-667.

Marras, W.S. and Sommerich, C.M., (1991a), A Three Dimensional Motion Model of Loads on the Lumbar Spine: I. Model Structure. *Human Factors*, , 33(2), 129-137.

Marras, W.S. and Sommerich, C.M.,(1991b), A Three Dimensional Motion Model of Loads on the Lumbar Spine: I. Model Validation. *Human Factors*, 33(2), 139-149

Mirka, G.A., and Marras, W.S., (1993), A Stochastic Model of Trunk Muscle Coactivation During Trunk Bending. *Spine*, 18(11), 1396-1409.

Reilly, C. H.,and Marras, W. S., (1989), SIMULIFT: A Simulation Model of Human Trunk Motion During Lifting, *Spine*, 14(1), 5-11.

Waters, T.R., Putz-Anderson, V., Garg, A., and L.J. Fine (1993) Revised NIOSH equation for the design and evaluation of manual lifting tasks. *Ergonomics*, 36(7): 749-776.

Author Index

Subject Index

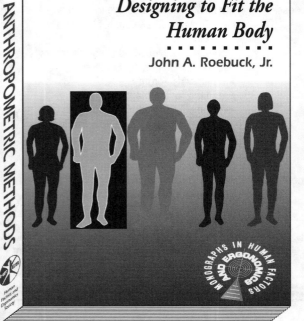

ERGONOMICS IN DESIGN serves all professionals who are concerned with workplace and system safety. If you require up-to-date demonstrations of the importance of ergonomics principles in design and implementation, *Ergonomics in Design* could prove to be the most valuable publication on your desk.

ERGONOMICS IN DESIGN contains articles describing how ergonomics improves the human-technology interface and thereby reduces risk and injury. Case studies, interviews, editorials, and debates focus on such workplace issues as workstation design, manual materials handling, visual displays and vision, and computerization of work processes.

ERGONOMICS IN DESIGN is published by the Human Factors and Ergonomics Society (HFES), the largest professional association for ergonomics researchers and practitioners in the world. HFES members receive *EID* with membership, and nonmembers may subscribe for $37 (outside North America, please add $7 postage if you would like expedited delivery). Issues appear in January, April, July, and October; subscriptions are for the calendar year.

Human Factors and Ergonomics Society
Membership Benefits

What Is HFES?

The Human Factors and Ergonomics Society (formerly the Human Factors Society) is the principal professional association in the United States that is concerned with the study of human characteristics and capabilities and with the application of that knowledge to the design of the products, systems, and environments that people use.

Since its formation in 1957, HFES has promoted the discovery and exchange of human factors and ergonomics knowledge, as well as education and training for students and practitioners.

Members Have Diverse Backgrounds

HFES has more than 5000 members located throughout the United States and in 43 other countries. They are employed in industry, universities and colleges, government, consulting, military, public utilities, and other settings.

Members have academic specialties in psychology (39%), engineering (22%), human factors/ergonomics (9%), industrial design (2%), medicine (4%), and many other fields.

HFES Publications

All members receive four regular publications as a benefit of membership.

Human Factors. This quarterly peer-reviewed journal presents reports of basic and applied research, advances in methods and applications, and reviews of the state of the art. *Human Factors* is an invaluable source of information for those who work in the human factors and ergonomics field and a service to researchers who wish to disseminate their findings.

Ergonomics in Design. The Society's quarterly magazine contains articles, case studies, debates, commentary, and book and product reviews. The focus of *Ergonomics in Design* is the application of human factors/ergonomics research to the design, development, test, and maintenance of human-machine systems and environments.

HFES Bulletin. This monthly newsletter covers news of Society events and committee activities, reviews of meetings and courses, job opportunities, ads for products and services, calls for papers, and issues of concern to human factors/ergonomics researchers and practitioners.

Directory and Yearbook. Each year the Society updates its directory of members. Included are descriptions of the previous year's activities within HFES committees, chapters, and technical groups; alphabetical and geographical member listings; the HFES Code of Ethics; and the Society's Bylaws.

Members also receive discounts on other publications.

Standards Development Activities

Members represent the field in the development of national and international ergonomics standards on computer workstation and software design, medical devices, safety, and a number of other areas.

Technical Areas

There are 20 technical interest groups within HFES, each organized to promote information exchange: Aerospace Systems, Aging, Cognitive Engineering and Decision Making, Communications, Computer Systems, Consumer Products, Educators' Professional, Environmental Design, Forensics Professional, Individual Differences in Performance, Industrial Ergonomics, Macroergonomics, Medical Systems and Rehabilitation, Safety, Surface Transportation, System Development, Test and Evaluation, Training, Virtual Environments, and Visual Performance.

Technical groups contribute to the HFES annual meeting program, distribute newsletters, and conduct periodic meetings.

Local Chapters

Chapters offer events featuring noted speakers, tours of local facilities, symposia on developments in human factors, and social activities. For the location of a chapter in your area, call HFES at the number below.

Technical Meetings

The five-day annual meeting of the Human Factors and Ergonomics Society is held each fall and includes an extensive program featuring the latest research discoveries; methods for research, design, and training; panel discussions and debates on important issues in the field and in the practice of human factors; hands-on workshops by technical specialists; tours of technical and research facilities in the host city; and technical group meetings.

More than 90 lecture, panel, debate, symposium, demonstration, and special sessions are offered. Published proceedings are available at the meeting, representing the work of more than 300 member and nonmember contributors.

HFES members receive substantial discounts on meeting registration. The proceedings are included in the registration fee.

Placement Service

A year-round job-matching database service assists companies and job seekers. One-time searches and subscriptions are available.

To receive information about joining HFES, contact the Society at:

Human Factors and Ergonomics Society
P.O. Box 1369
Santa Monica, CA 90406-1369 USA
310/394-1811   Fax 310/394-2410
hfes@compuserve.com
http://hfes.org